Protocols-Based Sliding Mode Control

This book discusses the Sliding Mode Control (SMC) problems of networked control systems (NCSs) under various communication protocols including static/dynamic/periodic event-triggered mechanism, and stochastic communication, Round-Robin, weighted try-once-discard, multiple-packet transmission, and the redundant channel transmission protocol. The super-twisting algorithm and the extended-state-observer-based SMC scheme are investigated in this book for suppressing chattering. Besides, the SMC designs for one-dimensional (1-D) and two-dimensional (2-D) NCSs are illustrated as well.

Features:

- Captures recent advances of theories, techniques, and applications of networked sliding mode control from an engineering-oriented perspective.

- Includes new design ideas and optimization techniques of networked sliding mode control theory.

- Provides advanced tools to apply networked sliding mode control techniques in the practical applications.

- Discusses some new tools to the engineering applications while dealing with the model uncertainties and external disturbances.

This book aims at Researchers and professionals in Control Systems, Computer Networks, Internet of Things, and Communication Systems.

Protocols-Based Sliding Mode Control

1-D and 2-D System Cases

Jun Song
Zidong Wang
Yugang Niu

CRC Press
Taylor & Francis Group
Boca Raton London New York

CRC Press is an imprint of the
Taylor & Francis Group, an **informa** business

First edition published 2023
by CRC Press
6000 Broken Sound Parkway NW, Suite 300, Boca Raton, FL 33487-2742

and by CRC Press
4 Park Square, Milton Park, Abingdon, Oxon, OX14 4RN

CRC Press is an imprint of Taylor & Francis Group, LLC

© 2023 Jun Song, Zidong Wang and Yugang Niu

Reasonable efforts have been made to publish reliable data and information, but the author and publisher cannot assume responsibility for the validity of all materials or the consequences of their use. The authors and publishers have attempted to trace the copyright holders of all material reproduced in this publication and apologize to copyright holders if permission to publish in this form has not been obtained. If any copyright material has not been acknowledged please write and let us know so we may rectify in any future reprint.

Except as permitted under U.S. Copyright Law, no part of this book may be reprinted, reproduced, transmitted, or utilized in any form by any electronic, mechanical, or other means, now known or hereafter invented, including photocopying, microfilming, and recording, or in any information storage or retrieval system, without written permission from the publishers.

For permission to photocopy or use material electronically from this work, access www.copyright. com or contact the Copyright Clearance Center, Inc. (CCC), 222 Rosewood Drive, Danvers, MA 01923, 978-750-8400. For works that are not available on CCC please contact mpkbookspermissions@tandf.co.uk

Trademark notice: Product or corporate names may be trademarks or registered trademarks and are used only for identification and explanation without intent to infringe.

ISBN: 9781032313870 (hbk)
ISBN: 9781032313887 (pbk)
ISBN: 9781003309499 (ebk)

DOI: 10.1201/9781003309499

Typeset in CMR10
by KnowledgeWorks Global Ltd.

Publisher's note: This book has been prepared from camera-ready copy provided by the authors.

To our families and our friends.

Contents

List of Figures   xi

List of Tables   xv

Preface   xvii

Author's Biography   xix

Acknowledgments   xxiii

Symbols   xxv

1 Introduction   1
    1.1 Sliding Mode Control   1
    1.2 Network Communication Protocols   10
    1.3 Outline   14

PART I: 1-D System Case   21

2 H_∞ Sliding Mode Control Under Stochastic Communication Protocol   23
    2.1 Problem Formulation   23
    2.2 Main Results   25
       2.2.1 Analysis of reachability   27
       2.2.2 Analysis of stochastic stability with H_∞ performance   29
       2.2.3 Solving algorithm   32
    2.3 Example   35
    2.4 Conclusion   39

3 Static Output-Feedback Sliding Mode Control Under Round-Robin   41
    3.1 Problem Formulation   41
    3.2 Main Results   43
       3.2.1 Token-dependent static output-feedback SMC law   44
       3.2.2 Analysis of the asymptotic stability   45
       3.2.3 Analysis of the reachability   49
       3.2.4 Solving algorithm   52

vii

viii   *Contents*

3.3	Example	56
3.4	Conclusion	60

4 Observer-Based Sliding Mode Control Under Weighted Try-Once-Discard Protocol   **61**

4.1	Problem Formulation	62
4.2	Main Results	64
	4.2.1 Token-dependent state-saturated observer	65
	4.2.2 Token-dependent sliding mode controller	65
	4.2.3 Analysis of the asymptotic stability	66
	4.2.4 Analysis of the reachability	69
	4.2.5 Solving algorithm	71
4.3	Example	74
4.4	Conclusion	77

5 Asynchronous Sliding Mode Control Under Static Event-Triggered Protocol   **79**

5.1	Problem Formulation	80
5.2	Main Results	82
	5.2.1 Designing of sliding surface and sliding mode controller	82
	5.2.2 Analysis of sliding mode dynamics	83
	5.2.3 Analysis of reachability	86
	5.2.4 Synthesis of SMC law	90
	5.2.5 Solving algorithm	93
5.3	Example	94
5.4	Conclusion	100

6 Sliding Mode Control Under Dynamic Event-Triggered Protocol   **101**

6.1	Problem Formulation	101
6.2	Main Results	103
	6.2.1 A novel sliding surface	103
	6.2.2 Dynamic event-triggered SMC law	103
	6.2.3 The reachability of sliding surface	104
	6.2.4 The stability of sliding mode dynamics	106
	6.2.5 Further discussions	109
	6.2.5.1 Special case: Static event-triggered SMC of slow-sampling SPSs	109
	6.2.5.2 Convergence of the quasi-sliding motion	112
	6.2.6 Solving algorithm	112
6.3	Example	113
6.4	Conclusion	116

Contents

ix

7 Reliable Sliding Mode Control Under Redundant Channel Transmission Protocol   **117**
   7.1   Problem Formulation 117
   7.2   Main Results 120
       7.2.1   Sliding function and sliding mode controller 120
       7.2.2   MSEUB of closed-loop system 123
       7.2.3   The reachability of sliding surface 126
       7.2.4   Solving algorithm 130
   7.3   Example 131
   7.4   Conclusion 138

8 State-Saturated Sliding Mode Control Under Multiple-Packet Transmission Protocol   **139**
   8.1   Problem Formulation 140
   8.2   Main Results 143
       8.2.1   Sliding function and sliding mode controller 143
       8.2.2   Analysis of sliding mode dynamics 145
       8.2.3   Analysis of reachability 147
       8.2.4   Synthesis of SMC law 151
       8.2.5   Solving algorithm 153
   8.3   Example 156
   8.4   Conclusion 160

9 ESO-Based Terminal Sliding Mode Control Under Periodic Event-Triggered Protocol   **161**
   9.1   Problem Formulation 162
   9.2   Main Results 163
       9.2.1   The design of ESO 163
       9.2.2   Design of periodic event-triggered TSMC based on ESO   165
       9.2.3   Estimation of actual bound for $\|e(t)\|$ 167
       9.2.4   Selection criterion for periodic sampling period λ ... 169
       9.2.5   Reachability and stability 170
       9.2.6   Solving algorithm 173
   9.3   Simulation and Experiment 177
       9.3.1   Simulation 177
       9.3.2   Experiment 183
   9.4   Conclusion 190

PART II: 2-D System Case   **191**

10 2-D Sliding Mode Control Under Event-Triggered Protocol   **193**
   10.1   Problem Formulation 194
   10.2   Main Results 195
       10.2.1   2-D sliding surface 195
       10.2.2   Design of 2-D event generator 195
       10.2.3   Stability of sliding mode dynamics 198

10.2.4 Solving algorithm	200
10.3 Example	204
10.4 Conclusion	210

11 2-D Sliding Mode Control Under Round-Robin Protocol — 213

11.1 Problem Formulation	214
11.2 Main Results	217
11.2.1 A novel 2-D common sliding function	217
11.2.2 First-order sliding mode case	218
11.2.3 Second-order sliding mode case	227
11.3 Solving Algorithms and Examples	236
11.3.1 Solving algorithms	237
11.3.2 Example 1: 2-D SMC without Round-Robin protocol	237
11.3.3 Example 2: 2-D SMC with Round-Robin protocol	242
11.4 Conclusion	247

12 Conclusions and Future Topics — 249

12.1 Conclusions	249
12.2 Future Topics	250

Bibliography — **251**

Index — **271**

List of Figures

1.1	Schematic diagram of PMSM speed regulation system based on vector control.	2
1.2	Response of the system trajectories $x_1(t)$ and $x_2(t)$.	5
1.3	Response of the SMC input $u(t)$.	6
1.4	Response of the sliding motion in phase plane.	7
1.5	The organization structure of this book.	15
2.1	The control systems under SCP scheduling.	24
2.2	Simulation results in Case 1.	37
2.3	A possible SCP scheduling sequence.	38
2.4	Simulation results in Case 2.	38
3.1	The control systems under Round-Robin protocol scheduling.	42
3.2	State trajectories of the uncontrolled system.	57
3.3	State trajectories of the closed-loop systems in Cases 1 & 2.	58
3.4	Sliding variables $s(k)$ of the closed-loop systems in Cases 1 & 2.	59
3.5	Actual control input $u(k)$ in Cases 1 & 2.	59
4.1	The control systems under the WTOD protocol scheduling.	63
4.2	State trajectories of the uncontrolled state-saturated system.	74
4.3	The selected sensor node $\xi(k)$ obtaining access to the network under the WTOD protocol.	75
4.4	System state $x(k)$ and the observer state $\hat{x}(k)$ in the closed-loop case.	76
4.5	Sliding variable $s(k)$ in the closed-loop case.	76
4.6	SMC input $u(k)$ in the closed-loop case.	77
5.1	The event-triggered asynchronous SMC for the networked Markovian jump Lur'e systems.	80
5.2	A possible sequence of system and controller modes.	97
5.3	State trajectories $x(k)$ in closed-loop case.	97
5.4	Release instants and release intervals for the event-triggered condition (5.6).	98
5.5	Sliding variable $s(k)$.	98
5.6	Event-triggered asynchronous SMC input $u(k)$.	99
5.7	Time response of the function $\gamma_D(k)$.	99

xi

xii *List of Figures*

6.1	θ vs. $\tilde{\varrho}$.	114
6.2	The simulation results under $\theta = 40$.	115
6.3	The simulation results under $\theta = 100$.	116
7.1	Reliable SMC of fast-sampling SPSs under RCTP.	119
7.2	Remote SMC of an operational amplifier circuit under the RCTP.	131
7.3	State trajectories $x(k)$ in open-loop case.	134
7.4	Random packet dropouts in three channels.	134
7.5	Measured outputs $y(k)$ in four cases.	135
7.6	State trajectories $x(k)$ in closed-loop cases.	136
7.7	Sliding variables $s(k)$ in four cases.	137
7.8	SMC input $u(k)$ in four cases.	138
8.1	The saturation function $\sigma_i(v_i)$.	141
8.2	Reliable SMC over hidden Markov fading channels.	143
8.3	Fitness value of each generation in solving Algorithm 2.	157
8.4	State trajectories $x(k)$ in open- and closed-loop cases.	158
8.5	A possible fading state and detection mode sequence.	158
8.6	The channel fading variable $\vartheta_{\varsigma(k)}(k)$.	159
8.7	Illustration of the reachability to the sliding region $\mathcal{O}$ in (8.23).	159
8.8	The SMC input $v(k)$ and the actual control input $u(k)$.	160
9.1	Schematic diagram of PMSM speed regulation system based on vector control.	162
9.2	The framework of the ESO-based periodic event-triggered TSMC.	167
9.3	The design and theoretical analysis procedure of the proposed periodic event-triggered TSMC.	176
9.4	The best and average fitness values in each generation.	178
9.5	Simulation results for the periodic event-triggered TSMC with ESO under $T_L(t) = 0$: (a) Response of the state trajectories $x_1(t)$ and $x_2(t)$; (b) Response of the terminal sliding function $s(t)$; (c) Response of the speed $\omega(t)$; (d) Response of the reference q-axis current $i_q^*(t)$.	179
9.6	Simulation comparison of the periodic event-triggered TSMC with and without ESO: (a) Response of the speed $\omega(t)$; (b) Response of the reference q-axis current $i_q^*(t)$.	180
9.7	Simulation comparison between static and periodic event-triggered TSMC schemes: (a) Response of the speed $\omega(t)$; (b) Response of the reference q-axis current $i_q^*(t)$.	181
9.8	Inter-event time between static and periodic event-triggered TSMC cases (simulation)	182
9.9	Schematic diagram of the considered PMSM speed regulation problem.	184

List of Figures

xiii

9.10 Experiment results in time-triggered TSMC. 185

9.11 Experiment results in periodic event-triggered TSMC. 186

9.12 Inter-event time between periodic event-triggered and time-triggered TSMC cases (experiment) 187

9.13 Experimental comparison in three different cases when the load torque $T_L(t) = 0.3$ N $\cdot$ m is added suddenly. 188

9.14 Experimental comparison of the proposed periodic event-triggered TSMC with and without ESO cases when the load torque $T_L(t) = 0.3$ N $\cdot$ m is added suddenly. 189

10.1 State trajectories $x^h(i,j)$ and $x^v(i,j)$ in open-loop case. . . 205

10.2 Fitness value of each generation. 206

10.3 State trajectories $x^h(i,j)$ and $x^v(i,j)$ in closed-loop case. . . 207

10.4 Evolutions of sliding variables $s^h(i,j)$ and $s^v(i,j)$. 208

10.5 Evolutions of event-based SMC inputs $u^h(i,j)$ and $u^v(i,j)$. . 209

10.6 Horizontal and vertical scheduling instants under 2-D event generators. 209

10.7 Auxiliary functions in horizontal and vertical directions. . . 210

11.1 2-D SMC subject to Round-Robin scheduling protocol . . . 215

11.2 $\hat{\rho}$ vs. n (First-Order SMC Case in Example 1). 239

11.3 Horizontal and vertical state trajectories in closed-loop case of Example 1 (First-Order SMC Case). 239

11.4 Evolutions of 2-D common sliding variable and 2-D first-order SMC input in Example 1 (First-Order SMC Case). 240

11.5 $\hat{\rho}$ vs. n (Second-Order SMC Case in Example 1). 240

11.6 Horizontal and vertical state trajectories in closed-loop case of Example 1 (Second-Order SMC Case). 241

11.7 Evolutions of 2-D common sliding variable and 2-D second-order SMC input in Example 1 (Second-Order SMC Case). . 241

11.8 Evolution of fitness value in each generation (First-Order SMC Case). 243

11.9 The state trajectories of the closed-loop system in Example 2 (First-Order SMC Case). 244

11.10 Evolutions of sliding variables in Example 2 (First-Order SMC Case). 244

11.11 Evolutions of actuators in Example 2 (First-Order SMC Case). 245

11.12 Evolution of fitness value in each generation (Second-Order SMC Case). 245

11.13 The state trajectories of the closed-loop system in Example 2 (Second-Order SMC Case). 246

11.14 Evolutions of sliding variables in Example 2 (Second-Order SMC Case). 246

11.15 Evolutions of actuators in Example 2 (Second-Order SMC Case). 247

List of Tables

5.1	Parameters in the networked DC Motor Device.	95
6.1	θ vs. $\bar{\varepsilon}$.	116
7.1	RCTP ($N = 3$) in the example.	132
7.2	Four different RCTP cases.	133
9.1	Parameters in the concerned PMSM.	177
10.1	Comparing results for five sets of δ^h and δ^v.	210

Preface

Practical plants are always affected by parameter uncertainties and external disturbances. The existence of these perturbations may worsen the performance or even cause system instability of the concerned system dynamics. To deal with these perturbations, many control theories and approaches have been developed. Among the existing methodologies to reject perturbations, the Sliding Mode Control (SMC) technique which was initially proposed by Emel'yanov and his colleagues in 1960s turns out to be characterized by high simplicity and robustness. Essentially, SMC utilizes discontinuous control laws to drive the system state trajectory onto a specified surface in the state space, the so-called sliding or switching surface, and to keep the system state on this manifold for all the subsequent times. The ability of compensating the *matched* model uncertainties and disturbances completely is the main advantage of SMC approach. Specifically, the dynamic of the system while in sliding mode shows *insensitive* (better than *robustness*) to model uncertainties and disturbances. Unfortunately, the real-life implementation of SMC techniques shows a major drawback: the so-called chattering effect, i.e., dangerous high-frequency vibrations of the controlled system. This phenomenon is due to the fact that, in real-life applications, it is not reasonable to assume that the control signal can switch at infinite frequency. Over the past decades, many efforts have been made in SMC community for the purpose of attenuating the chattering in the SMC input.

Nowadays, the rapid development in network technologies has triggered a revolution in engineering practices. More and more system information is delivered via communication networks such that the system components (e.g. controller, sensor, actuator, and sampler) are likely to be connected by communication networks, in which the sampled signals, sensor signals, and control signals are transmitted through a shared network medium. The usage of networks in control systems gives rise to the so-called networked control systems (NCSs) that facilitate the remote execution of certain tasks such as monitoring and control. The NCSs have many merits such as low cost, reduced weight and power, and easy manipulation. Nevertheless, the implementation of communication networks between system components has largely raised the level of complexities in the analysis and synthesis of the overall NCSs due mainly to the complex working conditions and limited bandwidth. Over the past decade, the SMC issues of NCSs have attracted a growing research attention due to the excellent feature of SMC to compensate external disturbance and parameter uncertainties. It should be mentioned that the available literature on this topic

xvii

is mainly focused on the design of SMC schemes for the one-dimensional (1-D) NCSs subject to various network-induced complexities (e.g., communication delays, packet dropout, packet disorder, quantization, etc.) in a *passive* way. Actually, in the practical engineering, an active way to prevent the data from collisions and improve the communication performance is always to introduce communication protocols in the control loop. Under the limited communication resource, the communication protocol is employed to determine which sensors or actuators should obtain the access to the communication network according to a prespecified scheduling rule.

In this book, we launch a systematic discussion to the SMC problems of NCSs under various communication protocols including static/dynamic/periodic event-triggered mechanism, stochastic communication protocol, Round-Robin protocol, weighted try-once-discard protocol, multiple-packet transmission protocol and the redundant channel transmission protocol. The super-twisting algorithm (a typical second-order SMC approach) and the extended-state-observer-based SMC scheme are involved in this book for suppressing chattering. Besides, the SMC designs for one-dimensional (1-D) and two-dimensional (2-D) NCSs are investigated, respectively, in Part I (Chapters 2–9) and Part II (Chapters 10–11) of this book. In each chapter of Parts I and II, the problem formulations are first considered with introducing the key idea of the used communication protocols. Then, some novel sliding functions and SMC schemes are properly designed by considering the scheduling effects from the communication protocols. Furthermore, by drawing on a variety of theories and methodologies such as Lyapunov function, finite-time stability and linear matrix inequalities, the sufficient conditions for guaranteeing the reachability to the ideal sliding surface and the desired stability of the sliding mode dynamics are derived simultaneously. It is shown that the communication protocols improve the usage of communication network at the cost of weakening some disturbance rejection performances of the SMC schemes. The satisfied SMC schemes are solved via some convex optimization algorithms as well as genetic algorithm. Finally, the proposed results are illustrated in some numerical examples and a practical experiment of permanent magnet synchronous motor speed regulation system. Chapter 1 introduces the main definitions related in SMC theory, the recent progress on SMC problems for 1-D NCSs as well as the outline of the book. Chapter 12 gives the conclusions and some possible future research topics on protocols-based SMC for 1-D and 2-D NCSs. This book is a research monograph whose intended audience is graduate and postgraduate students as well as researchers.

Jun Song
Hefei, China

Zidong Wang
London, U.K.

Yugang Niu
Shanghai, China

Author's Biography

Dr. Jun Song

Jun Song received his B.E. degree in electronic science and technology and the M.E. degree in pattern recognition and intelligent system from Anhui University, Hefei, China, in 2011 and 2014, respectively, and the Ph.D. in control science and engineering from East China University of Science and Technology, Shanghai, China, in 2018. He is currently Professor with Anhui University. From October 2016 to October 2017, he was a visiting Ph.D. student with the Department of Computer Science, Brunel University London, Uxbridge, U.K. He also was Research Assistant or Research Fellow with The University of Hong Kong, Hong Kong, China; City University of Hong Kong, Hong Kong, China and Western Sydney University, Penrith, NSW, Australia. His research interests include networked sliding mode control theory and its applications, Markovian jump systems, large-scale systems, and finite-time stability. Dr.Song was a recipient of the Chinese Association of Automation Outstanding Ph.D. Dissertation Award in 2019.

Prof. Zidong Wang

Zidong Wang is Professor of Dynamical Systems and Computing at Brunel University London, West London, United Kingdom. He was born in 1966 in Yangzhou, Jiangsu, China. He received his BSc degree in Mathematics in 1986 from Suzhou University, Suzhou, the MSc degree in Applied Mathematics in 1990 and the PhD in Electrical and Computer Engineering in 1994, both from Nanjing University of Science and Technology, Nanjing.

He was appointed as Lecturer in 1990 and Associate Professor in 1994 at Nanjing University of Science and Technology. From January 1997 to December 1998, he was an Alexander von Humboldt research fellow at the Control Engineering Laboratory, Ruhr-University Bochum, Germany. From January 1999 to February 2001, he was a Lecturer with the Department of Mathematics, University of Kaiserslautern, Germany. From March 2001 to July 2002, he was University Senior Research Fellow at the School of Mathematical and Information Sciences, Coventry University, U.K. In August 2002, he joined the Department of Computer Science, Brunel University London, U.K., as a Lecturer, and was then promoted to a Reader in September 2003, and to a Chair Professor in July 2007.

Professor Wang's research interests include dynamical systems, signal processing, bioinformatics, and control theory and applications. He has published more than 600 papers in refereed international journals. He was awarded the Humboldt research fellowship in 1996 from Alexander von Humboldt Foundation, the JSPS Research Fellowship in 1998 from Japan Society for the Promotion of Science, and the William Mong Visiting Research Fellowship in 2002 from the University of Hong Kong. He was a recipient of the State Natural Science Award from the State Council of China in 2014 and the Outstanding Science and Technology Development Awards (once in 2005 and twice in 1997) from the National Education Committee of China.

Professor Wang is currently serving or has served as the Editor-in-Chief for International Journal of Systems Science, the Editor-in-Chief for Neurocomputing, Executive Editor for Systems Science and Control Engineering, Subject Editor for Journal of The Franklin Institute, an Associate Editor for IEEE Transactions on Automatic Control, IEEE Transactions on Control Systems Technology, IEEE Transactions on Systems, Man, and Cybernetics - Systems, Asian Journal of Control, Science China Information Sciences, IEEE/CAA Journal of Automatica Sinica, Control Theory and Technology, an Action Editor for Neural Networks, an Editorial Board Member for Information Fusion, IET Control Theory & Applications, Complexity, International Journal of Systems Science, Neurocomputing, International Journal of General Systems, Studies in Autonomic, Data-driven and Industrial Computing, and a member of the Conference Editorial Board for the IEEE Control Systems Society. He served as Associate Editor for IEEE Transactions on Neural Networks, IEEE Transactions on Systems, Man, and Cybernetics - Part C, IEEE Transactions on Signal Processing, Circuits, Systems & Signal Processing, and an Editorial Board Member for International Journal of Computer Mathematics.

Professor Wang is a Member of the Academia Europaea (section of Physics and Engineering Sciences), a Fellow of the IEEE (for contributions to networked control and complex networks), a Fellow of the Chinese Association of Automation, a Member of the IEEE Press Editorial Board, a Member of the EPSRC Peer Review College of the UK, a Fellow of the Royal Statistical Society, a member of program committee for many international conferences, and a very active reviewer for many international journals. He was nominated an appreciated reviewer for IEEE Transactions on Signal Processing in 2006-2008 and 2011, an appreciated reviewer for IEEE Transactions on Intelligent Transportation Systems in 2008, an outstanding reviewer for IEEE Transactions on Automatic Control in 2004 and for the journal Automatica in 2000.

Author's Biography

Prof. Yugang Niu

Yugang Niu received his M.Sc. and Ph.D. degrees in control engineering from the Nanjing University of Science and Technology, Nanjing, China, in 1992 and 2001, respectively. In 2003, he joined the School of Information Science and Engineering, East China University of Science and Technology, Shanghai, China, where he is currently a Professor. His current research interests include stochastic systems, sliding mode control, Markovian jump systems, networked control systems, wireless sensor networks, and smart grid. Prof. Niu is currently an Associate Editor of several international journals, including *Information Sciences*, *Neurocomputing*, *IET Control Theory & Applications*, *Journal of The Franklin Institute*, and *International Journal of System Sciences*. He is also a member of Conference Editorial Board of IEEE Control Systems Society.

Acknowledgments

We would like to acknowledge the help of many people who have been directly involved in various aspects of the research leading to this book. Special thanks go to Professor Hak-Keung Lam from King's College London, London, the United Kingdom, Professor Hongli Dong from Northeast Petroleum University, Daqing, China, Professor Jun Hu from Harbin University of Science and Technology, Harbin, China, Professor Shuping He from Anhui University, Hefei, China, and Professor Hongjian Liu from Anhui Polytechnic University, Wuhu, China. Special thanks to Dr. Bei Chen, Dr. Haijuan Zhao, Dr. Zhina Zhang, Dr. Zhiru Cao, Mr. Yekai Yang, Ms. Jiarui Li, Ms. Xinyu Lv, Mr. Long-Yang Huang, Mr. Yu-Kun Wang, and Mr. Sai Zhou for their tremendous help in the editorial and proofreading work.

This book was supported in part by the National Natural Science Foundation of China (Grant Nos. 61903143, and 62073139), in part by the Shanghai Chenguang Program (Grant No. 19CG33), in part by the Shanghai Sailing Program (Grant No. 19YF1412100), the Engineering and Physical Sciences Research Council (EPSRC) of the UK, the Royal Society of the UK, and the Alexander von Humboldt Foundation of Germany. The support of these organizations is gratefully acknowledged.

Symbols

$\mathbb{R}^n$ The n-dimensional Euclidean space.

$\mathbb{N}^+$ The set of nonnegative integers.

$\mathbb{R}^{n \times m}$ The set of all $n \times m$ real matrices.

$\| \cdot \|$ The Euclidean norm in $\mathbb{R}^n$.

$A \circ B$ The Hadamard product $A \circ B \triangleq [a_{ij}b_{ij}]_{m \times n}$, where the real matrices $A = [a_{ij}]_{m \times n}$ and $B = [b_{ij}]_{m \times n}$.

$\delta(a)$ The Kronecker delta function.

$\otimes$ The Kronecker product of matrices.

$\mod(a, b)$ The unique nonnegative remainder on division of the integer a by the positive integer b.

$\Pr\{\cdot\}$ The occurrence probability of the event "$\cdot$".

$\mathbb{E}\{x\}$ The expectation of stochastic variable x.

$\mathbb{E}\{x|y\}$ The expectation of x conditional on y, x and y are all stochastic variables.

$X > Y$ The $X - Y$ is positive definite, where X and Y are symmetric matrices.

$X \geq Y$ The $X - Y$ is positive semi-definite, where X and Y are symmetric matrices.

M^{T} The transpose matrix of M.

$\mathrm{diag}\{M_1, ..., M_n\}$ The block diagonal matrix with diagonal blocks being the matrices $M_1, ..., M_n$.

$\star$ The ellipsis for terms induced by symmetry, in symmetric block matrices.

$\mathrm{sgn}(s(t))$ The sign function which is defined as $\mathrm{sgn}(s(t)) \triangleq$
$$\begin{cases} 1 & \text{if } s(t) > 0 \\ [-1, 1] & \text{if } s(t) = 0 \\ -1 & \text{if } s(t) < 0. \end{cases}$$

1

Introduction

1.1 Sliding Mode Control

The initial idea of Sliding Mode Control (SMC) was developed in some Russian papers by Emel'yanov and his colleagues in 1960s [41–43]. Motivated by a paper reported by Utkin [154] and a book by Itkis [67] in English, SMC theory and applications have attracted more and more research attentions in the control community. As a particular type of Variable Structure Control approach, the key advantage of SMC is that its insensitivity to the *matched* disturbances and parameter variations, i.e., the so-called *invariance property* of SMC (better than *robustness*).

For the conventional SMC [138, 154, 158, 195], its design process contains two phases, that is, the design of the sliding surface and the design of a control law to steer the system states onto the chosen sliding surface in finite time. There exists two kinds of stability related to the conventional SMC, i.e., the *asymptotical stability* in the sliding motion phase and the *finite-time stability* [11] in the reaching phase. For the reader's convenience, a simple example of using SMC theory to control permanent magnet synchronous motor (PMSM) speed regulation system is given as follows, and some main definitions and notions of SMC theory will be introduced.

Assume that the rotor of the motor has no damping winding and the permanent magnet has no damping effect, magnetic circuit saturation and the effect of eddy current and hysteresis are neglected, and the distribution of permanent magnetic field in air gap space is sinusoidal. The model of PMSM on d-q coordinates can be formulated as follows:

$$\begin{aligned}
\frac{\mathrm{d}i_d(t)}{\mathrm{d}t} &= \frac{1}{L_d}\left(u_d(t) - R_s i_d(t) + n_p\omega(t)L_q i_q(t)\right) \\
\frac{\mathrm{d}i_q(t)}{\mathrm{d}t} &= \frac{1}{L_q}\left(u_q(t) - R_s i_q(t) - n_p\omega(t)L_d i_d(t) - n_p\omega(t)\psi_f\right) \\
\frac{\mathrm{d}\omega(t)}{\mathrm{d}t} &= \frac{1}{J}\left(1.5n_p\psi_f i_q(t) - T_L(t) - B_v\omega(t)\right),
\end{aligned} \tag{1.1}$$

where L_d and L_q with $L_d = L_q$ are the stator inductances of d and q axes, respectively; R_s is the stator resistance; $u_d(t)$, $u_q(t)$ and $i_d(t)$, $i_q(t)$ are the stator voltages and currents of d, q axes, respectively; $\omega(t)$, n_p, and ψ_f are the rotor angular velocity, number of pole pairs, and permanent magnet flux

DOI: 10.1201/9781003309499-1

linkage, respectively; and J, $T_L(t)$, and B_v denote the moment of inertia, load torque, and viscous friction coefficient, respectively.

The vector control, which includes a speed loop and two current loops, is applied to the PMSM, and proportional-integral (PI) controllers are adopted to eliminate the tracking error of the two current loops. In vector control scheme, the output of the speed loop is used as the reference current $i_q^*(t)$ of the current loop $i_q(t)$. The principle of vector control of PMSM is shown in Fig. 1.1.

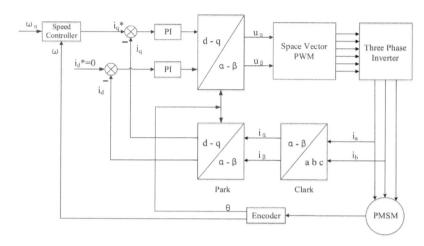

FIGURE 1.1: Schematic diagram of PMSM speed regulation system based on vector control.

Here, we assume that the dynamic response speed of the current loop is faster than the speed loop. Now, we can use $i_q(t)$ to replace reference current $i_q^*(t)$ approximately. By using the vector control strategy of $i_d^*(t) \equiv 0$, the model can be simplified as follows:

$$\begin{cases} \dfrac{\mathrm{d}i_q(t)}{\mathrm{d}t} = -\dfrac{R_s}{L_q}i_q(t) - \dfrac{n_p\psi_f}{L_q}\omega(t) + \dfrac{u_q(t)}{L_q}, \\ \dfrac{\mathrm{d}\omega(t)}{\mathrm{d}t} = ai_q(t) - b\omega(t) - \dfrac{T_L(t)}{J}, \end{cases} \quad (1.2)$$

where $a \triangleq \dfrac{1.5 n_p \psi_f}{J}$, $b \triangleq \dfrac{B_v}{J}$.

Define the following state variables:

$$x_1(t) \triangleq \omega_n(t) - \omega(t), x_2(t) \triangleq \dot{x}_1(t) = \dot{\omega}_n(t) - \dot{\omega}(t), \quad (1.3)$$

where $\omega_n(t)$ is the given reference speed. Then, we can rewrite the PMSM speed regulation system (1.2) as follows:

$$\begin{cases} \dot{x}_1(t) = x_2(t), \\ \dot{x}_2(t) = -au(t) + d(t), \end{cases} \quad (1.4)$$

Sliding Mode Control 3

where $u(t) \triangleq \dot{i}_q(t)$ is the control input, and $d(t) \triangleq \ddot{w}_n(t) + b\dot{w}(t) + \frac{\dot{T}_L(t)}{J}$, which is handled as the matched disturbance.

Assumption 1.1 *The disturbance $d(t)$ is bounded with known upper bound, which means that there exists a parameter $d_0 \geq 0$ such that $\sup_{t \geq 0} |d(t)| \leq d_0$.*

An output of the system is given as:

$$s(t) = c_1 x_1(t) + x_2(t), \tag{1.5}$$

where c_1 is a positive constant.

Definition 1.1 (Sliding Variable) *The variable $s(t)$ in (1.5) is defined as a **sliding variable**.*

Definition 1.2 (Sliding Surface) *The straight line in the state space $s(t) = 0$ is referred to as the **sliding surface**.*

Definition 1.3 (Ideal Sliding Motion) *Suppose that there exists a finite time t_s such that the solution to (1.4) represented by $x(t)$ satisfies*

$$s(t) = 0, \tag{1.6}$$

*for all $t \geq t_s$, then an **ideal sliding motion** is said to be taking place for all $t > t_s$.*

Now, a control law can be designed to make the point $(x_1(t), x_2(t))$ in phase plane arrive the sliding surface $s(t) = 0$ in a finite time. Then, the control law forces $(x_1(t), x_2(t))$ to $(0,0)$ along the sliding manifold asymptotically. The controlled plant, namely PMSM speed regulation system (1.4), is asymptotically stable. Suppose that at the instant t_s, the systems states lie on the surface $s(t) = 0$, and an ideal sliding motion takes place, which can be expressed as $s(t) = 0$ and $\dot{s}(t) = 0$ for all $t \geq t_s$ mathematically. Substituting (1.4) into $\dot{s}(t)$ gives

$$\dot{s}(t) = c_1 \dot{x}_1(t) + \dot{x}_2(t) = c_1 x_2(t) - au(t) + d(t) = 0. \tag{1.7}$$

Definition 1.4 (Equivalent Control Law) *The **equivalent control law** $u_{eq}(t)$ is defined as a unique solution to the algebraic equation (1.7), that is,*

$$u_{eq}(t) = a^{-1}\left(c_1 x_2(t) + d(t)\right). \tag{1.8}$$

In order to ensure the system states reaching the sliding surface within a finite time, a so-called η-reachability condition $s(t)\dot{s}(t) \leq -\eta|s|$ should be satisfied, where η is a positive scalar. The switching control law is chosen as $u_{sw}(t) = a^{-1}k\text{sgn}(s(t))$, where k is the switching gain to be designed. Without

4 *Introduction*

considering the external disturbance, the SMC law consisting of the equivalent control law $u_{eq}(t)$ and switching control law $u_{sw}(t)$ is developed as

$$
\begin{aligned}
u(t) &= u_{eq}(t) + u_{sw}(t) \\
&= a^{-1}c_1 x_2(t) + a^{-1}k\mathrm{sgn}(s(t)).
\end{aligned} \tag{1.9}
$$

In the sequel, the reachability of the sliding surface $s(t) = 0$ and the stability of the sliding mode dynamics (1.4) are analyzed, respectively.

- **Analysis of Reachability**. Choose a suitable Lyapunov function candidate $V(t) \triangleq \frac{1}{2}s^2(t)$. By applying the SMC law (1.9) into (1.4), it yields

$$
\begin{aligned}
\dot{V}(t) &= s(t)\dot{s}(t) \\
&= s(t)[-k\mathrm{sgn}(s(t)) + d(t)] \\
&\leq -k|s(t)| + d_0|s(t)| \\
&= -|s(t)|(k - d_0).
\end{aligned}
$$

Clearly, if the parameter k is chosen such that $k > d_0$, i.e., $\eta \triangleq k - d_0$, then the η-reachability is guaranteed. That is to say, the prespecified sliding surface $s(t) = 0$ can be arrived in a finite time.

- **Analysis of Stability**. When the ideal sliding motion occurs, it has $s(t) = c_1 x_1(t) + x_2(t) = 0$, which implies $x_2(t) = -c_1 x_1(t)$. From (1.4), one can get the following sliding motion dynamics on the sliding surface $s(t) = 0$:

$$
\begin{cases}
\dot{x}_1(t) = -c_1 x_1(t), \\
x_2(t) = -c_1 x_1(t).
\end{cases} \tag{1.10}
$$

Now, it can obtain the following solutions to (1.10):

$$
\begin{cases}
x_1(t) = x_1(t_s)e^{-c_1(t-t_s)}, \\
x_2(t) = -c_1 x_1(t_s)e^{-c_1(t-t_s)},
\end{cases} \tag{1.11}
$$

where $t_s > 0$ is the reaching instant to the sliding surface and the parameter c_1 is chosen as $c_1 > 0$. Obviously, the state trajectories $x_1(t)$ and $x_2(t)$ converge to the origin *asymptotically* in the sliding surface $s(t) = 0$.

One can observe from (1.10) that the sliding mode dynamics is independent of external disturbance $d(t)$, that is, it processes *insensitive* (better than *robustness*) to the disturbance. Besides, it is worth stressing that the design of the conventional SMC [138, 154, 158, 195] is related to two kinds of stability concept, that is, the finite-time stability [11] for the sliding variable and the asymptotical stability for the sliding mode dynamics.

A simple simulation is conducted to illustrate the above results. Suppose that the PMSM speed regulation system (1.4) is given as

$$
\begin{cases}
\dot{x}_1(t) = x_2(t), \\
\dot{x}_2(t) = -54.39u(t) + d(t),
\end{cases} \tag{1.12}
$$

Sliding Mode Control

where $d(t) = 2 + 0.5\sin(t)$ and the SMC law is designed as

$$u(t) = \frac{1}{54.39}x_2(t) + \frac{30}{54.39}\text{sgn}(s(t)),$$

with $c_1 = 1$ and $k = 30$.

The simulation results are shown in Figs. 1.2–1.4. It is seen from Fig. 1.2 that the state trajectories $x_1(t)$ and $x_2(t)$ converge to origin asymptotically. The SMC input $u(t)$ is depicted in Fig. 1.3. Besides, Fig. 1.4 shows the sliding motion in the phase plane.

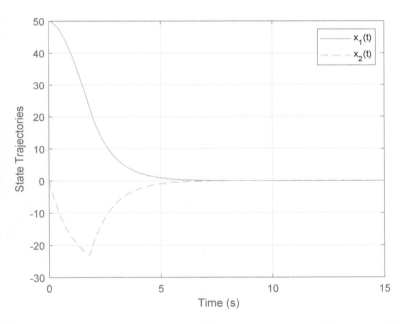

FIGURE 1.2: Response of the system trajectories $x_1(t)$ and $x_2(t)$.

With the development of SMC theory, various SMC approaches have been developed to meet different requirements in the engineering applications. In what follows, some SMC approaches related to this book will be reviewed briefly.

Terminal SMC

It is found that for the conventional SMC [7, 8, 154, 195], the convergence of sliding motion is *asymptotically stable* as shown in (1.11). That is to say, the system states may need *infinite* time to converge to the origin, which not necessarily satisfies the requirements of the practical applications. To achieve a higher control performance, a new approach of terminal SMC was developed

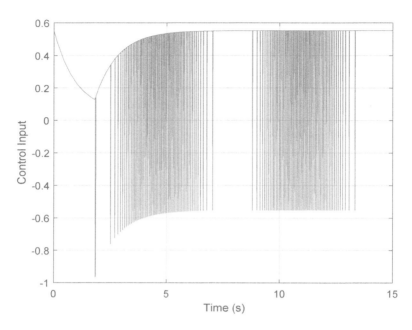

FIGURE 1.3: Response of the SMC input $u(t)$.

in [103, 160, 198]. The sliding motion of terminal SMC is based on terminal attractors [201], which guarantees the system convergence to the steady state within a finite time. For example, the following terminal sliding functions can be designed for the proposed PMSM speed regulation system (1.4):

- The terminal sliding function [18, 103, 104, 160]:

$$s(t) = x_2(t) + \beta x_1^{\frac{q}{p}}(t), \qquad (1.13)$$

where $\beta > 0$ and p, q are two odd integers satisfying $0 < \frac{q}{p} < 1$.

- In view of the slow convergence rate of terminal SMC based on the terminal sliding function (1.13) in the region far from the equilibrium point, the following fast terminal sliding function was developed [199]:

$$s(t) = x_2(t) + \alpha x_1(t) + \beta x_1^{\frac{q}{p}}(t), \qquad (1.14)$$

where $\alpha > 0$, $\beta > 0$, and p, q are two odd integers satisfying $0 < \frac{q}{p} < 1$. However, the terminal SMC schemes based on both terminal sliding function (1.13) and fast terminal sliding function (1.14) require the differential of the sliding function $s(t)$, which may result in terms $\frac{\beta q}{p} x_1^{\frac{q}{p}-1}(t)$. The singularity will happens when $x_1(t) = 0$ but $x_2(t) \neq 0$.

Sliding Mode Control

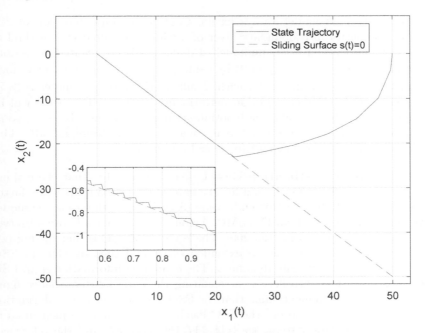

FIGURE 1.4: Response of the sliding motion in phase plane.

- To overcome the above singularity problem, the following nonsingular terminal sliding function was proposed [48]:

$$s(t) = x_1(t) + \beta x_2^{\frac{p}{q}}(t), \qquad (1.15)$$

where $\beta > 0$ and p, q are two odd integers satisfying $1 < \frac{p}{q} < 2$.

- To achieve the faster convergence rate of terminal SMC as well as avoid the singularity problem, the following nonsingular fast terminal sliding function was proposed in [189]:

$$s(t) = x_1(t) + \alpha x_1^{\frac{g}{h}}(t) + \beta x_2^{\frac{p}{q}}(t), \qquad (1.16)$$

where $\alpha > 0$, $\beta > 0$ and the odd integers p, q, g, h satisfying $1 < \frac{p}{q} < 2$ and $\frac{g}{h} > \frac{p}{q}$.

Chattering in SMC

Due to the existence of high-frequently switching of SMC input, a harmful phenomenon, i.e., *chattering* as shown in Fig. 1.3, is always occurred in the conventional SMC [156]. Chattering is considered as the main reason which hinders application of sliding mode control since dynamic mismatch between real plants and its model always exists. Continuous approximation of discontinuous control in a boundary layer of a sliding manifold is a method of chattering suppression if the unmodelled dynamics is not excited, for example, replacing sgn(s(t)) in Eq. (1.9) by using $\frac{s(t)}{\|s(t)\|+\rho}$ with a sufficiently small parameter $\rho > 0$ or $\tanh(\beta s(t))$ with a sufficiently large parameter $\beta \gg 1$. However, the disturbance rejection performance of the SMC scheme will be reduced inevitably by using the continuous approximation. To date, some other theories have been proposed to attenuating the chattering in SMC. This book will employ the following two typical approaches:

- **Disturbance-Estimation-Based SMC.** It is noted that the estimation precision of the upper-bound for the external disturbance is a key factor in using SMC/TSMC. To this end, some disturbance estimation techniques have been combined with SMC/TSMC. Among them, extended-state-observer (ESO) shows a popular one [120, 200]. The key idea of ESO is that the lumped disturbances are regarded as an extended state, and then develop an ESO based on the extended state dynamics. The estimated information from ESO will be sent to the controller for attenuating the external disturbance more effectively. It is worth mentioning that the ESO does not need any information about the disturbance. So far, the ESO has been successfully implemented in PMSM speed control system, see [116, 117, 166, 178, 187] and the references therein.

- **Higher-Order SMC.** It is widely recognized that the application of the *first-order* SMC law (i.e., conventional SMC as in [157,195] with discontinuous control input) possesses the undesirable chattering effect. In order to overcome the above shortcoming, many higher-order SMC approaches with designing continuous control input have been developed as in [73] for the purpose of reducing the chattering and improving the convergence of the sliding motion. To date, the studies of higher-order SMC have attracted considerable interest in both theoretical research [26, 74] and engineering application [94, 119, 148] of SMC theory.

Discrete-Time SMC

Because of the flexibility of implementation, a large class of continuous systems are controlled by digital signal processors (DSPs) and high end microcontrollers. To analyze the effect of sampling time, discrete-time SMC is well studied in the literature [51, 149]. In case of discrete-time SMC design, the control input is changed only at each sampling instant and the control effect is constant over the entire sampling period. Moreover, when the SMC law is based on a switching function, similar to its continuous-time counterpart, the subsequent control would generally be unable to keep the states confined to the sliding surface. As a result, the system states would approach the sliding surface but would generally be unable to stay on it, that is, switching based

Sliding Mode Control 9

discrete-time SMC can undergo only quasi-sliding mode (QSM) defined as follows.

Definition 1.5 *(Quasi-Sliding Mode, QSM). The QSM is the motion in a predefined ε vicinity of the sliding surface $s(k) = 0$ such that the system trajectory, after entering this band, never abandons it, i.e., $\|s(k)\| \leq \varepsilon$, where ε is the QSM band width.*

Over the past four decades, various discrete-time SMC algorithms have been reported in the literature for improving the disturbance rejection performance, see [6,158] and references therein. Specifically, Gao et al. in [51] elaborated the analytical characterization of an QSM, and designed a discrete-time reaching law such that the system was driven to the vicinity of the switching surface, in which the effect of sampling time to the QSM can be evaluated readily. Furthermore, Niu et al. in [110] developed a new reaching law such that a QSM can can be attained in finite time. Recently, a few research results have been reported for discrete-time higher-order SMC, see [123–125] and the references therein.

Discrete-Time SMC of 2-D Systems

Over the past few decades, two-dimensional (2-D) systems have gained considerable attention since a large number of practical systems can be modelled as 2-D systems such as those in repetitive processes, image data processing and transmission, thermal process, gas absorption, water stream heating and iterative learning control [14,52,183]. It is well known that 2-D systems can be represented by different models such as the Roesser model, Fornasini-Marchesini (FM) model, and Attasi model. So far, the stability analysis of 2-D systems described by these models has attracted a great deal of interest and some significant results have been obtained, see [3,30,38,75,121,173] and references therein.

As is well known, the 2-D system possesses more complicated dynamics than its 1-D counterpart since the former evolves along two independent directions but the latter only propagates along a single direction. By resorting to the reduced-order sliding mode dynamics via a model transformation, the problem of *first-order* SMC for 2-D systems has been addressed in [171] for Roesser model and in [191] for FM model. Following the pioneering work in [171], some preliminary results have been reported in the literature concerning the 2-D *first-order* SMC problems for discrete Roesser model in the past decade, see [2,4,192] for more details.

1.2 Network Communication Protocols

With rapid advances in communication network nowadays, more and more system information is delivered via communication networks. Accordingly, in control engineering practice, the system components (e.g. controller, sensor, actuator, and sampler) are likely to be connected by communication networks on a digital platform where the sampled signals, sensor signals, and control signals are transmitted through a shared network medium. The usage of networks in control systems gives rise to the so-called networked control systems (NCSs) that facilitate the remote execution of certain tasks such as monitoring and control. The NCSs have many merits such as low cost, reduced weight and power, and easy manipulation. Nevertheless, the implementation of communication networks between system components has largely raised the level of complexities in the analysis and synthesis of the overall NCSs due mainly to the inherent network constraints (e.g. limited bandwidth). The network-induced phenomena include, but are not limited to, packet dropouts, communication delays, delivery disorders, data collisions, network congestions and signal quantizations [16,53,203]. These phenomena, if not adequately dealt with, could lead to rather serious side effects (i.e. degradation or even destabilization) to the overall NCSs. Subsequently, in the past decade, the problems of eliminating/compensating these effects caused by network-induced phenomena have been extensively investigated, see [59,194,202] and references therein.

Nowadays, due to its excellent feature to compensate external disturbance and parameter uncertainties, the SMC problems of NCSs under network-induced complexities have also received a great deal of research attention and a rich body of literature has been available. For instance, by introducing a compensation strategy to the packet dropout, Niu & Ho in [109] addressed the SMC design problem subject to packet losses. Following this excellent work, some interesting results have been reported in the literature concerning the SMC problems subject to missing measurements, see [5,101,131,210] and the references therein. Other representative results on SMC subject to network-induced complexities can be found in [97,184,213,214] for quantization, [62,140] for randomly occurring phenomena, [89,90] for packet disorder, and [20,208] for cyber-attacks.

It should be pointed out that in all aforementioned literature, the network-induced complexities are handled in a *passive* way. Specifically, the proposed networked SMC schemes in [5,20,62,65,89,90,97–99,101,131,184,210,214] have been designed properly by considering the effects from the network-induced complexities to the transmission of control signal or measurement output. Actually, there always utilizes an *active* way in industry applications to prevent the data from collisions, that is, introducing communication protocols to the control loop for coordinating the network activities with the aim to

Network Communication Protocols 11

mitigate the traffic resulting from the limited communication resource. Unfortunately, to date, the corresponding study of networked SMC under various communication protocols has been relatively scattered despite certain limited results, such as [9] for networked *first-order* SMC under event-triggered protocols which just represents one specific communication protocol, and there still exists a huge gap towards the investigation on the networked SMC for different kinds of communication protocols, let alone the 2-D systems case with employing terminal or higher-order sliding modes.

In the sequel, we will introduce several communication protocols that have been widely applied in industry under different scheduling mechanisms.

Event-Triggered Protocols

The practical communication network always possesses the limited bandwidth constraint. Therefore, an important issue in NCSs is how to transmit signals more effectively by utilizing the available limited network bandwidth. To alleviate the unnecessary waste of communication/computation resources, a recently popular communication schedule called *event-triggered* strategy has been proposed in [34, 81, 86, 102, 152, 204]. The main idea of an event-triggered scheme is to transmit certain information only when a predesigned event happens. In comparison with the conventional time-triggered communication, a notable advantage of the event-triggering scheme is its capability of saving communication resource while preserving the guaranteed system performance. In view of this, particular interest has been poured to various event-triggered control problems, and quite a few results have been published for developing many improved event-triggered protocols with respect to the static one under the purpose of enhancing the scheduling performance as well as for constructing some new control strategies by incorporating the event-triggered mechanism with some existing control approaches.

• **Dynamic Event-Triggered Protocol.** Recently, a *dynamic* event-triggering mechanism was proposed in [54] by introducing an additional internal dynamical variable, which will adaptively adjust the event-triggering condition and thus may reduce the triggering times significantly as compared to the static event-triggering protocols. Following this excellent result, the dynamic event-triggering mechanism has been recently extended to multi-agent system [190], discrete-time system [64], and complex dynamical networks [78].

• **Periodic Event-Triggered Protocol.** It should be noted that for the *continuous*-time systems, the classical event-triggered protocol itself has inherent weakness in practical applications since the continuous monitoring of system states is required for determining the event-triggered condition [206]. To address the drawback of the classical event-triggered scheme, various schemes have been proposed, such as, periodic event-triggered scheme [56],

12 *Introduction*

self-triggered scheme [57,175], and co-design approach [106]. In periodic event-triggered method, the continuous state measurement is no longer required due to the periodic evaluation of the triggering rule. This just means that the triggering time is always an integral multiple of the sampling period which avoids the Zeno phenomenon. Recently, the control and filtering issues under *periodic event-triggered protocol* have gained increasingly research interesting, see [150, 188] for example. It is worth stressing that the main challenge in designing periodic event-triggered protocol is how to determine a proper periodic sampling period with guaranteeing the satisfied performances.

● **Networked SMC Under Event-Triggered Protocol.** It has been widely recognized that event-triggered protocol is one of the effective solutions via generating some specific events to determine whether transmitting the information in a sampling instant. By incorporating the event-triggered mechanism into SMC, an insightful idea, namely, *event-triggered SMC* with static event-triggering condition has been reported in [9,45,66,88,132,172] for first-order sliding mode, in [28] for second-order sliding mode, and in [47] for terminal sliding mode. More recently, by considering different event-triggering conditions, various event-triggered SMC techniques have been reported in the existing literature, see [19, 133] for dynamic event-triggered SMC, [76] for periodic event-triggered SMC, and [145] for self-triggered SMC. It should be pointed out that the reachability of the sliding surface will be affected by introducing event-triggered mechanism into SMC, which can be found in [9, 28, 88, 132] clearly that only the *practical* sliding motion but not the *ideal* one would be attained by the proposed event-triggered SMC. Besides, for the continuous-time systems, the analysis of excluding Zeno phenomena also is one of the mainly difficulties in designing event-triggered SMC.

Stochastic Communication Protocol (SCP)

In the context of SCP implementation, the Markov chain has been employed in [35,69] to characterize the SCP by considering the "random switch" behaviour of the node scheduling procedure. Following the pioneering results in [35], many researchers have investigated the system analysis and control synthesis problems subject to the SCP modelled as a discrete-time Markov chain, see [12, 92, 163, 217], where the zero-order holders (ZOHs) have been utilized to keep the received values unchanged until the renewed data is received.

Round-Robin Protocol (RRP)

In industry applications, the Round-Robin protocol, also known as the time-division multiple access (TDMA) protocol or Token Ring protocol, has been widely employed in communication and signal processing communities [25,72].

Network Communication Protocols

At each transmission instant, the Round-Robin protocol determines whether or not a node gets the access depends on a predetermined circular order (i.e., a periodic sequence). Up to now, by examining the effect from the "periodic assignment" behaviour of the Round-Robin protocol, some preliminary results have been reported in the literature concerning the robust control and state estimation problems, see [58, 85, 93, 153, 179, 218, 220] for more details.

Weighted Try-Once-Discard Protocol (WTODP)

The so-called WTOD protocol is categorized to the sort of quadratic protocols. The key idea of WTOD protocol is to utilize dynamic quadratic functions to select most-needed-devices for accessing the resource which, in turn, seamlessly incorporates the corresponding scheduling behaviours into Lyapunov-theory-based designs [151, 161]. It has been shown in [24] that the WTOD protocol could be implemented directly in wireless networks. So far, some pioneering results have recently been reported on WTOD-protocol-based control problems [36, 91, 221] and filtering problems [169].

Redundant Channel Transmission Protocol (RCTP)

As a *proactive* way to face packet dropouts, the RCTP has been widely utilized in IEC 62439-3-based industrial Ethernet [22] and some industrial systems and critical infrastructures such as power systems [212]. The key idea in RCTP is that if the primary channel suffers certain communication failure, which can be detected by means of software or hardware devices, other channels (i.e. redundant channels) will be automatically activated to protect the key data and thus improve the reliability of the communication network in a great sense. In the recent years, some particular research interests have been focused on the control/filtering under RCTP, see [107, 146, 209, 216] for example.

Multiple-Packet Transmission Protocol (MPTP)

In distributed NCSs (such as the systems based on DeviceNet), the MPTP is often introduced to transmit individual sensor or controller data in separate network packets for overcoming the packet size constraint or improving the reliability of communication network as opposed to a single-packet transmission policy. Obviously, for the NCSs under MPTP, the multiple separate packets may not arrive at the destination simultaneously due to network-induced complexities. Some efforts have been made in [63, 174, 215] to address the stabilization issues of NCSs under MPTP, in which the random packet dropout to MPTP was concerned. Furthermore, [82] investigated the sliding-mode predictive control problem for a class of NCSs under MPTP with packet losses.

14   *Introduction*

It is noteworthy that in wireless communication networks, the channel fading phenomenon is occurred inevitably due mainly to the physical fluctuation of wireless signals. In fact, the packet dropout can be treated as a special case of channel fading [23, 32, 44, 136]. Nonetheless, the networked SMC problem for NCSs under MPTP with channel fading remains an open yet interesting research topic.

1.3   Outline

As depicted in Fig. 1.5, the organization structure of this book contains two parts. In the first part, Chapters 2–9 focus on the networked SMC issues for uncertain 1-D systems under various communication protocols, in which Chapter 9 investigates periodic event-triggered terminal sliding mode speed control problem for networked PMSM system and a practical experiment is conducted to illustrated the proposed results. In the second part, Chapter 10–11 study the networked 2-D SMC issues for uncertain Roesser model under novel 2-D event-triggered and Round-Robin protocols, respectively. It is worth mentioning that a novel 2-D *second-order* SMC scheme is proposed in Chapter 11 by following the idea of discrete 1-D super-twisting-like algorithm.

The outline of this book is listed as follows:

- In Chapter 1, the research background is firstly introduced, which mainly involves the SMC, 1-D, and 2-D SMC cases, various network communication protocols, networked SMC. Then, the outline of the book is listed.

- **Part I focuses on the networked SMC issues for uncertain 1-D systems under various communication protocols. (Chapter 2– Chapter 9)**

    - Chapter 2 is concerned with the SMC problem for a class of uncertain discrete-time systems subject to unmatched external disturbances and communication constraints. In order to reduce the bandwidth usage between the controller and the actuators, the SCP is utilized to determine which actuator should be given the access to the network at a certain instant. A key issue of the addressed problem is to design both the sliding surface and the sliding mode controller under the SCP scheduling. An updating rule on actuator input is first introduced and then a token-dependent SMC law is designed. Sufficient conditions are established for the resultant SMC systems such that not only the reachability with a sliding domain around the specified sliding surface is ensured but also the stochastic stability with a prescribed H_∞ performance level is guaranteed. Based on these conditions, a set of coupled matrix inequalities is given to acquire the

Outline

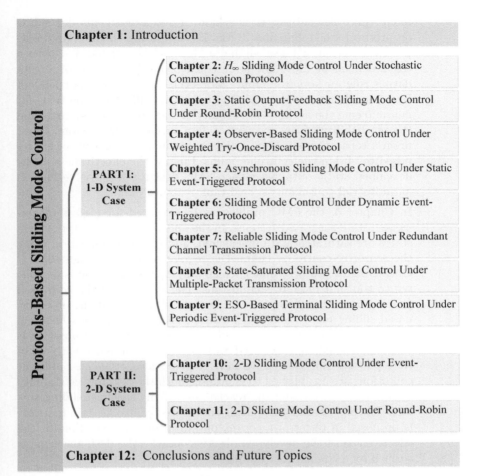

FIGURE 1.5: The organization structure of this book.

token-dependent parameter matrices in the proposed SMC law. Finally, a numerical example is presented to illustrate the effectiveness of the proposed H_∞ SMC scheme under the SCP scheduling.

— Chapter 3 addresses the static output-feedback SMC problem for a class of uncertain control systems subject to the Round-Robin protocol scheduling, in which the communication between the controller and the actuators is regulated by the Round-Robin protocol, that is, only one actuator node gets the access to the transmission network at each instant and the other actuators utilize the values stored in the zero-order holders (ZOHs). A key issue of the addressed problem is how to design both the sliding surface and the sliding mode controller subject to the kind of actuator signals depending on protocol scheduling and ZOHs. By only using the measured output information, a linear sliding surface is constructed and then a token-dependent SMC law is designed properly with aid of an updating rule on actuator input. Moreover, suitable token-dependent Lyapunov functions are exploited to obtain sufficient conditions for guaranteeing the asymptotic stability of the closed-loop systems and the reachability of the specified sliding surface. Furthermore, based upon a separation strategy, a convex optimization algorithm is proposed to obtain the controller gains. Finally, the effectiveness of the developed static output-feedback SMC design scheme is verified in an industrial continuous-stirred tank reactor system.

- In Chapter 4, the SMC problem is investigated for a class of networked state-saturated systems. In the underlying system, the system states are saturated and unmeasured, and the communication along the sensor-to-controller channel is regulated by the Weighted Try-Once-Discard protocol under which only one sensor node is permitted to transmit data at each instant. The key issue of the addressed problem is how to design both the state observer and the observer-based sliding mode controller in the presence of protocol scheduling and state saturation. A token-dependent state-saturated observer is first constructed, and a desired SMC law is then designed with aid of the updating rule for the actually received measurement. Sufficient conditions are established to guarantee both the asymptotic stability of the SMC system and the reachability of the specified sliding surface. The algorithm for acquiring the desired controller gain matrices is proposed in terms of a convex optimization problem. Finally, the effectiveness of the developed observer-based SMC design scheme is verified by a simulation example.

- In Chapter 5, the asynchronous SMC problem is investigated for networked Markovian jump Lur'e systems, in which the information of system modes is unavailable to the sliding mode controller but could be estimated by a mode detector via a hidden Markov model. In order to mitigate the burden of data communication, an event-triggered protocol is proposed to determine whether the system state should be released to the controller at certain time-point according to a specific triggering condition. By constructing a novel *common* sliding surface, this chapter designs an event-triggered asynchronous SMC law, which just depends on the hidden mode information. A combination of the stochastic Lur'e-type Lyapunov functional and the hidden Markov model approach is exploited to establish the sufficient conditions of the mean square stability with a prescribed H_∞ performance and the reachability of a sliding region around the specified sliding surface. Moreover, the solving algorithm for the control gain matrices is given

Outline 17

via a convex optimization problem. Finally, an example from the DC Motor device system is provided.

- Chapter 6 endeavours to investigate the SMC issue of the networked singularly perturbed systems (SPSs) under slow sampling. For the energy saving purpose in network communication, a *dynamic* event-triggering mechanism is introduced to SMC design. By considering the structure characteristics of the controlled system, a novel sliding function is constructed with taking the singular perturbed matrix E_ε into account properly. With the aid of some appropriate Lyapunov functionals, the sufficient conditions are derived to ensure the asymptotic stability of the sliding mode dynamics and the reachability of the specified sliding surface. Besides, it is shown that the quasi-sliding motion is dependent on the dynamic event-triggering parameters and its bound converges to a constant. Moreover, a convex optimization algorithm is formulated to solve the dynamic event-triggering SMC law with searching the upper bound of the singularly perturbed parameter. Finally, an numerical example is provided to illustrate the effectiveness of the proposed results.

- Chapter 7 endeavours to investigate the output-feedback SMC issue of the networked SPSs under fast sampling. In order to improve the reliability of the network communication, a redundant channel transmission protocol is introduced in the SMC design. Based on the measurement outputs, a sliding function is constructed with the consideration of the transmission protocol. With the aid of some appropriate Lyapunov functions, the sufficient conditions are derived to ensure the mean-square exponentially ultimately boundedness of the sliding mode dynamics and the reachability of the specified sliding surface. Moreover, a convex optimization algorithm is formulated to solve the output-feedback SMC law via searching the available upper bound of the singularly perturbed parameter. Finally, an operational amplifier circuit is exploited to explore the influences from the redundant channel transmission protocol to the output-feedback SMC performance and the estimated ε-bound.

- In Chapter 8, the SMC problem is studied for state-saturated systems under multiple-packet transmission protocol, in which a generalized *time-varying* fading channel is concerned. The underlying fading channels, whose channel fading amplitudes (characterized by the expectation and variance) are allowed to be different, are modelled as a finite-state Markov process. A key feature of the problem addressed is to use a hidden Markov mode detector to estimate the actual network mode. The novel model of *hidden Markov fading channels* is shown to be more general yet practical than the existing fading channel models. Based on a linear sliding surface, a switching-type SMC law is dedicatedly constructed by just using the estimated network mode.

By exploiting the concept of stochastic Lyapunov stability and the approach of hidden Markov models, sufficient conditions are obtained for the resultant SMC systems that ensure both the mean-square stability and the reachability with a sliding region. With the aid of the Hadamard product, a binary genetic algorithm (GA) is developed to solve the proposed SMC design problem subject to some nonconvex constraints induced by the state saturations and the fading channels, where the proposed GA is based on the objective function for optimal reachability. Finally, a numerical example is employed to verify the proposed GA-assisted SMC scheme under multiple-packet transmission protocol with the hidden Markov fading channels.

- Chapter 9 proposes a novel periodic event-triggered terminal sliding mode speed control scheme for networked PMSM. The novel speed control strategy combines advantages of both TSMC and periodic event-triggered mechanism. TSMC guarantees the satisfied speed tracking performance with compensating the external disturbances and parameter uncertainties. The communication burden between the sensors and the controller is reduced by periodic event-triggered protocol. In order to improve the tracking performance, an ESO is introduced to estimate all possible perturbations, in which the upperbound of the estimation error is analyzed explicitly. Then, a binary-based GA is adopted to find the optimal observer parameters. A key challenge here is how to estimate actual bound of the triggering error under the periodic event-triggered mechanism. To this end, an explicit selection criterion of the periodic sampling period is developed by utilizing the known constrained information of PMSM. It is proven that the proposed novel periodic event-triggered TSMC with a proper selection of the control gain can guarantee the reachability of sliding variable to a sliding region and the ultimate boundedness of the tracking errors simultaneously. Finally, the effectiveness of the proposed novel periodic event-triggered TSMC via GA-optimized ESO is demonstrated in both simulation and experiment results for a real PMSM platform.

- **Part II studies the networked 2-D SMC issues for uncertain Roesser model under event-triggered and Round-Robin protocols. (Chapter 10–11)**

  - In Chapter 10, the co-design problem of 2-D event generators and SMC is studied for the discrete-time 2-D Roesser model for reducing the communication usage between the plant and the controller. First, by resorting to the 2-D reaching law conditions, the event-based 2-D SMC schemes are proposed and a kind of horizontal/vertical independent event generators is designed. Then, the stability of the sliding mode dynamics is analyzed and a *nonlinear* matrix inequality condition associating with the horizontal/vertical sliding matrices is

Outline 19

obtained. In order to solve the derived *nonlinear* matrix inequality condition without introducing conservatism, a binary GA is applied by considering a multi-objective optimization problem, which reflects the trade-off between the convergence of the sliding mode dynamics and the scheduling performance of the designed horizontal/vertical event generators. Finally, a simulation example is exploited to demonstrate the effectiveness of the proposed co-design approach with GA.

– Chapter 11 investigates the SMC problem for a class of uncertain 2-D systems described by the Roesser models with a bounded disturbance. In order to reduce the communication usage between the controller and the actuators, it is supposed that only one actuator node can gain the access to the network at each sampling time along horizontal or vertical direction, where a proper 2-D Round-Robin protocol is designed to periodically regulate the access token and a set of ZOHs is employed to keep the other actuator nodes unchanged until the next renewed signal arrives. Based on a novel 2-D common sliding function, a token-dependent 2-D SMC scheme with *first-order sliding mode* is appropriately constructed to cope with the impacts from the periodic scheduling signal and the ZOHs. Furthermore, a novel super-twisting-like 2-D SMC scheme with *second-order sliding mode* is designed to improve the robustness against the bounded disturbance. By resorting to token-dependent Lyapunov-like function, sufficient conditions are obtained to guarantee the ultimate boundedness of the horizontal and vertical states as well as the 2-D common sliding function. For acquiring the optimized gain matrices, two searching algorithms are formulated to solve two optimization problems arising from finding optimized control performance. Finally, two comparative examples are exploited to demonstrate the effectiveness and the advantageous of the proposed first- and second-order 2-D SMC design schemes under Round-Robin scheduling mechanism.

• Chapter 12 gives the conclusions of this book and then discusses some potential directions for future research work.

PART I: 1-D System Case

2

H_∞ Sliding Mode Control Under Stochastic Communication Protocol

In this chapter, the sliding mode control (SMC) problem is investigated for a class of uncertain discrete-time systems with unmatched external disturbance and stochastic communication protocol (SCP) scheduling. At each transmission instant, only one actuator is allowed to obtain access to the communication network, and the SCP modelled as a discrete-time Markov chain is applied to determine which actuator should gain the access to the network at a certain instant. The zero-order holders (ZOHs) are utilized to keep the received values unchanged until the renewed data is received. By taking the impacts from the SCP scheduling into account, a suitable SMC law is constructed and the corresponding design approach is then established in terms of a convex optimization problem. *The main contributions of this chapter are highlighted as follows: 1) the SCP protocol is applied to reduce the bandwidth usage between the sliding mode controller and the actuators; 2) a token-dependent SMC law is designed to cope with the Markovian jumping scheduling and the reachability of the sliding mode dynamics is proven by using a new token-dependent stochastic Lyapunov function; 3) to handle the effects from the ZOHs, some time-varying bounds are specified in the SMC scheme such that the closed-loop system is stochastically stable with a given H_∞ disturbance attenuation level; and 4) a set of coupled linear matrix inequalities (LMIs) is established to compute the token-dependent parameter matrices in SMC law.*

2.1 Problem Formulation

Consider the following uncertain discrete-time systems:

$$\begin{cases} x(k+1) &= (A + \Delta A(k)) x(k) + Gw(k) + B \left(u(k) + f(x(k), k) \right) \\ z(k) &= Cx(k) + Lw(k) \end{cases} \tag{2.1}$$

where $x(k) \in \mathbb{R}^n$ is the state vector, $u(k) \in \mathbb{R}^m$ is the control signal, $z(k) \in \mathbb{R}^p$ is the controlled output, $f(x(k), k)$ is the unknown nonlinear function, and $w(k) \in \mathbb{R}^l$ is the *unmatched* external disturbance. A, B, C, L, and G are known constant matrices. Without loss of generality, it is assumed that the matrix

DOI: 10.1201/9781003309499-2

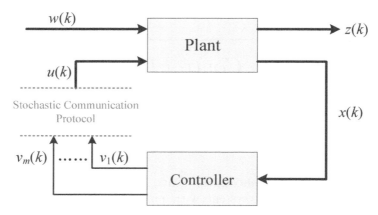

FIGURE 2.1: The control systems under SCP scheduling.

B has full column rank. The real-valued matrix $\Delta A(k)$ represents the norm-bounded parameter uncertainty in the form of $\Delta A(k) = EF(k)H$ with E and H known constant matrices and $F(k)$ an unknown matrix function satisfying $F^{\mathrm{T}}(k)F(k) \leq I$. The nonlinear function $f(x(k),k)$ satisfies $\|f(x(k),k)\| \leq \varepsilon \|x(k)\|$, where $\varepsilon > 0$ is a known scalar. The external disturbance $w(k)$ belongs to $\ell_2[0,\infty)$ and its peak value satisfies $\|w(k)\| \leq \bar{w}$, $\forall k \geq 0$, where $\bar{w} \geq 0$ is a known scalar.

In this chapter, the control signal $v(k)$ is transmitted to the actuators $u(k)$ via a shared communication channel as shown in Fig. 2.1. Moreover, the SCP scheduling is adopted for mitigating the network burden and reducing the risk of data collision in the controller-to-actuator (C/A) network. Without loss of generality, it is assumed that, at each instant, *only one actuator* obtains access to the communication network.

Let $\sigma(k) \in \mathcal{M} \triangleq \{1, 2, \ldots, m\}$ denote the selected actuator obtaining the *access token* to the C/A network at the time instant k. Under the SCP scheduling, the variable $\sigma(k)$ can be governed by a Markov chain with the transition probability matrix $\Pi = [\pi_{ij}]_{m \times m}$ where the transition probability π_{ij} is defined by

$$\pi_{ij} \triangleq \Pr\{\sigma(k+1) = j \mid \sigma(k) = i\}, \qquad (2.2)$$

with $0 \leq \pi_{ij} \leq 1$ for all $i, j \in \mathcal{M}$ and $\sum_{j=1}^{m} \pi_{ij} = 1$.

For technical analysis, we first write $v(k) \triangleq \begin{bmatrix} v_1^T(k) & v_2^T(k) & \cdots & v_m^T(k) \end{bmatrix}^T \in \mathbb{R}^m$ and $u(k) \triangleq \begin{bmatrix} u_1^{\mathrm{T}}(k) & u_2^{\mathrm{T}}(k) & \cdots & u_m^{\mathrm{T}}(k) \end{bmatrix}^{\mathrm{T}} \in \mathbb{R}^m$, where $v_i(k)$ is the ith control signal and $u_j(k)$ is the received signal by the jth actuator. In the C/A network, a set of ZOHs is used to store the received values at the actuators' end. Thus, the updating rule for the ith actuator $u_i(k)$ ($i \in \mathcal{M}$) is set

Main Results 25

to be

$$u_i(k) \;=\; \left\{ \begin{array}{ll} v_i(k), & \text{if } \sigma(k) = i, \\ u_i(k-1), & \text{otherwise.} \end{array} \right. \tag{2.3}$$

The updating rule (2.3) implies that, at the transmission instant k, the ith actuator is updated *if and only if* the SCP scheduling signal $\sigma(k) = i$; otherwise, the ith actuator is unchanged and the last value $u_i(k-1)$ stored by ZOH will be used. Therefore, the *actual* control input $u(k)$ at the plant side can be expressed as

$$u(k) = \Phi_{\sigma(k)} v(k) + \left(I - \Phi_{\sigma(k)} \right) u(k-1), \tag{2.4}$$

where $\Phi_{\sigma(k)} \triangleq \text{diag} \left\{ \tilde{\delta}^1_{\sigma(k)}, \tilde{\delta}^2_{\sigma(k)}, \cdots, \tilde{\delta}^m_{\sigma(k)} \right\}$, $\tilde{\delta}^b_a \triangleq \delta(a-b)$ with $\delta(\cdot)$ being the Kronecker delta function. Without loss of generality, we set $u(-1) = 0$.

The main objective of this chapter is to design a sliding mode controller $v(k)$ such that the uncertain system (2.1) under SCP scheduling is stochastically stable with a specified H_∞ performance. It is noted that there are two factors that complicate the SMC design problem due mainly to the utilization of the SCP scheduling scheme (2.2)–(2.4): one is the Markovian jumping scheduling mode $\sigma(k)$ and the other is the input component $u(k-1)$ caused by the use of ZOHs. To this end, the SMC law (2.9) will be designed later with some token-dependent parameter matrices $K_{\sigma(k)}$, which are the solutions of some coupled LMIs concerning the transition probabilities π_{ij} of scheduling modes.

2.2 Main Results

First, let us choose the following sliding function:

$$s(k) = B^{\mathrm{T}} X x(k) - B^{\mathrm{T}} X A x(k-1), \tag{2.5}$$

where $X > 0$ is a given matrix. Clearly, the non-singularity of $B^{\mathrm{T}} X B$ is ensured due to the full column rank of B.

By means of (2.1) and (2.4), we can obtain from (2.5):

$$\begin{aligned} s(k+1) =& B^{\mathrm{T}} X G w(k) + D_1(k) + D_2(k) + B^{\mathrm{T}} X B \Phi_{\sigma(k)} v(k) \\ & + B^{\mathrm{T}} X B \left(I - \Phi_{\sigma(k)} \right) u(k-1), \end{aligned} \tag{2.6}$$

where $D_1(k) \triangleq B^{\mathrm{T}} X \Delta A x(k)$ and $D_2(k) \triangleq B^{\mathrm{T}} X B f(x(k), k)$.

Assumption 2.1 *There exist known bounds $\underline{d}^i_1$, $\overline{d}^i_1$, $\underline{d}^i_2$, and $\overline{d}^i_2$ satisfying*

$$\underline{d}^i_1 \le d^i_1(k) \le \overline{d}^i_1, \quad \underline{d}^i_2 \le d^i_2(k) \le \overline{d}^i_2, \tag{2.7}$$

where $d_1^i(k)$ and $d_2^i(k)$ are the ith element in vectors $D_1(k)$ and $D_2(k)$, respectively.

Based on Assumption 2.1, we define $d_{1o}^i \triangleq \frac{\overline{d_1^i}+\underline{d_1^i}}{2}$, $d_{1s}^i \triangleq \frac{\overline{d_1^i}-\underline{d_1^i}}{2}$, $d_{2o}^i \triangleq \frac{\overline{d_2^i}+\underline{d_2^i}}{2}$, $d_{2s}^i \triangleq \frac{\overline{d_2^i}-\underline{d_2^i}}{2}$, $D_{1o} \triangleq \left[d_{1o}^1, d_{1o}^2, \ldots, d_{1o}^m\right]^{\mathrm{T}}$, $D_{1s} \triangleq \mathrm{diag}\{d_{1s}^1, d_{1s}^2, \ldots, d_{1s}^m\}$, $D_{2o} \triangleq \left[d_{2o}^1, d_{2o}^2, \ldots, d_{2o}^m\right]^{\mathrm{T}}$, and $D_{2s} \triangleq \mathrm{diag}\{d_{2s}^1, d_{2s}^2, \ldots, d_{2s}^m\}$. The definitions of $D_{1o}, D_{2o}, D_{1s}, D_{2s}$, and their bounds $\underline{d_1^i}, \overline{d_1^i}, \underline{d_2^i}, \overline{d_2^i}$ are standard in literature concerning discrete-time SMC, see e.g. [51, 62, 101, 109, 110]. These bounds, which are allowed to be *time-varying*, are obviously dependent on the norm-bounded uncertainty $\Delta A(k)$ and the nonlinear function $f(x(k), k)$.

Furthermore, denoting $D(k) \triangleq D_1(k) + D_2(k)$, $D_o \triangleq D_{1o} + D_{2o}$ and $D_s \triangleq D_{1s} + D_{2s}$, we have

$$\|D(k) - D_o - D_s\mathrm{sgn}(s(k))\| \leq 2\|D_s\|. \tag{2.8}$$

Now, a token-dependent SMC law $v(k)$ is designed as follows:

$$v(k) = -\left(B^{\mathrm{T}}XB\right)^{-1}\left[K_{\sigma(k)}x(k) + D_o + D_s\mathrm{sgn}(s(k))\right], \tag{2.9}$$

where the matrices $K_{\sigma(k)} \in \mathbb{R}^{m \times n}$ $(\sigma(k) \in \mathcal{M})$ will be designed later. Clearly, the effect of Markovian jumping scheduling from one actuator to another is reflected in the SMC law $v(k)$ just via the token-dependent matrices $K_{\sigma(k)}$.

With (2.1), (2.4), and (2.9), the resulting closed-loop system is obtained as:

$$\begin{aligned}
x(k+1) =& \left[A - B\Phi_{\sigma(k)}(B^{\mathrm{T}}XB)^{-1}K_{\sigma(k)} + \tilde{X}_{\sigma(k)}\Delta A(k)\right]x(k) + Gw(k) \\
&+ B(I - \Phi_{\sigma(k)})u(k-1) + B\left(I - \Phi_{\sigma(k)}\right)f(x(k), k) \\
&+ B\Phi_{\sigma(k)}(B^{\mathrm{T}}XB)^{-1}\left[D(k) - D_o - D_s\mathrm{sgn}(s(k))\right], \tag{2.10}
\end{aligned}$$

where $\tilde{X}_{\sigma(k)} \triangleq I - B\Phi_{\sigma(k)}(B^{\mathrm{T}}XB)^{-1}B^{\mathrm{T}}X$.

Definition 2.1 *The closed-loop system (2.10) with $w(k) = 0$ is said to be stochastically stable if, for every initial condition $\{x(0) \neq 0, u(-1) = 0, \sigma(0) \in \mathcal{M}\}$, the relationship $\mathbb{E}\left\{\sum_{k=0}^{\infty}\|x(k)\|^2 \mid x(0) \neq 0, u(-1) = 0, \sigma(0) \in \mathcal{M}\right\} < \infty$ holds.*

Definition 2.2 *Given a scalar $\gamma > 0$, the closed-loop system (2.10) is said to be stochastically stable with the H_∞ disturbance attenuation level γ if, under the zero initial conditions $x(0) = 0$ and $u(-1) = 0$, the condition $\sum_{k=0}^{\infty}\mathbb{E}\left\{\|z(k)\|^2\right\} < \gamma^2\sum_{k=0}^{\infty}\|w(k)\|^2$ holds for all nonzero $w(k) \in \ell_2[0, \infty)$.*

In the sequel, both of the reachability and the stochastic stability issues will be analyzed.

Main Results 27

2.2.1 Analysis of reachability

Substituting (2.9) into (2.6), we have

$$
\begin{aligned}
s(k+1) =& B^{\mathrm{T}}XB\left(I - \Phi_{\sigma(k)}\right)u(k-1) + \left[\left(I - \Phi_{\sigma(k)}\right)B^{\mathrm{T}}X\Delta A(k)\right.\\
& - \Phi_{\sigma(k)}K_{\sigma(k)}]x(k) + (I - \Phi_{\sigma(k)})B^{\mathrm{T}}XBf(x(k),k) + B^{\mathrm{T}}XGw(k)\\
& + \Phi_{\sigma(k)}\left[D(k) - D_o - D_s\mathrm{sgn}(s(k))\right].
\end{aligned}
\tag{2.11}
$$

In the following, a stochastic Lyapunov function method is utilized to analyze the reachability of the sliding mode dynamics.

Theorem 2.1 *Consider the uncertain system (2.1) with the updating rule (2.4) and the token-dependent SMC law (2.9). For $i \in \mathcal{M}$, if there exist matrices $K_i \in \mathbb{R}^{m \times n}$, $P_i > 0$, $Q_i > 0$, $W_i > 0$, and scalars $\kappa_i > 0$, $\theta_i > 0$, $\mu_i > 0$, $\rho_i > 0$ satisfying the following matrix inequalities:*

$$
(I - \Phi_i)B^{\mathrm{T}}\bar{P}_iB(I - \Phi_i) \le \kappa_i I,
\tag{2.12}
$$

$$
B^{\mathrm{T}}XB(I - \Phi_i)\bar{W}_i(I - \Phi_i)B^{\mathrm{T}}XB \le \mu_i I,
\tag{2.13}
$$

$$
\Omega_i \triangleq \left[\begin{array}{cc} \Omega_i^{11} & \Omega_i^{12} \\ \star & \Omega_i^{22} \end{array}\right] < 0,
\tag{2.14}
$$

where

$$
\bar{P}_i \triangleq \sum_{j=1}^m \pi_{ij}P_j, \bar{Q}_i \triangleq \sum_{j=1}^m \pi_{ij}Q_j, \bar{W}_i \triangleq \sum_{j=1}^m \pi_{ij}W_j,
$$

$$
\begin{aligned}
\Omega_i^{11} \triangleq & -P_i + 4[A - B\Phi_i(B^{\mathrm{T}}XB)^{-1}K_i + \tilde{X}_i\Delta A(k)]^{\mathrm{T}}\bar{P}_i[A - B\Phi_i(B^{\mathrm{T}}XB)^{-1}K_i\\
& + \tilde{X}_i\Delta A(k)] + 5\varepsilon^2\kappa_i I + 5\varepsilon^2\mu_i I + 2K_i^{\mathrm{T}}(B^{\mathrm{T}}XB)^{-1}\Phi_i\bar{Q}_i\Phi_i(B^{\mathrm{T}}XB)^{-1}K_i\\
& + 4\left[(I - \Phi_i)B^{\mathrm{T}}X\Delta A(k) - \Phi_iK_i\right]^{\mathrm{T}}\bar{W}_i\left[(I - \Phi_i)B^{\mathrm{T}}X\Delta A(k) - \Phi_iK_i\right],
\end{aligned}
$$

$$
\begin{aligned}
\Omega_i^{12} \triangleq & \left[A - B\Phi_i(B^{\mathrm{T}}XB)^{-1}K_i + \tilde{X}_i\Delta A(k)\right]^{\mathrm{T}}\bar{P}_iB(I - \Phi_i) - K_i^{\mathrm{T}}(B^{\mathrm{T}}XB)^{-1}\Phi_i\\
& \times \bar{Q}_i(I - \Phi_i) + \left[(I - \Phi_i)B^{\mathrm{T}}X\Delta A(k) - \Phi_iK_i\right]^{\mathrm{T}}\bar{W}_iB^{\mathrm{T}}XB(I - \Phi_i),
\end{aligned}
$$

$$
\begin{aligned}
\Omega_i^{22} \triangleq & -Q_i + 4(I - \Phi_i)B^{\mathrm{T}}\bar{P}_iB(I - \Phi_i) + 2(I - \Phi_i)\bar{Q}_i(I - \Phi_i) + 4(I - \Phi_i)B^{\mathrm{T}}\\
& \times XB\bar{W}_iB^{\mathrm{T}}XB(I - \Phi_i),
\end{aligned}
$$

then the state trajectories of the closed-loop system (2.10) will be driven (in mean square) into the following domain Ξ around the specified sliding surface (2.5):

$$
\Xi \triangleq \left\{s(k) \;\middle|\; \|s(k)\| \le \max_{i \in \mathcal{M}}\left\{\sqrt{\frac{\Delta_i}{\lambda_{\min}(W_i)}}\right\}\right\},
\tag{2.15}
$$

where $\Delta_i \triangleq 5\|G^{\mathrm{T}}\bar{P}_iG\|\bar{w}^2 + 5\|\left(B^{\mathrm{T}}XG\right)^{\mathrm{T}}\bar{W}_i\left(B^{\mathrm{T}}XG\right)\|\bar{w}^2 + 20\|(B^{\mathrm{T}}XB)^{-1}\Phi_i \times B^{\mathrm{T}}\bar{P}_iB\Phi_i(B^{\mathrm{T}}XB)^{-1}\| \cdot \|D_s\|^2 + 20\|\Phi_i\bar{W}_i\Phi_i\| \cdot \|D_s\|^2 + 6\|(B^{\mathrm{T}}XB)^{-1}\Phi_i\bar{Q}_i\Phi_i \times (B^{\mathrm{T}}XB)^{-1}\|\left(\|D_o\|^2 + m\|D_s\|^2\right).$

Proof *Define* $\eta(k) \triangleq \begin{bmatrix} x^{\mathrm{T}}(k) & u^{\mathrm{T}}(k-1) & s^{\mathrm{T}}(k) \end{bmatrix}^{\mathrm{T}}$ *and choose the token-dependent Lyapunov function candidate as*

$$V(\eta(k), \sigma(k)) = x^{\mathrm{T}}(k)P_{\sigma(k)}x(k) + u^{\mathrm{T}}(k-1)Q_{\sigma(k)}u(k-1) + s^{\mathrm{T}}(k)W_{\sigma(k)}s(k).$$

Letting $\sigma(k) = i$ *and* $\sigma(k+1) = j$ ($i, j \in \mathcal{M}$), *we compute the conditional expectation of* $V(\eta(k+1), \sigma(k+1))$ *[139] as follows:*

$$
\begin{aligned}
&\mathbb{E}\left\{V(\eta(k+1), \sigma(k+1)) \mid \eta(k), \sigma(k)\right\} \\
=&\mathbb{E}\left\{x^{\mathrm{T}}(k+1)P_{\sigma(k+1)}x(k+1) \mid \eta(k), \sigma(k)\right\} \\
&+ \mathbb{E}\left\{u^{\mathrm{T}}(k)Q_{\sigma(k+1)}u(k) \mid \eta(k), \sigma(k)\right\} \\
&+ \mathbb{E}\left\{s^{\mathrm{T}}(k+1)W_{\sigma(k+1)}s(k+1) \mid \eta(k), \sigma(k)\right\}.
\end{aligned}
\tag{2.16}
$$

First, let us consider the term $x^{\mathrm{T}}(k+1)P_{\sigma(k+1)}x(k+1)$ *in (2.16). For* $\kappa_i > 0$ *and* $\theta_i > 0$, *it follows from (2.10) and (2.12) that*

$$
\begin{aligned}
&\mathbb{E}\left\{x^{\mathrm{T}}(k+1)P_{\sigma(k+1)}x(k+1) \mid \eta(k), \sigma(k)\right\} \\
\leq& x^{\mathrm{T}}(k)\left\{4\left[A - B\Phi_i(B^{\mathrm{T}}XB)^{-1}K_i + \tilde{X}_i\Delta A(k)\right]^{\mathrm{T}}\bar{P}_i \right. \\
&\times \left[A - B\Phi_i(B^{\mathrm{T}}XB)^{-1}K_i + \tilde{X}_i\Delta A(k)\right] \\
&\left. + 5\varepsilon^2\kappa_i I\right\}x(k) + 4u^{\mathrm{T}}(k-1)(I - \Phi_i)B^{\mathrm{T}}\bar{P}_iB(I - \Phi_i)u(k-1) \\
&+ 2x^{\mathrm{T}}(k)\left[A - B\Phi_i(B^{\mathrm{T}}XB)^{-1}K_i + \tilde{X}_i\Delta A(k)\right]^{\mathrm{T}}\bar{P}_iB(I - \Phi_i)u(k-1) \\
&+ 5\|G^{\mathrm{T}}\bar{P}_iG\|\bar{w}^2 + 20\|(B^{\mathrm{T}}XB)^{-1}\Phi_iB^{\mathrm{T}}\bar{P}_iB\Phi_i(B^{\mathrm{T}}XB)^{-1}\| \cdot \|D_s\|^2.
\end{aligned}
\tag{2.17}
$$

Secondly, let us tackle the term $u^{\mathrm{T}}(k)Q_{\sigma(k+1)}u(k)$ *in (2.16). By means of* $\|D_o + D_s\mathrm{sgn}(s(k))\|^2 \leq 2\|D_o\|^2 + 2m\|D_s\|^2$, *it obtains that*

$$
\begin{aligned}
&\mathbb{E}\left\{u^{\mathrm{T}}(k)Q_{\sigma(k+1)}u(k) \mid \eta(k), \sigma(k)\right\} \\
\leq& 2x^{\mathrm{T}}(k)K_i^{\mathrm{T}}(B^{\mathrm{T}}XB)^{-1}\Phi_i\bar{Q}_i\Phi_i(B^{\mathrm{T}}XB)^{-1}K_ix(k) \\
&+ 2u^{\mathrm{T}}(k-1)(I - \Phi_i)\bar{Q}_i(I - \Phi_i)u(k-1) \\
&- 2x^{\mathrm{T}}(k)K_i^{\mathrm{T}}(B^{\mathrm{T}}XB)^{-1}\Phi_i\bar{Q}_i(I - \Phi_i)u(k-1) \\
&+ 6\|(B^{\mathrm{T}}XB)^{-1}\Phi_i\bar{Q}_i\Phi_i(B^{\mathrm{T}}XB)^{-1}\| \left(\|D_o\|^2 + m\|D_s\|^2\right).
\end{aligned}
\tag{2.18}
$$

Next, let us focus on the third term $s^{\mathrm{T}}(k+1)W_{\sigma(k+1)}s(k+1)$ *in (2.16).*

Main Results 29

From (2.13), it gets

$$\mathbb{E}\left\{ s^{\mathrm{T}}(k+1) W_{\sigma(k+1)} s(k+1) \mid \eta(k), \sigma(k) \right\}$$

$$\leq x^{\mathrm{T}}(k) \Big\{ 4 \left[(I - \Phi_i) B^{\mathrm{T}} X \Delta A(k) - \Phi_i K_i \right]^{\mathrm{T}} \bar{W}_i \left[(I - \Phi_i) B^{\mathrm{T}} X \Delta A(k) - \Phi_i K_i \right]$$

$$+ 5\varepsilon^2 \mu_i I \Big\} x(k) + 4u^{\mathrm{T}}(k-1)(I - \Phi_i) B^{\mathrm{T}} X B \bar{W}_i B^{\mathrm{T}} X B (I - \Phi_i) u(k-1)$$

$$+ 2x^{\mathrm{T}}(k) \left[(I - \Phi_i) B^{\mathrm{T}} X \Delta A(k) - \Phi_i K_i \right]^{\mathrm{T}} \bar{W}_i B^{\mathrm{T}} X B (I - \Phi_i) u(k-1)$$

$$+ 20 \|\Phi_i \bar{W}_i \Phi_i\| \cdot \|D_s\|^2 + 5 \| \left(B^{\mathrm{T}} X G \right)^{\mathrm{T}} \bar{W}_i \left(B^{\mathrm{T}} X G \right) \| \bar{w}^2. \tag{2.19}$$

Combining (2.17), (2.18), and (2.19) together, we have

$$\mathbb{E}\left\{ V(\eta(k+1), \sigma(k+1)) \mid \eta(k), \sigma(k) \right\} - V(\eta(k), \sigma(k))$$

$$\leq \left[\begin{array}{c} x(k) \\ u(k-1) \end{array} \right]^{\mathrm{T}} \Omega_i \left[\begin{array}{c} x(k) \\ u(k-1) \end{array} \right] - s^{\mathrm{T}}(k) W_i s(k) + \Delta_i. \tag{2.20}$$

Notice that, if the state trajectories escape from the band Ξ around the specified sliding surface (2.5), that is, $\|s(k)\| \geq \max_{i \in \mathcal{M}} \left\{ \sqrt{\frac{\Delta_i}{\lambda_{\min}(W_i)}} \right\}$, we will have for any $i \in \mathcal{M}$, $\Delta_i \leq \lambda_{\min}(W_i) s^{\mathrm{T}}(k) s(k)$, which implies $-s^{\mathrm{T}}(k) W_i s(k) + \Delta_i \leq 0$. Therefore, for all $x(k) \neq 0$, it follows from the condition (2.14) that $\mathbb{E}\left\{ V(\eta(k+1), \sigma(k+1)) \mid \eta(k), \sigma(k) \right\} - V(\eta(k), \sigma(k)) \leq$

$$\left[\begin{array}{c} x(k) \\ u(k-1) \end{array} \right]^{\mathrm{T}} \Omega_i \left[\begin{array}{c} x(k) \\ u(k-1) \end{array} \right] < 0, \text{ which means that the state trajectories of}$$

the closed-loop system (2.10) will be driven (in mean square) into the band Ξ around the sliding surface (2.5) by the SMC law (2.9).

Notice that Theorem 2.1 ensures that the state trajectories enter (in mean square) into the band Ξ around the specified sliding surface (2.5). In the following subsection, it will be shown that the sliding motion within the sliding domain Ξ is stochastically stable.

2.2.2 Analysis of stochastic stability with H_∞ performance

In this subsection, we shall firstly analyze the stochastic stability of the closed-loop system (2.10) under $w(k) = 0$. Notice that, due to the usage of SCP (2.4) with a set of ZOHs, there exists an input term $u(k-1)$ in the expression (2.10). This implies that, under the stochastic stability, both the state $x(k)$ and the input term $u(k-1)$ should converge to zero in mean square. Obviously, the existing discrete-time SMC approaches (as in [51, 110]) cannot attain such convergence because of the nonzero constant coefficients of switching terms, that is, $D_s \neq 0$ in (2.9). In order to solve this problem, we are going to design a time-varying D_s in the following lemma such that not only the relationships in (2.7) are ensured but also $D_s \to 0$ holds in mean square.

Lemma 2.1 *The bounds $\{\underline{d_n^i}\}$ and $\{\overline{d_n^i}\}$ $(i = 1, 2, \ldots, m;\ n = 1, 2)$ in (2.7) can be specified as:*

$$\overline{d_1^i} = -\underline{d_1^i} = \alpha\|x(k)\|,\ \overline{d_2^i} = -\underline{d_2^i} = \epsilon\|x(k)\|, \tag{2.21}$$

with $\alpha \triangleq \|B^T X E\| \cdot \|H\|$, $\epsilon \triangleq \varepsilon\|B^T X B\|$.

Proof *According to the definition of Euclidean vector norm, it is easily found that for any $i \in \mathcal{M}$, one has $\left(d_1^i(k)\right)^2 \leq \sum_{i=1}^m \left(d_1^i(k)\right)^2 = \|D_1(k)\|^2$ and $\left(d_2^i(k)\right)^2 \leq \sum_{i=1}^m \left(d_2^i(k)\right)^2 = \|D_2(k)\|^2$. Clearly, the relationships in (2.7) are satisfied for the time-varying bounds in (2.21), which completes the proof.*

Based on the above lemma, we rewrite D_o and D_s as $D_o = 0$, $D_s = (\alpha + \epsilon)\|x(k)\|I$. Then, the SMC law (2.9) is further rewritten as

$$v(k) = -\left(B^T X B\right)^{-1}\left[K_{\sigma(k)}x(k) + (\alpha + \epsilon)\|x(k)\| \cdot \text{sgn}(s(k))\right]. \tag{2.22}$$

It is easily observed that, if $x(k) \to 0$ in mean square, then we have $D_s \to 0$ and the control signal $v(k) \to 0$, which implies that, under the SCP scheduling mechanism (2.4), the input term $u(k-1)$ in (2.10) could also converge to zero along with $x(k) \to 0$, and thus the stochastic stability of the closed-loop system (2.10) is attained by the SMC law (2.22). In the following theorem, a sufficient condition for guaranteeing the stochastic stability with H_∞ performance of the closed-loop system (2.10) is obtained.

Theorem 2.2 *Consider the system (2.1) with the SCP (2.4) and the token-dependent SMC law (2.22). For a given $\gamma > 0$, if there exist matrices $K_i \in \mathbb{R}^{m \times n}$, $P_i > 0$, $Q_i > 0$, and scalars $\kappa_i > 0$, $\omega_i > 0$, $\nu_i > 0$ satisfying the condition (2.12) and the following matrix inequalities:*

$$(B^T X B)^{-1}\Phi_i B^T \bar{P}_i B \Phi_i (B^T X B)^{-1} \leq \omega_i I, \tag{2.23}$$

$$(B^T X B)^{-1}\Phi_i \bar{Q}_i \Phi_i (B^T X B)^{-1} \leq \nu_i I, \tag{2.24}$$

$$\bar{\Omega}_i \triangleq \begin{bmatrix} \bar{\Omega}_i^{11} & \bar{\Omega}_i^{12} & \bar{\Omega}_i^{13} \\ \star & \bar{\Omega}_i^{22} & 0 \\ \star & \star & \bar{\Omega}_i^{33} \end{bmatrix} < 0, \tag{2.25}$$

where

$$\bar{\Omega}_i^{11} \triangleq -P_i + 4\left[A - B\Phi_i(B^T X B)^{-1}K_i + \tilde{X}_i\Delta A(k)\right]^T \bar{P}_i\left[A - B\Phi_i(B^T \right.$$
$$\left. \times X B)^{-1}K_i + \tilde{X}_i\Delta A(k)\right] + 5\varepsilon^2\kappa_i I + 20(\alpha + \epsilon)^2\omega_i I + 3m(\alpha + \epsilon)^2\nu_i I$$
$$+ C^T C + 2K_i^T(B^T X B)^{-1}\Phi_i\bar{Q}_i\Phi_i(B^T X B)^{-1}K_i,$$

$$\bar{\Omega}_i^{12} \triangleq \left[A - B\Phi_i(B^T X B)^{-1}K_i + \tilde{X}_i\Delta A(k)\right]^T \bar{P}_i B(I - \Phi_i) - K_i^T(B^T X B)^{-1}$$
$$\times \Phi_i\bar{Q}_i(I - \Phi_i),$$

$$\bar{\Omega}_i^{13} \triangleq C^T L, \bar{\Omega}_i^{33} \triangleq -\gamma^2 I + L^T L + 5G^T \bar{P}_i G,$$

$$\bar{\Omega}_i^{22} \triangleq 4(I - \Phi_i)B^T \bar{P}_i B(I - \Phi_i) - Q_i + 2(I - \Phi_i)\bar{Q}_i(I - \Phi_i),$$

Main Results   31

then the closed-loop system (2.10) is stochastically stable with the H_∞ distur-bance attenuation level γ.

Proof *Define* $\tilde{\eta}(k) \triangleq \begin{bmatrix} x^{\mathrm{T}}(k) & u^{\mathrm{T}}(k-1) \end{bmatrix}^{\mathrm{T}}$ *and choose the token-dependent Lyapunov function candidate as* $\tilde{V}(\tilde{\eta}(k), i) = x^{\mathrm{T}}(k)P_i x(k) + u^{\mathrm{T}}(k-1)Q_i u(k-1)$, $i \in \mathcal{M}$. *Then, for* $\sigma(k) = i$ *and* $\sigma(k+1) = j$ $(i, j \in \mathcal{M})$, *it easily obtains that*

$$\mathbb{E}\left\{ \tilde{V}(\tilde{\eta}(k+1), \sigma(k+1)) \mid \tilde{\eta}(k), \sigma(k) \right\}$$

$$\leq 4x^{\mathrm{T}}(k) \left[A - B\Phi_i(B^{\mathrm{T}}XB)^{-1}K_i + \tilde{X}_i \Delta A(k) \right]^{\mathrm{T}} \bar{P}_i$$

$$\times \left[A - B\Phi_i(B^{\mathrm{T}}XB)^{-1}K_i + \tilde{X}_i \Delta A(k) \right] x(k)$$

$$+ 4u^{\mathrm{T}}(k-1)(I - \Phi_i)B^{\mathrm{T}}\bar{P}_i B(I - \Phi_i)u(k-1)$$

$$+ 5\varepsilon^2 \kappa_i x^{\mathrm{T}}(k)x(k) + 5w^{\mathrm{T}}(k)G^{\mathrm{T}}\bar{P}_i Gw(k)$$

$$+ 5\left[D(k) - D_o - D_s \mathrm{sgn}(s(k)) \right]^{\mathrm{T}} (B^{\mathrm{T}}XB)^{-1}\Phi_i B^{\mathrm{T}}\bar{P}_i B\Phi_i(B^{\mathrm{T}}XB)^{-1}$$

$$\times \left[D(k) - D_o - D_s \mathrm{sgn}(s(k)) \right] + 2x^{\mathrm{T}}(k)[A - B\Phi_i(B^{\mathrm{T}}XB)^{-1}K_i$$

$$+ \tilde{X}_i \Delta A(k)]^{\mathrm{T}}\bar{P}_i B(I - \Phi_i)u(k-1) + 2x^{\mathrm{T}}(k)K_i^{\mathrm{T}}(B^{\mathrm{T}}XB)^{-1}\Phi_i\bar{Q}_i\Phi_i$$

$$\times (B^{\mathrm{T}}XB)^{-1}K_i x(k) + 2u^{\mathrm{T}}(k-1)(I - \Phi_i)\bar{Q}_i(I - \Phi_i)u(k-1)$$

$$+ 3\left(D_o + D_s \mathrm{sgn}(s(k)) \right)^{\mathrm{T}} (B^{\mathrm{T}}XB)^{-1}\Phi_i\bar{Q}_i\Phi_i(B^{\mathrm{T}}XB)^{-1}\left(D_o + D_s \mathrm{sgn}(s(k)) \right)$$

$$- 2x^{\mathrm{T}}(k)K_i^{\mathrm{T}}(B^{\mathrm{T}}XB)^{-1}\Phi_i\bar{Q}_i(I - \Phi_i)u(k-1). \qquad (2.26)$$

Furthermore, it has $\|D_s\|^2 = (\alpha + \epsilon)^2 x^{\mathrm{T}}(k)x(k)$, $\|D(k) - D_o - D_s \mathrm{sgn}(s(k))\|^2 \leq 4(\alpha + \epsilon)^2 x^{\mathrm{T}}(k)x(k)$ *and* $\|D_o + D_s \mathrm{sgn}(s(k))\|^2 \leq m(\alpha + \epsilon)^2 x^{\mathrm{T}}(k)x(k)$, *which render the following relationships from conditions (2.23) and (2.24):*

$$\left[D(k) - D_o - D_s \mathrm{sgn}(s(k)) \right]^{\mathrm{T}} (B^{\mathrm{T}}XB)^{-1}\Phi_i B^{\mathrm{T}}\bar{P}_i B\Phi_i(B^{\mathrm{T}}XB)^{-1}$$

$$\times \left[D(k) - D_o - D_s \mathrm{sgn}(s(k)) \right] \leq 4\omega_i(\alpha + \epsilon)^2 x^{\mathrm{T}}(k)x(k), \qquad (2.27)$$

$$\left(D_o + D_s \mathrm{sgn}(s(k)) \right)^{\mathrm{T}} (B^{\mathrm{T}}XB)^{-1}\Phi_i\bar{Q}_i\Phi_i(B^{\mathrm{T}}XB)^{-1}\left(D_o + D_s \mathrm{sgn}(s(k)) \right)$$

$$\leq m\nu_i(\alpha + \epsilon)^2 x^{\mathrm{T}}(k)x(k). \qquad (2.28)$$

Substituting (2.27)–(2.28) into (2.26), we have

$$\mathbb{E}\left\{ \tilde{V}(\tilde{\eta}(k+1), \sigma(k+1)) \mid \tilde{\eta}(k), \sigma(k) \right\} - \tilde{V}(\tilde{\eta}(k), \sigma(k))$$

$$\leq \tilde{\eta}^{\mathrm{T}}(k)\tilde{\Omega}_i\tilde{\eta}(k) + 5w^{\mathrm{T}}(k)G^{\mathrm{T}}\bar{P}_i Gw(k), \qquad (2.29)$$

where

$$\tilde{\Omega}_i \triangleq \begin{bmatrix} \tilde{\Omega}_i^{11} & \bar{\Omega}_i^{12} \\ \star & \bar{\Omega}_i^{22} \end{bmatrix}$$

with
$$\tilde{\Omega}_i^{11} \triangleq -P_i + 4\left[A - B\Phi_i(B^{\mathrm{T}}XB)^{-1}K_i + \tilde{X}_i\Delta A(k)\right]^{\mathrm{T}}\bar{P}_i \times \left[A - B\Phi_i(B^{\mathrm{T}}XB)^{-1}K_i\right.$$
$$\left. + \tilde{X}_i\Delta A(k)\right] + 5\varepsilon^2\kappa_i I + 20(\alpha+\epsilon)^2\omega_i I + 3m(\alpha+\epsilon)^2\nu_i I + 2K_i^{\mathrm{T}}(B^{\mathrm{T}}XB)^{-1}\Phi_i\bar{Q}_i\Phi_i$$
$$(B^{\mathrm{T}}XB)^{-1}K_i.$$

It is easily seen that the condition (2.25) implies $\tilde{\Omega}_i < 0$. Therefore, in case of $w(k) = 0$ and $x(k) \neq 0$, there exists a sufficiently small $c > 0$ such that the inequality (2.29) satisfies $\mathbb{E}\left\{\tilde{V}(\tilde{\eta}(k+1), \sigma(k+1)) \mid \tilde{\eta}(k), \sigma(k)\right\} - \tilde{V}(\tilde{\eta}(k), \sigma(k)) \leq \tilde{\eta}^{\mathrm{T}}(k)\tilde{\Omega}_i\tilde{\eta}(k) < -c\|\tilde{\eta}(k)\|^2$. Following a similar line in the proof of Theorem 1 in [15], it can be obtained that $\mathbb{E}\{\sum_{k=0}^{\infty}[\|x(k)\|^2 + \|u(k-1)\|^2]\} < \infty$, which implies $\mathbb{E}\left\{\sum_{k=0}^{\infty}\|x(k)\|^2\right\} < \infty$. Thus, the closed-loop system (2.10) is stochastically stable.

Next, we shall prove that, under the zero initial condition $\tilde{\eta}(0) = 0$, the closed-loop system (2.10) satisfies $\mathbb{E}\left\{\sum_{k=0}^{\infty}\|z(k)\|^2\right\} < \gamma^2 \sum_{k=0}^{\infty}\|w(k)\|^2$ for all nonzero $w(k) \in \ell_2[0, \infty)$. Again, it follows from Schur complement and (2.29) that the condition (2.25) guarantees

$$\mathbb{E}\left\{\tilde{V}(\tilde{\eta}(k+1), \sigma(k+1)) \mid \tilde{\eta}(k), \sigma(k)\right\} - \tilde{V}(\tilde{\eta}(k), \sigma(k))$$
$$+ \mathbb{E}\left\{\|z(k)\|^2\right\} - \gamma^2\|w(k)\|^2$$
$$= \begin{bmatrix} x(k) \\ u(k-1) \\ w(k) \end{bmatrix}^{\mathrm{T}} \bar{\Omega}_i \begin{bmatrix} x(k) \\ u(k-1) \\ w(k) \end{bmatrix} < 0. \tag{2.30}$$

By taking the sum on both sides of (2.30) from 0 to ∞ with respect to k and noting the fact that $\tilde{\eta}(0) = 0$ and $\mathbb{E}\left\{\tilde{V}(\tilde{\eta}(\infty))\right\} \geq 0$, we have $\sum_{k=0}^{\infty}\mathbb{E}\left\{\|z(k)\|^2\right\} < \gamma^2 \sum_{k=0}^{\infty}\|w(k)\|^2$, which completes the proof.

2.2.3 Solving algorithm

In order to guarantee both the reachability and the stochastic stability with a specified H_∞ performance level of the SMC system (2.10), the matrices K_i ($i \in \mathcal{M}$) in (2.22) should be designed by Theorems 2.1 and 2.2 simultaneously.

Lemma 2.2 *The conditions (2.14) and (2.25) are guaranteed simultaneously by the following inequality:*

$$\hat{\Omega}_i \triangleq \begin{bmatrix} \hat{\Omega}_i^{11} & \bar{\Omega}_i^{12} & \bar{\Omega}_i^{13} & \hat{\Omega}_i^{14} \\ \star & \hat{\Omega}_i^{22} & 0 & \hat{\Omega}_i^{24} \\ \star & \star & \bar{\Omega}_i^{33} & 0 \\ \star & \star & \star & -\bar{W}_i \end{bmatrix} < 0, \ \forall i \in \mathcal{M}, \tag{2.31}$$

Main Results 33

where

$$\hat{\Omega}_i^{11} \triangleq -P_i + 2K_i^{\mathrm{T}}(B^{\mathrm{T}}XB)^{-1}\Phi_i\bar{Q}_i\Phi_i(B^{\mathrm{T}}XB)^{-1}K_i + 4[A - B\Phi_i(B^{\mathrm{T}}XB)^{-1}$$
$$\times K_i + \tilde{X}_i\Delta A(k)]^{\mathrm{T}}\bar{P}_i[A - B\Phi_i(B^{\mathrm{T}}XB)^{-1}K_i + \tilde{X}_i\Delta A(k)]$$
$$+ 20(\alpha + \epsilon)^2\omega_i I + 3m(\alpha + \epsilon)^2\nu_i I + 5\varepsilon^2\kappa_i I + C^{\mathrm{T}}C + 5\varepsilon^2\mu_i I$$
$$+ 3[(I - \Phi_i)B^{\mathrm{T}}X\Delta A(k) - \Phi_iK_i]^{\mathrm{T}}\bar{W}_i[(I - \Phi_i)B^{\mathrm{T}}X\Delta A(k) - \Phi_iK_i],$$
$$\hat{\Omega}_i^{22} \triangleq -Q_i + 3(I - \Phi_i)B^{\mathrm{T}}XB\bar{W}_iB^{\mathrm{T}}XB(I - \Phi_i) + 4(I - \Phi_i)B^{\mathrm{T}}$$
$$\times \bar{P}_iB(I - \Phi_i) + 2(I - \Phi_i)\bar{Q}_i(I - \Phi_i),$$
$$\hat{\Omega}_i^{14} \triangleq [(I - \Phi_i)B^{\mathrm{T}}X\Delta A(k) - \Phi_iK_i]^{\mathrm{T}}\bar{W}_i,$$
$$\hat{\Omega}_i^{24} \triangleq (I - \Phi_i)B^{\mathrm{T}}XB\bar{W}_i.$$

Proof *On one hand, since* $20(\alpha + \epsilon)^2\omega_i I + 3m(\alpha + \epsilon)^2\nu_i I + C^{\mathrm{T}}C > 0$, *it follows from Schur complement that* $\hat{\Omega}_i < 0 \implies \Omega_i < 0$. *On the other hand, it is easily found that* $\hat{\Omega}_i^{11} > \bar{\Omega}_i^{11}$ *and* $\hat{\Omega}_i^{22} > \bar{\Omega}_i^{22}$. *Thus, it can be concluded that* $\hat{\Omega}_i < 0 \implies \bar{\Omega}_i < 0$. *The proof is complete.*

By means of Lemma 2.2, we shall give an effective design scheme for the token-dependent SMC law (2.22), which can simultaneously ensures the reachability and the stochastic stability with a specified H_∞ performance level of the SMC system (2.10).

Theorem 2.3 *Consider the system (2.1) with the SCP (2.4) and the token-dependent SMC law (2.22). For a given* $\gamma > 0$, *if there exist matrices* $\check{K}_i \in \mathbb{R}^{m\times n}$, $\check{P}_i > 0$, $\check{Q}_i > 0$, $\check{W}_i > 0$, *and scalars* $\check{\kappa}_i > 0$, $\check{\mu}_i > 0$, $\check{\omega}_i > 0$, $\check{\nu}_i > 0$, $\zeta_i > 0$ *such that the following LMIs hold:*

$$\begin{bmatrix} -\check{\kappa}_i I & \Lambda_i \\ \star & -\check{\mathbb{P}} \end{bmatrix} \leq 0, \quad \begin{bmatrix} -\check{\mu}_i I & \Upsilon_i \\ \star & -\check{\mathcal{W}} \end{bmatrix} \leq 0, \tag{2.32}$$

$$\begin{bmatrix} -\check{\omega}_i I & \Gamma_i \\ \star & -\check{\mathbb{P}} \end{bmatrix} \leq 0, \quad \begin{bmatrix} -\check{\nu}_i I & \Theta_i \\ \star & -\check{\mathcal{Q}} \end{bmatrix} \leq 0, \tag{2.33}$$

$$\check{\Omega}_i \triangleq \begin{bmatrix} -\mathrm{diag}\left\{\check{P}_i, \check{Q}_i, \gamma^2 I\right\} & \check{\Sigma}_i & 0 \\ \star & -\check{F}_i & \daleth_i \\ \star & \star & -\zeta_i I \end{bmatrix} < 0, \tag{2.34}$$

where

$$\check{\mathbb{P}} \triangleq \mathrm{diag}\{\check{P}_1, \ldots, \check{P}_m\}, \check{\mathcal{Q}} \triangleq \mathrm{diag}\{\check{Q}_1, \ldots, \check{Q}_m\}, \check{\mathcal{W}} \triangleq \mathrm{diag}\{\check{W}_1, \ldots, \check{W}_m\},$$
$$\check{\Sigma}_i \triangleq [\mathbb{K}_i^1 \; \mathbb{K}_i^2 \; \mathbb{K}_i^3], \Lambda_i \triangleq [\sqrt{\pi_{i1}}\mathbb{B}_i \; \cdots \; \sqrt{\pi_{im}}\mathbb{B}_i]], \Upsilon_i \triangleq [\sqrt{\pi_{i1}}\mathbb{X}_i \; \cdots \; \sqrt{\pi_{im}}\mathbb{X}_i],$$
$$\Gamma_i \triangleq [\sqrt{\pi_{i1}}\check{\mathbb{X}}_i \; \cdots \; \sqrt{\pi_{im}}\check{\mathbb{X}}_i], \Theta_i \triangleq [\sqrt{\pi_{i1}}\hat{\mathbb{X}}_i \; \cdots \; \sqrt{\pi_{im}}\hat{\mathbb{X}}_i],$$
$$\check{F}_i \triangleq \mathrm{diag}\{\check{\mathcal{W}}, \check{\mathbb{P}}, \check{\mathcal{Q}}, I, \check{\kappa}_i I, \check{\mathcal{Q}}, \check{\mathbb{P}}, \check{\omega}_i I, \check{\nu}_i I, \check{\mu}_i I, \check{\mathcal{W}}, \check{\mathbb{P}}, \check{\mathcal{Q}}, \check{\mathcal{W}}, \check{\mathbb{P}}, \zeta_i I\},$$

$$\mathbb{K}_i^1 \triangleq \left[\breve{\Sigma}_i^{1,1} \ \breve{\Sigma}_i^{1,2} \ \breve{\Sigma}_i^{1,3} \ \breve{P}_i C^{\mathrm{T}} \ \sqrt{5}\varepsilon\breve{P}_i \ \breve{\Sigma}_i^{1,6}; \breve{\Sigma}_i^{2,1}\breve{\Sigma}_i^{2,2} \ \breve{\Sigma}_i^{2,3} \ 0 \ 0 \ 0; 0 \ 0 \ 0 \ L^{\mathrm{T}} \ 0 \ 0\right],$$

$$\mathbb{K}_i^2 \triangleq \left[\breve{\Sigma}_i^{1,7} \ \breve{\Sigma}_i^{1,8} \ \breve{\Sigma}_i^{1,9} \ \sqrt{5}\varepsilon\breve{P}_i \ \breve{\Sigma}_i^{1,11}; 0 \ 0 \ 0 \ 0 \ 0; 0 \ 0 \ 0 \ 0 \ 0\right],$$

$$\mathbb{K}_i^3 \triangleq \left[0 \ 0 \ 0 \ 0 \ \breve{P}_i H^{\mathrm{T}}; \breve{\Sigma}_i^{2,12} \ \breve{\Sigma}_i^{2,13} \ \breve{\Sigma}_i^{2,14} \ 0 \ 0; 0 \ 0 \ 0 \ \breve{\Sigma}_i^{3,15} \ 0\right],$$

$$\breve{\Sigma}_i^{1,1} \triangleq \left[-\sqrt{\pi_{i1}}\breve{K}_i^{\mathrm{T}}\Phi_i \ \cdots \ -\sqrt{\pi_{im}}\breve{K}_i^{\mathrm{T}}\Phi_i\right], \breve{\Sigma}_i^{1,2} \triangleq \left[\sqrt{\pi_{i1}}\mathbb{A}_i \ \cdots \ \sqrt{\pi_{im}}\mathbb{A}_i\right],$$

$$\breve{\Sigma}_i^{1,3} \triangleq \left[-\sqrt{\pi_{i1}}\bar{\mathbb{K}}_i \ \cdots \ -\sqrt{\pi_{im}}\bar{\mathbb{K}}_i\right], \breve{\Sigma}_i^{1,6} \triangleq \left[\sqrt{\pi_{i1}}\bar{\mathbb{K}}_i \ \cdots \ \sqrt{\pi_{im}}\bar{\mathbb{K}}_i\right],$$

$$\breve{\Sigma}_i^{1,7} \triangleq \left[\sqrt{3\pi_{i1}}\mathbb{A}_i \ \cdots \ \sqrt{3\pi_{im}}\mathbb{A}_i\right], \breve{\Sigma}_i^{1,8} \triangleq 2\sqrt{5}(\alpha + \epsilon)\breve{P}_i,$$

$$\breve{\Sigma}_i^{1,9} \triangleq \sqrt{3m}(\alpha + \epsilon)\breve{P}_i, \breve{\Sigma}_i^{1,11} \triangleq \left[-\sqrt{3\pi_{i1}}\breve{K}_i^{\mathrm{T}}\Phi_i \ \cdots \ -\sqrt{3\pi_{im}}\breve{K}_i^{\mathrm{T}}\Phi_i\right],$$

$$\breve{\Sigma}_i^{2,1} \triangleq \left[\sqrt{\pi_{i1}}\mathbb{Q}_i \ \cdots \ \sqrt{\pi_{im}}\mathbb{Q}_i\right], \breve{\Sigma}_i^{2,2} \triangleq \left[\sqrt{\pi_{i1}}\bar{\mathbb{Q}}_i \ \cdots \ \sqrt{\pi_{im}}\bar{\mathbb{Q}}_i\right],$$

$$\breve{\Sigma}_i^{2,3} \triangleq \left[\sqrt{\pi_{i1}}\hat{\mathbb{Q}}_i \ \cdots \ \sqrt{\pi_{im}}\hat{\mathbb{Q}}_i\right], \breve{\Sigma}_i^{2,12} \triangleq \left[\sqrt{3\pi_{i1}}\bar{\mathbb{Q}}_i \ \cdots \ \sqrt{3\pi_{im}}\bar{\mathbb{Q}}_i\right],$$

$$\breve{\Sigma}_i^{2,13} \triangleq \left[\sqrt{\pi_{i1}}\hat{\mathbb{Q}}_i \ \cdots \ \sqrt{\pi_{im}}\hat{\mathbb{Q}}_i\right], \breve{\Sigma}_i^{2,14} \triangleq \left[\sqrt{3\pi_{i1}}\mathbb{Q}_i \cdots \ \sqrt{3\pi_{im}}\mathbb{Q}_i\right],$$

$$\breve{\Sigma}_i^{3,15} \triangleq \left[\sqrt{5\pi_{i1}}G^{\mathrm{T}} \ \cdots \ \sqrt{5\pi_{im}}G^{\mathrm{T}}\right], \daleth_i \triangleq \left[\mathbb{E}_i^1 \ \mathbb{E}_i^2 \ \mathbb{E}^3 \ \mathbb{E}_i^4 \ \mathbb{E}_i^5 \ \mathbb{E}^6\right]^{\mathrm{T}},$$

$$\mathbb{E}_i^1 \triangleq \left[\sqrt{\pi_{i1}}\tilde{\mathbb{X}}_i \ \cdots \ \sqrt{\pi_{im}}\tilde{\mathbb{X}}_i\right], \mathbb{E}_i^2 \triangleq \left[\sqrt{\pi_{i1}}E^{\mathrm{T}}\tilde{X}_i^{\mathrm{T}}\zeta_i \ \cdots \ \sqrt{\pi_{im}}E^{\mathrm{T}}\tilde{X}_i^{\mathrm{T}}\zeta_i\right],$$

$$\mathbb{E}^3 \triangleq \left[\mathbf{0}_{1 \times m} \text{ blocks } 0 \ 0\mathbf{0}_{1 \times m} \text{ blocks}\right],$$

$$\mathbb{E}_i^4 \triangleq \left[\sqrt{3\pi_{i1}}E^{\mathrm{T}}\tilde{X}_i^{\mathrm{T}}\zeta_i \ \cdots \ \sqrt{3\pi_{im}}E^{\mathrm{T}}\tilde{X}_i^{\mathrm{T}}\zeta_i \ 0 \ 0 \ 0\right],$$

$$\mathbb{E}_i^5 \triangleq \left[\sqrt{3\pi_{i1}}\tilde{\mathbb{X}}_i \ \cdots \ \sqrt{3\pi_{im}}\tilde{\mathbb{X}}_i\right],$$

$$\mathbb{E}^6 \triangleq \left[\mathbf{0}_{1 \times m} \text{ blocks } \mathbf{0}_{1 \times m} \text{ blocks } \mathbf{0}_{1 \times m} \text{ blocks } \mathbf{0}_{1 \times m} \text{ blocks } 0\right],$$

$$\mathbb{B}_i \triangleq (I - \Phi_i)B^{\mathrm{T}}\breve{\kappa}_i, \mathbb{X}_i \triangleq B^{\mathrm{T}}XB(I - \Phi_i)\breve{\mu}_i,$$

$$\bar{\mathbb{X}}_i \triangleq (B^{\mathrm{T}}XB)^{-1}\Phi_iB^{\mathrm{T}}\breve{\omega}_i, \hat{\mathbb{X}}_i \triangleq (B^{\mathrm{T}}XB)^{-1}\Phi_i\breve{\nu}_i,$$

$$\mathbb{A}_i \triangleq \left[A\breve{P}_i - B\Phi_i(B^{\mathrm{T}}XB)^{-1}\breve{K}_i\right]^{\mathrm{T}}, \bar{\mathbb{K}}_i \triangleq \breve{K}_i^{\mathrm{T}}(B^{\mathrm{T}}XB)^{-1}\Phi_i,$$

$$\mathbb{Q}_i \triangleq \breve{Q}_i(I - \Phi_i)B^{\mathrm{T}}XB, \bar{\mathbb{Q}}_i \triangleq \breve{Q}_i(I - \Phi_i)B^{\mathrm{T}},$$

$$\hat{\mathbb{Q}}_i \triangleq \breve{Q}_i(I - \Phi_i), \tilde{\mathbb{X}}_i \triangleq E^{\mathrm{T}}XB(I - \Phi_i)\zeta_i,$$

then the reachability, the stochastic stability, and the specified H_∞ disturbance attention performance γ of the closed-loop system (2.10) can be ensured simultaneously by the SMC law $v(k)$ (2.22) with the matrices $K_i = \breve{K}_i\breve{P}_i^{-1}$, $i \in \mathcal{M}$.

Proof *For any $i \in \mathcal{M}$, denote $\breve{\kappa}_i \triangleq \kappa_i^{-1}$, $\breve{\mu}_i \triangleq \mu_i^{-1}$, $\breve{\omega}_i \triangleq \omega_i^{-1}$, $\breve{\nu}_i \triangleq \nu_i^{-1}$, $\breve{P}_i \triangleq P_i^{-1}$, $\breve{Q}_i \triangleq Q_i^{-1}$, $\breve{W}_i \triangleq W_i^{-1}$, and $\breve{K}_i \triangleq K_i\breve{P}_i$. Consider the inequalities $(I - \Phi_i)B^{\mathrm{T}}\bar{P}_iB(I - \Phi_i) \leq \kappa_iI$, $B^{\mathrm{T}}XB(I - \Phi_i)\bar{W}_i(I - \Phi_i)B^{\mathrm{T}}XB \leq \mu_iI$, $(B^{\mathrm{T}}XB)^{-1}\Phi_iB^{\mathrm{T}}\bar{P}_iB\Phi_i(B^{\mathrm{T}}XB)^{-1} \leq \omega_iI$, $(B^{\mathrm{T}}XB)^{-1}\Phi_i\bar{Q}_i\Phi_i(B^{\mathrm{T}}XB)^{-1} \leq \nu_iI$ and $\hat{\Omega}_i < 0$, it follows from Lemma 2.2 that the conditions in Theorems 2.1 and 2.2 are guaranteed by the above matrix inequalities, which are equivalent to LMIs (2.32)–(2.34). This completes the proof.*

Example 35

To obtain an optimized H_∞ sliding mode controller for system (2.1) under SCP, the attenuation level γ can be made as small as possible such that LMIs (2.32)–(2.34) are satisfied. The optimization problem is formulated as follows:

$$\min_{\breve{P}_i,\breve{Q}_i,\breve{W}_i,\breve{K}_i,\breve{\kappa}_i,\breve{\mu}_i,\breve{\omega}_i,\breve{\nu}_i,\breve{\zeta}_i,\breve{\gamma}} \breve{\gamma}$$

$$\text{s.t. LMIs (2.32)–(2.34) with } \breve{\gamma} \triangleq \gamma^2. \tag{2.35}$$

Remark 2.1 *It is noted that the selection of the matrix X in (2.5) would largely affects the feasibility of the LMIs (2.32)–(2.34). For practicality, one can simply choose $X = \varsigma I$, where the scalar $\varsigma > 0$ is determined by checking the feasibility of the LMIs (2.32)–(2.34) by using the linear search algorithm.*

Remark 2.2 *In this chapter, the SMC problem is investigated for a class of uncertain discrete-time systems with unmatched external disturbance and SCP scheduling. Our main results in Theorems 2.1—2.3 exhibit the following distinct features: 1) the designed SMC law is token-dependent as a reflection of the introduction of the Markovian jumping scheduling; 2) the stochastic Lyapunov function is also token-dependent for ensuring the reachability of the sliding mode dynamics; and 3) some time-varying bounds are purposely exploited to handle the effects from the ZOHs.*

2.3 Example

Consider system (2.1) with the following parameters:

$$A = \begin{bmatrix} 1.0307 & 0 & 0.0557 \\ 0.0333 & 0.2466 & -0.0091 \\ 0.0071 & -0.01 & 1.0130 \end{bmatrix}, B = \begin{bmatrix} 0.1817 & 0.4286 \\ 0.1597 & 0.0793 \\ 0.1138 & 0.0581 \end{bmatrix},$$

$$C = \begin{bmatrix} 0.2 & 0 & -0.1 \\ 0.1 & 0.15 & 0 \end{bmatrix}, G = \begin{bmatrix} 0.015 \\ 0.01 \\ 0.02 \end{bmatrix},$$

$$L = \begin{bmatrix} 0.015 \\ 0.02 \end{bmatrix}, E = \begin{bmatrix} 0.01 \\ 0.02 \\ 0 \end{bmatrix}, H = \begin{bmatrix} 0 & 0.01 & 0 \end{bmatrix},$$

$$F(k) = \sin(0.6k), f(x(k), k) = \begin{bmatrix} 0.4\sqrt{|x_1(k)x_2(k)|} \\ 0.3x_3(k) \end{bmatrix}.$$

It is easily to verify that the uncontrolled system (2.1) (when $u(k) = 0$ and $w(k) = 0$) is unstable under an initial condition $x(0) = \begin{bmatrix} 0.5 & -1.2 & 1 \end{bmatrix}^T$. Choose the scalar $\varepsilon = 0.3$ and the constant matrix $X = 0.093I_{3\times3}$ in sliding function (2.5). To examine the impacts of the SCP scheduling, we design SMC law in the cases of SMC without and with SCP.

36 H_∞ *SMC Under SCP*

▶ *Case 1 (SMC without SCP).* It is observed from the actual control law (2.4) that the SCP constraints are actually reflected in the *update matrix* $\Phi_{\sigma(k)}$, $\sigma(k) \in \mathcal{M}$. In this example, if we let the update matrix Φ_1 be the identity matrix $I_{2\times2}$ and the transition probability matrix $\Pi = [1\ 0; 1\ 0]$, then the underlying communication network reduces to the common type without SCP, and the actual control input (2.4) in plant side becomes $u(k) = v(k)$. By solving the optimization problem (2.35), we have the optimization H_∞ disturbance attenuation level $\gamma^* = 3.0667$ and the matrix in (2.9) as $K_1 = \begin{bmatrix} 0.0175 & -0.0003 & 0.0331 \\ 0.0412 & 0.0001 & 0.0168 \end{bmatrix}$. Thus, the SMC law $v(k)$ in form of (2.22) is designed as:

$$v(k) = - \begin{bmatrix} -1.4961 & -0.1891 & 11.7008 \\ 3.0415 & 0.1009 & -4.9453 \end{bmatrix} x(k)$$

$$- \begin{bmatrix} 3.2754 & -1.6457 \\ -1.6457 & 1.2106 \end{bmatrix} \|x(k)\| \cdot \text{sgn}(s(k)). \qquad (2.36)$$

▶ *Case 2 (SMC with SCP).* For this case, by solving the optimization problem (2.35) with the transition probability matrix $\Pi = [0.55\ 0.45; 0.6\ 0.4]$ in the SCP (2.4), we have the optimization H_∞ disturbance attenuation level $\gamma^\diamond = 10.6282$ and the parameter matrices in (2.9) as $K_1 = \begin{bmatrix} 0.0018 & 0.0000 & 0.0001 \\ -0.0029 & -0.0000 & -0.0046 \end{bmatrix}$ and $K_2 = \begin{bmatrix} -0.0041 & -0.0000 & -0.0000 \\ 0.0015 & 0.0000 & 0.0014 \end{bmatrix}$. Then, the SMC law with SCP in (2.9) is obtained as:

$$v(k) = \begin{cases} - \begin{bmatrix} 1.5524 & 0.0026 & 1.1625 \\ -0.9433 & -0.0019 & -0.8405 \end{bmatrix} x(k) \\ \qquad - \begin{bmatrix} 3.2754 & -1.6457 \\ -1.6457 & 1.2106 \end{bmatrix} \|x(k)\| \cdot \text{sgn}(s(k)), i = 1; \\ - \begin{bmatrix} -2.3175 & -0.0148 & -0.3668 \\ 1.2462 & 0.0076 & 0.2643 \end{bmatrix} x(k) \\ \qquad - \begin{bmatrix} 3.2754 & -1.6457 \\ -1.6457 & 1.2106 \end{bmatrix} \|x(k)\| \cdot \text{sgn}(s(k)), i = 2. \end{cases}$$

$$\qquad (2.37)$$

Under the aforementioned initial condition $x(0)$ and the external disturbance $w(k) = 0$, the state trajectories $x(k)$, the sliding function $s(k)$ and the actual control input $u(k)$ of the controlled system are depicted in Figs. 2.2(a)–(c) and 2.4(a)–(c), respectively, for Cases 1 and 2. Figure 2.3 gives the realization of the SCP scheduling sequence in Case 2. Furthermore,

Example 37

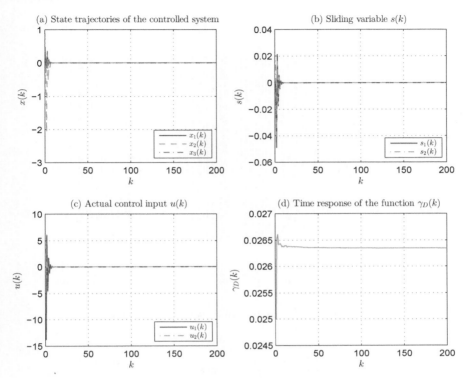

FIGURE 2.2: Simulation results in Case 1.

under the initial condition $x(0) = 0$ and the external disturbance $w(k) = \sin^2(k)/(1+k)$, Figs. 2.2(d) and 2.4(d) show the evolutions of the function $\gamma_D(k) \triangleq \sqrt{\sum_{n=0}^{k} \|z(n)\|^2 / \sum_{n=0}^{k} \|w(n)\|^2}$ in Cases 1 and 2, respectively, which illustrate the H_∞ performance of the obtained SMC laws (2.36) and (2.37). By comparing the above two cases, it is concluded that the SCP mechanism alleviates the communication burden at the cost of sacrificing certain performances of the SMC systems, which include the convergence speed of the closed-loop systems (see Figs. 2.2(a) and 2.4(a)) and the H_∞ performances because $\gamma^* < \gamma^\circ$ (see Figs. 2.2(d) and 2.4(d)). Nevertheless, all the simulation results have illustrated the effectiveness of the proposed H_∞ SMC design approach under SCP scheduling and unmatched external disturbance.

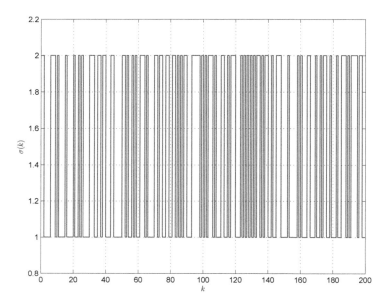

FIGURE 2.3: A possible SCP scheduling sequence.

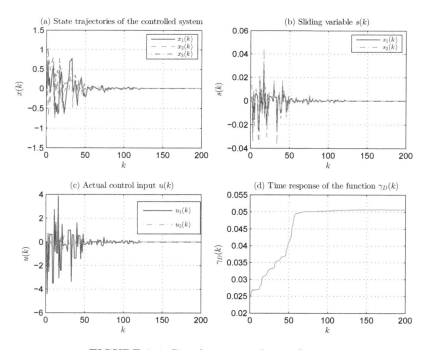

FIGURE 2.4: Simulation results in Case 2.

2.4 Conclusion

This chapter has investigated the H_∞ SMC problem for discrete-time uncertain systems under SCP scheduling. The SCP scheduling has been applied to determine which actuator obtains access to the communication network between sliding mode controller and actuators. To cope with the impacts from the SCP, a token-dependent SMC law with some specific bounds has been designed such that the reachability, the stochastic stability and H_∞ performance of the SMC systems are guaranteed simultaneously in mean square sense. Simulation results have demonstrated the effectiveness of the proposed H_∞ SMC approach under the SCP scheduling.

3

Static Output-Feedback Sliding Mode Control Under Round-Robin Protocol

In this chapter, we endeavour to design the static output-feedback sliding mode controller for a class of uncertain NCSs under the Round-Robin protocol. At each transmission instant, only one actuator obtains access to the communication network, and the other actuators use the values stored by the zero-order holders (ZOHs). By depending on a predetermined circular order (i.e. a periodic sequence), the Round-Robin scheduling scheme is applied to determine which actuator should be given the access to the network at a certain instant. *The main contributions of this chapter are highlighted as follows: 1) In order to reduce the communication burden, the Round-Robin protocol with a set of ZOHs is introduced to schedule the SMC signals between the controller and the actuators; 2) To cope with the impacts from the periodic scheduling signal and the ZOHs on actuators, a suitable token-dependent SMC law is constructed and the reachability of the specified sliding surface is proven by using a new token-dependent Lyapunov function; and 3) Sufficient conditions for guaranteeing the asymptotic stability of the resultant SMC system are derived by exploiting matrix inequality techniques, and a separation strategy based algorithm is proposed to compute the design matrices.*

3.1   Problem Formulation

Consider the following class of uncertain NCSs:

$$x(k+1) \;=\; (A + \Delta A(k)) \, x(k) + B \, (u(k) + g(x,y)) \qquad (3.1)$$
$$y(k) \;=\; Cx(k) \qquad (3.2)$$

where $x(k) \in \mathbb{R}^n$ is the state vector, $u(k) \in \mathbb{R}^m$ is the control input vector, and $y(k) \in \mathbb{R}^p$ is the measured output vector. The time-varying uncertainty $\Delta A(k)$ is norm-bounded by $\Delta A(k) = MZ(k)N$ where M, N are known constant matrices and the unknown time-varying matrix $Z(k)$ satisfies $Z^{\mathrm{T}}(k)Z(k) \leq I$. The nonlinear function $g(x,y)$ satisfies the following condition:

$$g^{\mathrm{T}}(x,y)g(x,y) \leq \alpha x^{\mathrm{T}}(k)x(k) + \beta y^{\mathrm{T}}(k)y(k) \qquad (3.3)$$

DOI: 10.1201/9781003309499-3

41

where α and β are known nonnegative scalars. $A \in \mathbb{R}^{n \times n}$, $B \in \mathbb{R}^{n \times m}$ and $C \in \mathbb{R}^{p \times n}$ are known constant matrices. In addition, we suppose that the uncertain NCS (3.1)–(3.2) possesses the following standard assumption as in [180, 211].

Assumption 3.1 *The input matrix B has a full column rank, the output matrix C has a full row rank, and* $\mathrm{rank}\,(CB) = m$.

As shown in Fig. 3.1, the control signal $\bar{u}(k)$ is transmitted to the actuators $u(k)$ via a shared communication channel subject to the Round-Robin protocol scheduling. At a certain transmission instant k, it is assumed that *only one actuator* obtains the token to access the controller-to-actuator (C/A) network. Under the Round-Robin scheduling scheme, all actuators are arranged in a particular sequence, by which they will get the token to access the channel one by one. When the last actuator completes the communication task, the channel token will be turned over to the first actuator. By doing so, the communication burden is significantly reduced at each instant, thereby relieving the pressure on the communication channel.

The scheduling signal $\sigma(k) \in \mathcal{M} \triangleq \{1, 2, \ldots, m\}$ denotes the selected actuator obtaining the *access token* to the C/A network at the transmission instant k. For technical convenience, we denote

$$\bar{u}(k) \triangleq \begin{bmatrix} \bar{u}_1^{\mathrm{T}}(k) & \bar{u}_2^{\mathrm{T}}(k) & \cdots & \bar{u}_m^{\mathrm{T}}(k) \end{bmatrix}^{\mathrm{T}} \in \mathbb{R}^m,$$
$$u(k) \triangleq \begin{bmatrix} u_1^{\mathrm{T}}(k) & u_2^{\mathrm{T}}(k) & \cdots & u_m^{\mathrm{T}}(k) \end{bmatrix}^{\mathrm{T}} \in \mathbb{R}^m,$$

where $\bar{u}_i(k)$ is the ith control signal and $u_i(k)$ is the received signal by the ith actuator.

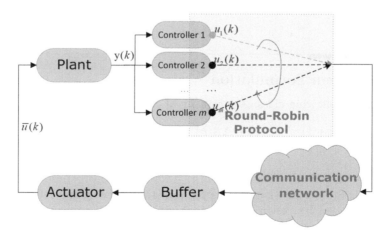

FIGURE 3.1: The control systems under Round-Robin protocol scheduling.

Main Results 43

According to the Round-Robin protocol, at the transmission instant k, the ith actuator is updated *if and only if* the scheduling signal $\sigma(k) = i$, i.e., $\mathrm{mod}(k - i, m) = 0$; otherwise, there is no any information received by the ith actuator and the last value $u_i(k - 1)$ stored by ZOH (i.e., a *buffer*) will be used. Thus, the updating rule for the ith actuator $u_i(k)$ ($i \in \mathcal{M}$) under the Round-Robin communication protocol is given as follows:

$$u_i(k) = \begin{cases} \bar{u}_i(k), & \text{if } \sigma(k) = i, \\ u_i(k - 1), & \text{otherwise,} \end{cases} \tag{3.4}$$

from which it is concluded that the value of $\sigma(k)$ satisfies $\sigma(k + m) = \sigma(k)$ for all $k \in \mathbb{N}^+$ and $\sigma(k)$ can be calculated by

$$\sigma(k) = \mathrm{mod}(k - 1, m) + 1. \tag{3.5}$$

Based upon the updating rule (3.4) on each actuator, we can further express the *actual* control input $u(k)$ at plant side as

$$u(k) = \Phi_{\sigma(k)} \bar{u}(k) + \left(I - \Phi_{\sigma(k)} \right) u(k - 1), \quad k \in \mathbb{N}^+, \tag{3.6}$$

where the update matrix is given by

$$\Phi_{\sigma(k)} \triangleq \mathrm{diag}\left\{ \tilde{\delta}^1_{\sigma(k)}, \tilde{\delta}^2_{\sigma(k)}, \cdots, \tilde{\delta}^m_{\sigma(k)} \right\}.$$

Here, $\tilde{\delta}^b_a \triangleq \delta(a - b)$ with $\delta(\cdot)$ being the Kronecker delta function. Without loss of generality, we set $u(-1) = 0$ here.

From (3.6), we rewrite the uncertain system (3.1) as follows:

$$\begin{aligned} x(k + 1) = &\left(A + \Delta A(k) \right) x(k) + B \left(I - \Phi_{\sigma(k)} \right) u(k - 1) \\ &+ Bg(x, y) + B\Phi_{\sigma(k)} \bar{u}(k). \end{aligned} \tag{3.7}$$

Definition 3.1 *The uncertain system (3.7) with $\bar{u}(k) = 0$ is asymptotically stable if, for every initial conditions $x(0) \neq 0$ and $u(-1) = 0$, $\lim_{k \to \infty} \|x(k)\|^2 = 0$ is true.*

Our objective of this chapter is to design an output-feedback sliding mode controller $\bar{u}(k)$ such that the resulting closed-loop system under Round-Robin protocol (3.4)–(3.5) is asymptotically stable.

3.2 Main Results

In this section, an output-feedback SMC law $\bar{u}(k)$ shall be designed to asymptotically stabilize the uncertain NCS (3.1)–(3.2) under Round-Robin protocol (3.4)–(3.5).

44 *Static Output-Feedback SMC under RRP*

It is observed from the closed-loop system (3.7) that the introduction of Round-Robin protocol (3.4)–(3.5) brings two key features to the controlled system, that is, the periodic scheduler $\sigma(k)$ and the input component $u(k-1)$ caused from ZOHs. Notice that the existence of the input component $u(k-1)$ in the closed-loop system (3.7) implies that, in order to achieve the asymptotic stability for $x(k)$, $u(k-1)$ must converge to *zero* along with $k \to \infty$. These observations have led to the following two questions in designing output-feedback SMC under the Round-Robin protocol (3.4)–(3.5):

Q-1) How to cope with the effects from the scheduling signal $\sigma(k)$ in the output-feedback SMC law?

Q-2) How to ensure that the input term $u(k-1)$ converges to zero as $k \to \infty$?

The above two issues will be kept in mind when designing the desired output-feedback SMC law.

3.2.1 Token-dependent static output-feedback SMC law

Firstly, based on the measured output $y(k)$, the following sliding function is exploited:

$$s(k) = Fy(k), \tag{3.8}$$

where F is a matrix to be designed.

It is well known that, in order to realize the static output-feedback SMC based on the specified sliding function (3.8), the non-singularity of FCB should be satisfied as pointed out in [40, 180, 181, 211]. In this chapter, the matrix F is designed to satisfy the following equality:

$$B^{\mathrm{T}}X = FC, \tag{3.9}$$

where $X > 0$ is a matrix to be determined later (as will be seen in Theorem 3.3). Clearly, the matrix FCB is nonsingular since B has full column rank.

Remark 3.1 *It is easy to see that, when the matrix X is known in the equality constraint (3.9), the matrix F can be obtained by solving the following convex optimization problem [62]:*

$$\begin{bmatrix} -\varrho I & B^{\mathrm{T}}X - FC \\ \star & -I \end{bmatrix} < 0, \tag{3.10}$$

where $\varrho > 0$ is a sufficiently small scalar.

Now, bearing the issues **Q-1** and **Q-2** in mind, we design a token-dependent output-feedback SMC law $\bar{u}(k)$ as follows:

$$\bar{u}(k) = -K_{\sigma(k)}y(k) - \vartheta\|y(k)\|\frac{s(k)}{\|s(k)\|}, \tag{3.11}$$

where the matrices $K_{\sigma(k)} \in \mathbb{R}^{m \times n}$ $(\sigma(k) \in \mathcal{M})$ are to be designed and $\vartheta > 0$ is a prescribed parameter.

Main Results 45

Remark 3.2 *It is worth stressing that the effect of periodic scheduling from one actuator to another is reflected in the proposed SMC law (3.11) just via the token-dependent matrices $K_{\sigma(k)}$, which are the solutions of a set of coupled linear matrix inequalities (LMIs) relating to the periodic scheduling signal $\sigma(k)$ (as will be seen in Theorem 3.3). Thus, the question **Q-1** has been answered.*

Remark 3.3 *Notice that the equivalent control lawindex Equivalent Control Law approach has been widely employed to analyze the stability of the SMC systems, see e.g. [62, 80, 97, 101, 109]. However, it follows from (3.7) and (3.8) that*

$$s(k+1) = FC\Big(A + \Delta A(k)\Big)x(k) + FCBg(x,y) + FCB\left(I - \Phi_{\sigma(k)}\right)u(k-1)$$
$$+ FCB\Phi_{\sigma(k)}\bar{u}(k), \tag{3.12}$$

from which one it is clear that the equivalent control law approach is no longer applicable here due to the singularity of the matrix $FCB\Phi_{\sigma(k)}$, and this makes the stability analysis of the SMC systems under the Round-Robin protocol more challenging yet interesting.

Substituting (3.11) into (3.7), we obtain the closed-loop system as follows:

$$x(k+1) = \bar{A}_{\sigma(k)}x(k) + B(I - \Phi_{\sigma(k)})u(k-1) + Bg(x,y) - B\Phi_{\sigma(k)}\vartheta_s(k) \tag{3.13}$$

where $\bar{A}_{\sigma(k)} \triangleq A + \Delta A(k) - B\Phi_{\sigma(k)}K_{\sigma(k)}C$ and $\vartheta_s(k) \triangleq \vartheta\|y(k)\|\frac{s(k)}{\|s(k)\|}$. It is straightforward to see that

$$\|\vartheta_s(k)\|^2 = \vartheta^2 x^{\mathrm{T}}(k)C^{\mathrm{T}}Cx(k). \tag{3.14}$$

In the sequel, the asymptotic stability of the closed-loop system (3.13) and the reachability of the specified sliding surface (3.8) will be analyzed, respectively. To this end, the following lemmas are introduced.

Lemma 3.1 *For any vectors x, y, a matrix $U > 0$, and a scalar $\varepsilon > 0$, the following matrix inequality holds:*

$$\pm 2x^{\mathrm{T}}Uy \leq \varepsilon x^{\mathrm{T}}Ux + \varepsilon^{-1}y^{\mathrm{T}}Uy.$$

Lemma 3.2 *For any matrices $X > 0$, Y, a scalar $\epsilon > 0$, the following matrix inequality holds:*

$$YX^{-1}Y^{\mathrm{T}} \geq \epsilon(Y + Y^{\mathrm{T}}) - \epsilon^2 X.$$

3.2.2 Analysis of the asymptotic stability

In this subsection, we aim to analyze the asymptotic stability of the closed-loop system (3.13). By resorting to the constructed SMC law (3.11), we first rewrite the updating rule (3.6) for the actual control input $u(k)$ as follows:

$$u(k) = -\Phi_{\sigma(k)}K_{\sigma(k)}Cx(k) - \Phi_{\sigma(k)}\vartheta_s(k) + \left(I - \Phi_{\sigma(k)}\right)u(k-1). \tag{3.15}$$

The following theorem presents a sufficient condition for guaranteeing the asymptotic stability of the closed-loop system (3.13) under the Round-Robin protocol (3.4)–(3.5).

Theorem 3.1 *Consider the uncertain NCS (3.1)–(3.2) with the Round-Robin protocol (3.4)–(3.5). Let the matrix F in sliding function (3.8) and the parameter ϑ in the token-dependent SMC law (3.11) be given. If there exist matrices $K_{\sigma(k)} \in \mathbb{R}^{m \times p}$, $P_{\sigma(k)} > 0$, $Q_{\sigma(k)} > 0$, and scalars $\eta_{\sigma(k)} > 0$, $\delta_{\sigma(k),\sigma(k+1)} > 0$, $\gamma_{\sigma(k),\sigma(k+1)} > 0$ such that the following matrix inequalities*

$$B^{\mathrm{T}} P_{\sigma(k+1)} B - \eta_{\sigma(k+1)} I \le 0, \tag{3.16}$$

$$\Phi_{\sigma(k)} B^{\mathrm{T}} P_{\sigma(k+1)} B \Phi_{\sigma(k)} - \delta_{\sigma(k),\sigma(k+1)} I \le 0, \tag{3.17}$$

$$\Phi_{\sigma(k)} Q_{\sigma(k+1)} \Phi_{\sigma(k)} - \gamma_{\sigma(k),\sigma(k+1)} I \le 0, \tag{3.18}$$

$$\Psi_{\sigma(k),\sigma(k+1)} \triangleq \begin{bmatrix} \Psi^{(1,1)}_{\sigma(k),\sigma(k+1)} & \Psi^{(1,2)}_{\sigma(k),\sigma(k+1)} \\ \star & \Psi^{(2,2)}_{\sigma(k),\sigma(k+1)} \end{bmatrix} < 0, \tag{3.19}$$

hold for any $\sigma(k), \sigma(k+1) \in \mathcal{M}$, where

$$\begin{aligned} \Psi^{(1,1)}_{\sigma(k),\sigma(k+1)} &\triangleq -P_{\sigma(k)} + 3\bar{A}^{\mathrm{T}}_{\sigma(k)} P_{\sigma(k+1)} \bar{A}_{\sigma(k)} + 4\vartheta^2 \delta_{\sigma(k),\sigma(k+1)} C^{\mathrm{T}} C \\ &\quad + 4\eta_{\sigma(k+1)}(\alpha I + \beta C^{\mathrm{T}} C) + 2C^{\mathrm{T}} K^{\mathrm{T}}_{\sigma(k)} \Phi_{\sigma(k)} Q_{\sigma(k+1)} \Phi_{\sigma(k)} K_{\sigma(k)} C \\ &\quad + 3\vartheta^2 \gamma_{\sigma(k),\sigma(k+1)} C^{\mathrm{T}} C, \end{aligned}$$

$$\Psi^{(1,2)}_{\sigma(k),\sigma(k+1)} \triangleq \bar{A}^{\mathrm{T}}_{\sigma(k)} P_{\sigma(k+1)} B \left(I - \Phi_{\sigma(k)} \right) - C^{\mathrm{T}} K^{\mathrm{T}}_{\sigma(k)} \Phi_{\sigma(k)} Q_{\sigma(k+1)} \left(I - \Phi_{\sigma(k)} \right),$$

$$\begin{aligned} \Psi^{(2,2)}_{\sigma(k),\sigma(k+1)} &\triangleq -Q_{\sigma(k)} + 2 \left(I - \Phi_{\sigma(k)} \right) Q_{\sigma(k+1)} \left(I - \Phi_{\sigma(k)} \right) \\ &\quad + 3 \left(I - \Phi_{\sigma(k)} \right) B^{\mathrm{T}} P_{\sigma(k+1)} B \left(I - \Phi_{\sigma(k)} \right), \end{aligned}$$

then the closed-loop system (3.13) is asymptotically stable.

Proof *Define the following token-dependent Lyapunov function for any $\sigma(k) \in \mathcal{M}$:*

$$V(x(k), u(k-1), \sigma(k)) \triangleq x^{\mathrm{T}}(k) P_{\sigma(k)} x(k) + u^{\mathrm{T}}(k-1) Q_{\sigma(k)} u(k-1). \tag{3.20}$$

It follows from (3.13) that

$$\begin{aligned} &x^{\mathrm{T}}(k+1) P_{\sigma(k+1)} x(k+1) \\ =\,&x^{\mathrm{T}}(k) \bar{A}^{\mathrm{T}}_{\sigma(k)} P_{\sigma(k+1)} \bar{A}_{\sigma(k)} x(k) + g^{\mathrm{T}}(x,y) B^{\mathrm{T}} P_{\sigma(k+1)} B g(x,y) \\ &+ u^{\mathrm{T}}(k-1) \left(I - \Phi_{\sigma(k)} \right) B^{\mathrm{T}} P_{\sigma(k+1)} B \left(I - \Phi_{\sigma(k)} \right) u(k-1) \\ &+ \vartheta^{\mathrm{T}}_s(k) \Phi_{\sigma(k)} B^{\mathrm{T}} P_{\sigma(k+1)} B \Phi_{\sigma(k)} \vartheta_s(k) \\ &+ 2x^{\mathrm{T}}(k) \bar{A}^{\mathrm{T}}_{\sigma(k)} P_{\sigma(k+1)} B \left(I - \Phi_{\sigma(k)} \right) u(k-1) \\ &+ 2x^{\mathrm{T}}(k) \bar{A}^{\mathrm{T}}_{\sigma(k)} P_{\sigma(k+1)} B g(x,y) \\ &- 2x^{\mathrm{T}}(k) \bar{A}^{\mathrm{T}}_{\sigma(k)} P_{\sigma(k+1)} B \Phi_{\sigma(k)} \vartheta_s(k) \end{aligned}$$

Main Results 47

$$+ 2u^{\mathrm{T}}(k-1)\left(I - \Phi_{\sigma(k)}\right)B^{\mathrm{T}}P_{\sigma(k+1)}Bg(x,y)$$

$$- 2u^{\mathrm{T}}(k-1)\left(I - \Phi_{\sigma(k)}\right)B^{\mathrm{T}}P_{\sigma(k+1)}B\Phi_{\sigma(k)}\vartheta_s(k)$$

$$- 2g^{\mathrm{T}}(x,y)B^{\mathrm{T}}P_{\sigma(k+1)}B\Phi_{\sigma(k)}\vartheta_s(k). \tag{3.21}$$

Keeping the relationships of (3.3), (3.14), and (3.16)–(3.17) in mind, we have from Lemma 3.1 that

$$g^{\mathrm{T}}(x,y)B^{\mathrm{T}}P_{\sigma(k+1)}Bg(x,y) \leq \eta_{\sigma(k+1)}\left[\alpha x^{\mathrm{T}}(k)x(k) + \beta x^{\mathrm{T}}(k)C^{\mathrm{T}}Cx(k)\right], \tag{3.22}$$

$$\vartheta_s^{\mathrm{T}}(k)\Phi_{\sigma(k)}B^{\mathrm{T}}P_{\sigma(k+1)}B\Phi_{\sigma(k)}\vartheta_s(k) \leq \vartheta^2\delta_{\sigma(k),\sigma(k+1)}x^{\mathrm{T}}(k)C^{\mathrm{T}}Cx(k), \tag{3.23}$$

$$2x^{\mathrm{T}}(k)\bar{A}_{\sigma(k)}^{\mathrm{T}}P_{\sigma(k+1)}Bg(x,y)$$

$$\leq x^{\mathrm{T}}(k)\bar{A}_{\sigma(k)}^{\mathrm{T}}P_{\sigma(k+1)}\bar{A}_{\sigma(k)}x(k) + \eta_{\sigma(k+1)}\left[\alpha x^{\mathrm{T}}(k)x(k) + \beta x^{\mathrm{T}}(k)C^{\mathrm{T}}Cx(k)\right], \tag{3.24}$$

$$- 2x^{\mathrm{T}}(k)\bar{A}_{\sigma(k)}^{\mathrm{T}}P_{\sigma(k+1)}B\Phi_{\sigma(k)}\vartheta_s(k)$$

$$\leq x^{\mathrm{T}}(k)\bar{A}_{\sigma(k)}^{\mathrm{T}}P_{\sigma(k+1)}\bar{A}_{\sigma(k)}x(k) + \vartheta^2\delta_{\sigma(k),\sigma(k+1)}x^{\mathrm{T}}(k)C^{\mathrm{T}}Cx(k), \tag{3.25}$$

$$2u^{\mathrm{T}}(k-1)\left(I - \Phi_{\sigma(k)}\right)B^{\mathrm{T}}P_{\sigma(k+1)}Bg(x,y)$$

$$\leq u^{\mathrm{T}}(k-1)\left(I - \Phi_{\sigma(k)}\right)B^{\mathrm{T}}P_{\sigma(k+1)}B\left(I - \Phi_{\sigma(k)}\right)u(k-1)$$

$$+ \eta_{\sigma(k+1)}\left[\alpha x^{\mathrm{T}}(k)x(k) + \beta x^{\mathrm{T}}(k)C^{\mathrm{T}}Cx(k)\right], \tag{3.26}$$

$$- 2u^{\mathrm{T}}(k-1)\left(I - \Phi_{\sigma(k)}\right)B^{\mathrm{T}}P_{\sigma(k+1)}B\Phi_{\sigma(k)}\vartheta_s(k)$$

$$\leq u^{\mathrm{T}}(k-1)\left(I - \Phi_{\sigma(k)}\right)B^{\mathrm{T}}P_{\sigma(k+1)}B\left(I - \Phi_{\sigma(k)}\right)u(k-1)$$

$$+ \vartheta^2\delta_{\sigma(k),\sigma(k+1)}x^{\mathrm{T}}(k)C^{\mathrm{T}}Cx(k), \tag{3.27}$$

$$- 2g^{\mathrm{T}}(x,y)B^{\mathrm{T}}P_{\sigma(k+1)}B\Phi_{\sigma(k)}\vartheta_s(k)$$

$$\leq \eta_{\sigma(k+1)}\left[\alpha x^{\mathrm{T}}(k)x(k) + \beta x^{\mathrm{T}}(k)C^{\mathrm{T}}Cx(k)\right] + \vartheta^2\delta_{\sigma(k),\sigma(k+1)}x^{\mathrm{T}}(k)C^{\mathrm{T}}Cx(k). \tag{3.28}$$

Substituting (3.22)–(3.28) into (3.21) gives

$$x^{\mathrm{T}}(k+1)P_{\sigma(k+1)}x(k+1)$$

$$\leq x^{\mathrm{T}}(k)\Big\{3\bar{A}_{\sigma(k)}^{\mathrm{T}}P_{\sigma(k+1)}\bar{A}_{\sigma(k)}$$

$$+ 4\vartheta^2\delta_{\sigma(k),\sigma(k+1)}C^{\mathrm{T}}C + 4\eta_{\sigma(k+1)}\left(\alpha I + \beta C^{\mathrm{T}}C\right)\Big\}x(k)$$

$$+ 3u^{\mathrm{T}}(k-1)\left(I - \Phi_{\sigma(k)}\right)B^{\mathrm{T}}P_{\sigma(k+1)}B\left(I - \Phi_{\sigma(k)}\right)u(k-1)$$

$$+ 2x^{\mathrm{T}}(k)\bar{A}_{\sigma(k)}^{\mathrm{T}}P_{\sigma(k+1)}B\left(I - \Phi_{\sigma(k)}\right)u(k-1). \tag{3.29}$$

On the other hand, it yields from (3.15) that

$$u^{\mathrm{T}}(k)Q_{\sigma(k+1)}u(k)$$

$$= x^{\mathrm{T}}(k)C^{\mathrm{T}}K_{\sigma(k)}^{\mathrm{T}}\Phi_{\sigma(k)}Q_{\sigma(k+1)}\Phi_{\sigma(k)}K_{\sigma(k)}Cx(k)$$

$$+ u^{\mathrm{T}}(k-1)\left(I - \Phi_{\sigma(k)}\right) Q_{\sigma(k+1)}\left(I - \Phi_{\sigma(k)}\right) u(k-1)$$
$$+ \vartheta_s^{\mathrm{T}}(k)\Phi_{\sigma(k)}Q_{\sigma(k+1)}\Phi_{\sigma(k)}\vartheta_s(k)$$
$$- 2x^{\mathrm{T}}(k)C^{\mathrm{T}}K_{\sigma(k)}^{\mathrm{T}}\Phi_{\sigma(k)}Q_{\sigma(k+1)}\left(I - \Phi_{\sigma(k)}\right) u(k-1)$$
$$+ 2x^{\mathrm{T}}(k)C^{\mathrm{T}}K_{\sigma(k)}^{\mathrm{T}}\Phi_{\sigma(k)}Q_{\sigma(k+1)}\Phi_{\sigma(k)}\vartheta_s(k)$$
$$- 2u^{\mathrm{T}}(k-1)\left(I - \Phi_{\sigma(k)}\right) Q_{\sigma(k+1)}\Phi_{\sigma(k)}\vartheta_s(k). \tag{3.30}$$

By resorting to Lemma 3.1 again, it obtains from (3.18) that

$$\vartheta_s^{\mathrm{T}}(k)\Phi_{\sigma(k)}Q_{\sigma(k+1)}\Phi_{\sigma(k)}\vartheta_s(k) \leq \vartheta^2\gamma_{\sigma(k),\sigma(k+1)}x^{\mathrm{T}}(k)C^{\mathrm{T}}Cx(k), \tag{3.31}$$
$$2x^{\mathrm{T}}(k)C^{\mathrm{T}}K_{\sigma(k)}^{\mathrm{T}}\Phi_{\sigma(k)}Q_{\sigma(k+1)}\Phi_{\sigma(k)}\vartheta_s(k)$$
$$\leq x^{\mathrm{T}}(k)C^{\mathrm{T}}K_{\sigma(k)}^{\mathrm{T}}\Phi_{\sigma(k)}Q_{\sigma(k+1)}\Phi_{\sigma(k)}K_{\sigma(k)}Cx(k)$$
$$+ \vartheta^2\gamma_{\sigma(k),\sigma(k+1)}x^{\mathrm{T}}(k)C^{\mathrm{T}}Cx(k), \tag{3.32}$$
$$- 2u^{\mathrm{T}}(k-1)\left(I - \Phi_{\sigma(k)}\right) Q_{\sigma(k+1)}\Phi_{\sigma(k)}\vartheta_s(k)$$
$$\leq u^{\mathrm{T}}(k-1)\left(I - \Phi_{\sigma(k)}\right) Q_{\sigma(k+1)}\left(I - \Phi_{\sigma(k)}\right) u(k-1)$$
$$+ \vartheta^2\gamma_{\sigma(k),\sigma(k+1)}x^{\mathrm{T}}(k)C^{\mathrm{T}}Cx(k). \tag{3.33}$$

Combining (3.30)–(3.33), we have

$$u^{\mathrm{T}}(k)Q_{\sigma(k+1)}u(k)$$
$$\leq x^{\mathrm{T}}(k)\left[2C^{\mathrm{T}}K_{\sigma(k)}^{\mathrm{T}}\Phi_{\sigma(k)}Q_{\sigma(k+1)}\Phi_{\sigma(k)}K_{\sigma(k)}C + 3\vartheta^2\gamma_{\sigma(k),\sigma(k+1)}C^{\mathrm{T}}C\right]x(k)$$
$$+ 2u^{\mathrm{T}}(k-1)\left(I - \Phi_{\sigma(k)}\right) Q_{\sigma(k+1)}\left(I - \Phi_{\sigma(k)}\right) u(k-1)$$
$$- 2x^{\mathrm{T}}(k)C^{\mathrm{T}}K_{\sigma(k)}^{\mathrm{T}}\Phi_{\sigma(k)}Q_{\sigma(k+1)}\left(I - \Phi_{\sigma(k)}\right) u(k-1). \tag{3.34}$$

Thus, we further obtain from (3.29) and (3.34) that

$$V(x(k+1), u(k), \sigma(k+1)) - V(x(k), u(k-1), \sigma(k))$$
$$= x^{\mathrm{T}}(k+1)P_{\sigma(k+1)}x(k+1) - x^{\mathrm{T}}(k)P_{\sigma(k)}x(k) + u^{\mathrm{T}}(k)Q_{\sigma(k+1)}u(k)$$
$$- u^{\mathrm{T}}(k-1)Q_{\sigma(k)}u(k-1)$$
$$\leq \xi^{\mathrm{T}}(k)\Psi_{\sigma(k),\sigma(k+1)}\xi(k), \tag{3.35}$$

where $\xi(k) \triangleq \begin{bmatrix} x^{\mathrm{T}}(k) & u^{\mathrm{T}}(k-1) \end{bmatrix}^{\mathrm{T}}$. *It is clear that the condition (3.19) guarantees*

$$V(x(k+1), u(k), \sigma(k+1)) < V(x(k), u(k-1), \sigma(k)), \tag{3.36}$$

which means that the asymptotic stability of the closed-loop system (3.13) is eventually attained. The proof is now complete.

Remark 3.4 *In Theorem 3.1, a sufficient condition is established to ensure that the dynamics of the SMC system (3.13) is asymptotically stable under the*

Main Results 49

Round-Robin protocol scheduling (3.6). It is shown from (3.36) that the control input signal $u(k-1)$ is also asymptotically stable. Actually, one can observe from the constructed output-feedback SMC law (3.11) that the control signal $\bar{u}(k) \to 0$ along with $x(k) \to 0$, which implies that the input term $u(k-1)$ could converge to zero as per the Round-Robin scheduling mechanism (3.6). This just responds to the question **Q-2.** *It should be pointed out that the existing discrete-time SMC approaches (as developed in [51, 62, 101, 109, 110, 180]) cannot be applied here since the coefficients of the switching term* $\text{sgn}(s(k))$ *in these proposed SMC laws are actually* nonzero *and, as such, the asymptotical convergence of the input signal $u(k-1)$ cannot be possessed.*

3.2.3 Analysis of the reachability

This subsection analyzes the reachability of the specified sliding surface (3.8). More specifically, by exploiting a token-dependent Lyapunov function, a sufficient condition will be derived to ensure that the state trajectories of the SMC system (3.13) enter into a sliding domain Ω around the specified sliding surface (3.8).

Theorem 3.2 *Consider the uncertain NCS (3.1)–(3.2) with the Round-Robin protocol (3.4)–(3.5). Let the parameter ϑ in the token-dependent SMC law (3.11) be given. If there exist matrices $K_{\sigma(k)} \in \mathbb{R}^{m \times p}$, $P_{\sigma(k)} > 0$, $Q_{\sigma(k)} > 0$, $X > 0$, and scalars $\eta_{\sigma(k)} > 0$, $\delta_{\sigma(k),\sigma(k+1)} > 0$, $\gamma_{\sigma(k),\sigma(k+1)} > 0$, $\upsilon > 0$, $\phi_{\sigma(k)} > 0$ such that the conditions (3.16)–(3.18) and the following matrix inequalities*

$$B^{\mathrm{T}} X B - \upsilon I \leq 0, \tag{3.37}$$

$$\Phi_{\sigma(k)} B^{\mathrm{T}} X B \Phi_{\sigma(k)} - \phi_{\sigma(k)} I \leq 0, \tag{3.38}$$

$$\tilde{\Psi}_{\sigma(k),\sigma(k+1)} \triangleq \begin{bmatrix} \tilde{\Psi}^{(1,1)}_{\sigma(k),\sigma(k+1)} & \Psi^{(1,2)}_{\sigma(k),\sigma(k+1)} \\ \star & \tilde{\Psi}^{(2,2)}_{\sigma(k),\sigma(k+1)} \end{bmatrix} < 0, \tag{3.39}$$

hold for any $\sigma(k), \sigma(k+1) \in \mathcal{M}$, where $\Psi^{(1,2)}_{\sigma(k),\sigma(k+1)}$ is defined as Theorem 3.1 and

$$\tilde{\Psi}^{(1,1)}_{\sigma(k),\sigma(k+1)} \triangleq \Psi^{(1,1)}_{\sigma(k),\sigma(k+1)} + 4\upsilon\alpha I + 4\tilde{A}^{\mathrm{T}}_{\sigma(k)} \left(B^{\mathrm{T}} X B\right)^{-1} \tilde{A}_{\sigma(k)},$$

$$\tilde{\Psi}^{(2,2)}_{\sigma(k),\sigma(k+1)} \triangleq \Psi^{(2,2)}_{\sigma(k),\sigma(k+1)} + 4 \left(I - \Phi_{\sigma(k)}\right) B^{\mathrm{T}} X B \left(I - \Phi_{\sigma(k)}\right),$$

$$\tilde{A}_{\sigma(k)} \triangleq B^{\mathrm{T}} X \left(A + \Delta A(k)\right) - B^{\mathrm{T}} X B \Phi_{\sigma(k)} K_{\sigma(k)} C,$$

then the state trajectories of the closed-loop system (3.13) will be forced into the following domain Ω around the specified sliding surface $s(k) = 0$:

$$\Omega \triangleq \left\{ s(k) \mid \|s(k)\| \leq \mu\|y(k)\| \right\}, \tag{3.40}$$

where $\mu \triangleq \max\limits_{\sigma(k) \in \mathcal{M}} \left\{ \sqrt{\dfrac{4\upsilon\beta + 4\phi_{\sigma(k)}\vartheta^2}{\lambda_{\min}\left[(B^{\mathrm{T}} X B)^{-1}\right]}} \right\}.$

Proof *Combining (3.9), (3.11) and (3.12), we have*

$$s(k+1) = \tilde{A}_{\sigma(k)} x(k) + B^{\mathrm{T}} X B \left(I - \Phi_{\sigma(k)} \right) u(k-1) + B^{\mathrm{T}} X B g(x, y)$$
$$- B^{\mathrm{T}} X B \Phi_{\sigma(k)} \vartheta_s(k). \tag{3.41}$$

Define the following token-dependent Lyapunov-like function as:

$$\tilde{V}(x(k), u(k-1), \sigma(k)) \triangleq x^{\mathrm{T}}(k) P_{\sigma(k)} x(k) + u^{\mathrm{T}}(k-1) Q_{\sigma(k)} u(k-1)$$
$$+ s^{\mathrm{T}}(k) \left(B^{\mathrm{T}} X B \right)^{-1} s(k), \qquad \sigma(k) \in \mathcal{M}. \tag{3.42}$$

From (3.41), we have

$$s^{\mathrm{T}}(k+1) \left(B^{\mathrm{T}} X B \right)^{-1} s(k+1)$$
$$= x^{\mathrm{T}}(k) \tilde{A}_{\sigma(k)}^{\mathrm{T}} \left(B^{\mathrm{T}} X B \right)^{-1} \tilde{A}_{\sigma(k)} x(k)$$
$$+ u^{\mathrm{T}}(k-1) \left(I - \Phi_{\sigma(k)} \right) B^{\mathrm{T}} X B \left(I - \Phi_{\sigma(k)} \right) u(k-1)$$
$$+ g^{\mathrm{T}}(x, y) B^{\mathrm{T}} X B g(x, y) + \vartheta_s^{\mathrm{T}}(k) \Phi_{\sigma(k)} B^{\mathrm{T}} X B \Phi_{\sigma(k)} \vartheta_s(k)$$
$$+ 2 x^{\mathrm{T}}(k) \tilde{A}_{\sigma(k)}^{\mathrm{T}} \left(I - \Phi_{\sigma(k)} \right) u(k-1) + 2 x^{\mathrm{T}}(k) \tilde{A}_{\sigma(k)}^{\mathrm{T}} g(x, y)$$
$$+ 2 u^{\mathrm{T}}(k-1) \left(I - \Phi_{\sigma(k)} \right) B^{\mathrm{T}} X B g(x, y)$$
$$- 2 u^{\mathrm{T}}(k-1) \left(I - \Phi_{\sigma(k)} \right) B^{\mathrm{T}} X B \Phi_{\sigma(k)} \vartheta_s(k)$$
$$- 2 x^{\mathrm{T}}(k) \tilde{A}_{\sigma(k)}^{\mathrm{T}} \Phi_{\sigma(k)} \vartheta_s(k) - 2 g^{\mathrm{T}}(x, y) B^{\mathrm{T}} X B \Phi_{\sigma(k)} \vartheta_s(k). \tag{3.43}$$

Bearing the conditions (3.37)–(3.38) in mind, we obtain

$$g^{\mathrm{T}}(x, y) B^{\mathrm{T}} X B g(x, y) \leq \upsilon \left[\alpha x^{\mathrm{T}}(k) x(k) + \beta \|y(k)\|^2 \right], \tag{3.44}$$
$$\vartheta_s^{\mathrm{T}}(k) \Phi_{\sigma(k)} B^{\mathrm{T}} X B \Phi_{\sigma(k)} \vartheta_s(k) \leq \phi_{\sigma(k)} \vartheta^2 \|y(k)\|^2, \tag{3.45}$$
$$2 x^{\mathrm{T}}(k) \tilde{A}_{\sigma(k)}^{\mathrm{T}} g(x, y)$$
$$\leq x^{\mathrm{T}}(k) \tilde{A}_{\sigma(k)}^{\mathrm{T}} \left(B^{\mathrm{T}} X B \right)^{-1} \tilde{A}_{\sigma(k)} x(k) + \upsilon \left[\alpha x^{\mathrm{T}}(k) x(k) + \beta \|y(k)\|^2 \right], \tag{3.46}$$
$$- 2 x^{\mathrm{T}}(k) \tilde{A}_{\sigma(k)}^{\mathrm{T}} \Phi_{\sigma(k)} \vartheta_s(k)$$
$$\leq x^{\mathrm{T}}(k) \tilde{A}_{\sigma(k)}^{\mathrm{T}} \left(B^{\mathrm{T}} X B \right)^{-1} \tilde{A}_{\sigma(k)} x(k) + \phi_{\sigma(k)} \vartheta^2 \|y(k)\|^2, \tag{3.47}$$
$$2 u^{\mathrm{T}}(k-1) \left(I - \Phi_{\sigma(k)} \right) B^{\mathrm{T}} X B g(x, y)$$
$$\leq u^{\mathrm{T}}(k-1) \left(I - \Phi_{\sigma(k)} \right) B^{\mathrm{T}} X B \left(I - \Phi_{\sigma(k)} \right) u(k-1)$$
$$+ \upsilon \left[\alpha x^{\mathrm{T}}(k) x(k) + \beta \|y(k)\|^2 \right], \tag{3.48}$$
$$- 2 u^{\mathrm{T}}(k-1) \left(I - \Phi_{\sigma(k)} \right) B^{\mathrm{T}} X B \Phi_{\sigma(k)} \vartheta_s(k)$$
$$\leq u^{\mathrm{T}}(k-1) \left(I - \Phi_{\sigma(k)} \right) B^{\mathrm{T}} X B \left(I - \Phi_{\sigma(k)} \right) u(k-1) + \phi_{\sigma(k)} \vartheta^2 \|y(k)\|^2, \tag{3.49}$$
$$- 2 g^{\mathrm{T}}(x, y) B^{\mathrm{T}} X B \Phi_{\sigma(k)} \vartheta_s(k)$$
$$\leq \upsilon \left[\alpha x^{\mathrm{T}}(k) x(k) + \beta \|y(k)\|^2 \right] + \phi_{\sigma(k)} \vartheta^2 \|y(k)\|^2, \tag{3.50}$$

Main Results 51

$$2x^{\mathrm{T}}(k)\tilde{A}_{\sigma(k)}^{\mathrm{T}}\left(I - \Phi_{\sigma(k)}\right)u(k-1)$$

$$\leq x^{\mathrm{T}}(k)\tilde{A}_{\sigma(k)}^{\mathrm{T}}\left(B^{\mathrm{T}}XB\right)^{-1}\tilde{A}_{\sigma(k)}x(k)$$

$$+ u^{\mathrm{T}}(k-1)\left(I - \Phi_{\sigma(k)}\right)B^{\mathrm{T}}XB\left(I - \Phi_{\sigma(k)}\right)u(k-1). \tag{3.51}$$

Thus, we have from (3.44)–(3.51) that

$$s^{\mathrm{T}}(k+1)\left(B^{\mathrm{T}}XB\right)^{-1}s(k+1)$$

$$\leq x^{\mathrm{T}}(k)\left[4\tilde{A}_{\sigma(k)}^{\mathrm{T}}\left(B^{\mathrm{T}}XB\right)^{-1}\tilde{A}_{\sigma(k)} + 4v\alpha I\right]x(k) + \left(4v\beta + 4\phi_{\sigma(k)}\vartheta^2\right)\|y(k)\|^2$$

$$+ 4u^{\mathrm{T}}(k-1)\left(I - \Phi_{\sigma(k)}\right)B^{\mathrm{T}}XB\left(I - \Phi_{\sigma(k)}\right)u(k-1) \tag{3.52}$$

Combining (3.29), (3.34), and (3.52) results in

$$\tilde{V}(x(k+1), u(k), \sigma(k+1)) - \tilde{V}(x(k), u(k-1), \sigma(k))$$

$$= x^{\mathrm{T}}(k+1)P_{\sigma(k+1)}x(k+1) - x^{\mathrm{T}}(k)P_{\sigma(k)}x(k) + u^{\mathrm{T}}(k)Q_{\sigma(k+1)}u(k)$$

$$- u^{\mathrm{T}}(k-1)Q_{\sigma(k)}u(k-1) + s^{\mathrm{T}}(k+1)\left(B^{\mathrm{T}}XB\right)^{-1}s(k+1)$$

$$- s^{\mathrm{T}}(k)\left(B^{\mathrm{T}}XB\right)^{-1}s(k)$$

$$\leq \xi^{\mathrm{T}}(k)\tilde{\Psi}_{\sigma(k),\sigma(k+1)}\xi(k) + \left(4v\beta + 4\phi_{\sigma(k)}\vartheta^2\right)\|y(k)\|^2$$

$$- \lambda_{\min}\left[\left(B^{\mathrm{T}}XB\right)^{-1}\right]\|s(k)\|^2. \tag{3.53}$$

If the state trajectories escape the sliding domain Ω, one has $\|s(k)\| \geq \mu\|y(k)\|$. Then, it is concluded from condition (3.39) that

$$\tilde{V}(x(k+1), u(k), \sigma(k+1)) - \tilde{V}(x(k), u(k-1), \sigma(k))$$

$$\leq \xi^{\mathrm{T}}(k)\tilde{\Psi}_{\sigma(k),\sigma(k+1)}\xi(k) < 0,$$

which implies that the state trajectories of the closed-loop system (3.13) are strictly deceasing outside the domain Ω defined in (3.40). Note that the proposed upper bound of the domain is a function of the measured output $y(k)$. The proof is now complete.

Remark 3.5 *It is noted that Theorem 3.2 can only guarantee the state trajectories of the closed-loop system (3.13) to be driven into a neighbourhood of the specified sliding surface (3.8) by the static output-feedback SMC law (3.11) with the Round-Robin protocol scheduling (3.6). This is just the characteristic of discrete-time SMC, see [6, 51, 62, 101, 109] for more details. In addition, it is seen from (3.40) that the upper bound of the obtained sliding domain Ω is related to the periodic scheduling signal $\sigma(k)$, which is another key feature of the proposed Round-Robin-protocol-based output-feedback SMC scheme.*

3.2.4 Solving algorithm

For any $\sigma(k), \sigma(k+1) \in \mathcal{M}$, the following notations are introduced:

$$\bar{P}_{\sigma(k)} \triangleq P^{-1}_{\sigma(k)}, \ \bar{Q}_{\sigma(k)} \triangleq Q^{-1}_{\sigma(k)}, \ \bar{X} \triangleq X^{-1}, \ \bar{\delta}_{\sigma(k),\sigma(k+1)} \triangleq \delta^{-1}_{\sigma(k),\sigma(k+1)},$$
$$\bar{\eta}_{\sigma(k+1)} \triangleq \eta^{-1}_{\sigma(k+1)}, \ \bar{\gamma}_{\sigma(k+1)} \triangleq \gamma^{-1}_{\sigma(k+1)}, \ \bar{\upsilon} \triangleq \upsilon^{-1}, \ \bar{\phi}_{\sigma(k)} \triangleq \phi^{-1}_{\sigma(k)}. \tag{3.54}$$

It is noted that, in order to guarantee both the reachability of the specified sliding surface (3.8) and the asymptotic stability of the SMC system (3.13), the conditions in Theorems 3.1 and 3.2 should be solved simultaneously, which is a fairly difficult task due primarily to some nonconvex terms $B\Phi_{\sigma(k)}K_{\sigma(k)}C\bar{P}_{\sigma(k)}$ in the conditions (3.19) and (3.39). As is well known, the coupling relationship between the controller gain and the Lyapunov variable is the key obstacle in addressing static output-feedback control issues, see [130] for more details. In this chapter, inspired by [105, 126], we introduce the following separation strategy:

$$B\Phi_{\sigma(k)}K_{\sigma(k)}C\bar{P}_{\sigma(k)} = B\Phi_{\sigma(k)}\bar{K}_{\sigma(k)} + B\Phi_{\sigma(k)}\bar{L}_{\sigma(k)}, \tag{3.55}$$

where $\bar{L}_{\sigma(k)} \triangleq K_{\sigma(k)}C\bar{P}_{\sigma(k)} - \bar{K}_{\sigma(k)}$.

Remark 3.6 *Along the similar line to Remark 3.1, if the matrices $\bar{K}_{\sigma(k)}$, $\bar{L}_{\sigma(k)}$ and $\bar{P}_{\sigma(k)}$ in (3.55) are known, the gain matrices $K_{\sigma(k)}$ can be obtained by solving the following convex optimization problem [62]:*

$$\begin{bmatrix} -\tilde{\varrho}I & \bar{K}_s + \bar{L}_s - K_sC\bar{P}_s \\ \star & -I \end{bmatrix} < 0, \ s \in \mathcal{M}, \tag{3.56}$$

where $\tilde{\varrho} > 0$ is a sufficiently small scalar. Although the two approximations (3.10) and (3.56) may lead to some extra conservatism of the obtained sufficient conditions, the sufficiently small ϱ and $\tilde{\varrho}$ prescribed in practical application can be used to render the algorithm feasible (see the example to be given later).

Remark 3.7 *It is worth mentioning that by means of the descriptor system representation, an effective separation approach to the Lyapunov matrices and the controller matrices was developed in [130], which may motivate us to consider how to solve the static output-feedback SMC problem with the system augmentation technique in future work.*

By resorting to the notations in (3.54) and the proposed separation strategy (3.55), we are going to establish a sufficient condition in terms of LMIs in the following theorem such that the conditions in Theorems 3.1 and 3.2 are ensured simultaneously.

Theorem 3.3 *Consider the uncertain NCS (3.1)–(3.2) with the Round-Robin protocol (3.4)–(3.5). Let the parameter ϑ in the token-dependent SMC law*

Main Results 53

(3.11) be given. If, for some prescribed positive scalars τ_1, τ_2, and κ, there exist matrices $\bar{K}_s \in \mathbb{R}^{m\times n}$, $\bar{L}_s \in \mathbb{R}^{m\times n}$, $\bar{P}_s > 0$, $\bar{Q}_s > 0$, $\bar{X} > 0$, and scalars $\bar{\eta}_s > 0$, $\bar{\delta}_{s,s+1} > 0$, $\bar{\gamma}_{s,s+1} > 0$, $\bar{\upsilon} > 0$, $\bar{\phi}_s > 0$, $\bar{\epsilon} > 0$, $\varepsilon_s > 0$ $(s \in \mathcal{M})$ such that the following LMIs hold:

$$\begin{bmatrix} -\bar{\upsilon}I & \bar{\upsilon}B^{\mathrm{T}} \\ \star & -\bar{X} \end{bmatrix} \leq 0, \tag{3.57}$$

$$\begin{bmatrix} -2\tau_1\bar{X} + \tau_1^2\bar{\epsilon}I & B \\ \star & -2\tau_2 B^{\mathrm{T}}B + \tau_2^2 B^{\mathrm{T}}\bar{X}B \end{bmatrix} < 0, \tag{3.58}$$

and for $s \in \{1, 2, \cdots, m-1\}$

$$\begin{bmatrix} -\bar{\eta}_{s+1}I & \bar{\eta}_{s+1}B^{\mathrm{T}} \\ \star & -\bar{P}_{s+1} \end{bmatrix} \leq 0, \tag{3.59}$$

$$\begin{bmatrix} -\bar{\delta}_{s,s+1}I & \bar{\delta}_{s,s+1}\Phi_s B^{\mathrm{T}} \\ \star & -\bar{P}_{s+1} \end{bmatrix} \leq 0, \tag{3.60}$$

$$\begin{bmatrix} -\bar{\gamma}_{s,s+1}I & \bar{\gamma}_{s,s+1}\Phi_s \\ \star & -\bar{Q}_{s+1} \end{bmatrix} \leq 0, \tag{3.61}$$

$$\begin{bmatrix} -\bar{\phi}_s I & \bar{\phi}_s\Phi_s B^{\mathrm{T}} \\ \star & -\bar{X} \end{bmatrix} \leq 0, \tag{3.62}$$

$$\begin{bmatrix} -\mathrm{diag}\left\{\bar{P}_s, \bar{Q}_s\right\} & \Gamma_s & \Xi_s \\ \star & -\Lambda_{s,s+1} & \Sigma_s \\ \star & \star & -\mathrm{diag}\left\{\varepsilon_s I, \varepsilon_s I\right\} \end{bmatrix} < 0, \tag{3.63}$$

and for $s = m$

$$\begin{bmatrix} -\bar{\eta}_1 I & B^{\mathrm{T}}\bar{\eta}_1 \\ \star & -\bar{P}_1 \end{bmatrix} \leq 0, \tag{3.64}$$

$$\begin{bmatrix} -\bar{\delta}_{m,1}I & \bar{\delta}_{m,1}\Phi_m B^{\mathrm{T}} \\ \star & -\bar{P}_1 \end{bmatrix} \leq 0, \tag{3.65}$$

$$\begin{bmatrix} -\bar{\gamma}_{m,1}I & \bar{\gamma}_{m,1}\Phi_m \\ \star & -\bar{Q}_1 \end{bmatrix} \leq 0, \tag{3.66}$$

$$\begin{bmatrix} -\bar{\phi}_m I & \bar{\phi}_m\Phi_m B^{\mathrm{T}} \\ \star & -\bar{X} \end{bmatrix} \leq 0, \tag{3.67}$$

$$\begin{bmatrix} -\mathrm{diag}\left\{\bar{P}_m, \bar{Q}_m\right\} & \Gamma_m & \Xi_m \\ \star & -\Lambda_{m,1} & \Sigma_m \\ \star & \star & -\mathrm{diag}\left\{\varepsilon_m I, \varepsilon_m I\right\} \end{bmatrix} < 0, \tag{3.68}$$

where $\tilde{A}_s \triangleq A\bar{P}_s - \tilde{K}_s$, $\tilde{K}_s \triangleq B\Phi_s\bar{K}_s + B\Phi_s\bar{L}_s$, $\Gamma_s \triangleq \begin{bmatrix} \Gamma_s^1 & \Gamma_s^2 & \Gamma_s^3 & \Gamma_s^4 \end{bmatrix}$, and

$$\Gamma_s^1 \triangleq \begin{bmatrix} \tilde{A}_s^{\mathrm{T}} & \sqrt{2}\tilde{A}_s^{\mathrm{T}} & \tilde{K}_s^{\mathrm{T}} & \tilde{K}_s^{\mathrm{T}} \\ \bar{Q}_s(I - \Phi_s)B^{\mathrm{T}} & 0 & -\bar{Q}_s(I - \Phi_s) & 0 \end{bmatrix},$$

$$\Gamma_s^2 \triangleq \begin{bmatrix} 2\vartheta\bar{P}_s C^{\mathrm{T}} & 2\sqrt{\alpha}\bar{P}_s & 2\sqrt{\beta}\bar{P}_s C^{\mathrm{T}} & \sqrt{3}\vartheta\bar{P}_s C^{\mathrm{T}} \\ 0 & 0 & 0 & 0 \end{bmatrix},$$

$$\Gamma_s^3 \triangleq \begin{bmatrix} 2\sqrt{\alpha}\bar{P}_s & 2\sqrt{1+\kappa^{-1}}\bar{P}_s A^{\mathrm{T}} & 2\sqrt{1+\kappa}\tilde{K}_s^{\mathrm{T}} \\ 0 & 0 & 0 \end{bmatrix},$$

$$\Gamma_s^4 \triangleq \begin{bmatrix} 0 & 0 & 0 \\ \sqrt{2}\bar{Q}_s(I-\Phi_s)B^{\mathrm{T}} & -\bar{Q}_s(I-\Phi_s) & 2\bar{Q}_s(I-\Phi_s)B^{\mathrm{T}} \end{bmatrix},$$

$$\Xi_s \triangleq \begin{bmatrix} \bar{P}_s N^{\mathrm{T}} & 0 \\ 0 & 0 \end{bmatrix},$$

$$\Sigma_s \triangleq \begin{bmatrix} \Sigma_s^1 & \Sigma_s^2 \end{bmatrix}^{\mathrm{T}},$$

$$\Sigma_s^1 \triangleq \begin{bmatrix} 0 & 0 & 0 & 0 & 0 & 0 & 0 & 0 & 0 \\ \varepsilon_s M^{\mathrm{T}} & \sqrt{2}\varepsilon_s M^{\mathrm{T}} & 0 & 0 & 0 & 0 & 0 & 0 & 0 \end{bmatrix},$$

$$\Sigma_s^2 \triangleq \begin{bmatrix} 0 & 0 & 0 & 0 & 0 \\ 2\varepsilon_s\sqrt{1+\kappa^{-1}}M^{\mathrm{T}} & 0 & 0 & 0 & 0 \end{bmatrix},$$

$s \in \mathcal{M};$

$$\Lambda_{s,s+1} \triangleq \mathrm{diag}\Big\{ \bar{P}_{s+1}, \bar{P}_{s+1}, \bar{Q}_{s+1}, \bar{Q}_{s+1}, \bar{\delta}_{s,s+1}I, \bar{\eta}_{s+1}I, \bar{\eta}_{s+1}I, \bar{\gamma}_{s,s+1}I, \bar{\upsilon}I, \bar{\phi}_s I,$$
$$\bar{\epsilon}I, \bar{X}, \bar{P}_{s+1}, \bar{Q}_{s+1}, \bar{X} \Big\}, \forall s \in \{1,2,\ldots,m-1\};$$

$$\Lambda_{m,1} \triangleq \mathrm{diag}\Big\{ \bar{P}_1, \bar{P}_1, \bar{Q}_1, \bar{Q}_1, \bar{\delta}_{m,1}I, \bar{\eta}_1 I, \bar{\eta}_1 I, \bar{\gamma}_{m,1}I, \bar{\upsilon}I, \bar{\epsilon}I, \bar{X}, \bar{P}_1, \bar{Q}_1, \bar{X} \Big\},$$

then the asymptotic stability of the closed-loop system (3.13) and the reachability of the sliding domain Ω defined in (3.40) can be guaranteed simultaneously by the token-dependent SMC law (3.11) with the matrix K_s ($s \in \mathcal{M}$) satisfying $\bar{L}_s + \bar{K}_s = K_s C\bar{P}_s$.

Proof *Considering the following conditions:*

$$B^{\mathrm{T}}P_{\sigma(k+1)}B \le \eta_{\sigma(k+1)}I, \tag{3.69}$$

$$\Phi_{\sigma(k)}B^{\mathrm{T}}P_{\sigma(k+1)}B\Phi_{\sigma(k)} \le \delta_{\sigma(k),\sigma(k+1)}I, \tag{3.70}$$

$$\Phi_{\sigma(k)}Q_{\sigma(k+1)}\Phi_{\sigma(k)} \le \gamma_{\sigma(k),\sigma(k+1)}I, \tag{3.71}$$

$$B^{\mathrm{T}}XB \le \upsilon I, \tag{3.72}$$

$$\Phi_{\sigma(k)}B^{\mathrm{T}}XB\Phi_{\sigma(k)} \le \phi_{\sigma(k)}I, \tag{3.73}$$

$$\bar{\Psi}_{\sigma(k),\sigma(k+1)} \triangleq \begin{bmatrix} \bar{\Psi}_{\sigma(k),\sigma(k+1)}^{(1,1)} & \Psi_{\sigma(k),\sigma(k+1)}^{(1,2)} \\ \star & \tilde{\Psi}_{\sigma(k),\sigma(k+1)}^{(2,2)} \end{bmatrix} < 0, \tag{3.74}$$

where

$$\bar{\Psi}_{\sigma(k),\sigma(k+1)}^{(1,1)} \triangleq \Psi_{\sigma(k),\sigma(k+1)}^{(1,1)} + 4\upsilon\alpha I + 4(1+\kappa)C^{\mathrm{T}}K_{\sigma(k)}^{\mathrm{T}}\Phi_{\sigma(k)}B^{\mathrm{T}}XB\Phi_{\sigma(k)}K_{\sigma(k)}C$$
$$+ 4(1+\kappa^{-1})(A+\Delta A(k))^{\mathrm{T}}XB\left(B^{\mathrm{T}}XB\right)^{-1}B^{\mathrm{T}}X(A+\Delta A(k)).$$

Since $\bar{\Psi}_{\sigma(k),\sigma(k+1)}^{(1,1)} > \Psi_{\sigma(k),\sigma(k+1)}^{(1,1)}$, the conditions (3.16)–(3.19) in Theorem 3.1 are ensured by the inequalities (3.69)–(3.71) and (3.74). On the other

Main Results   55

hand, for an adjustable scalar $\kappa > 0$, one has

$$\tilde{A}_{\sigma(k)}^{\mathrm{T}} \left(B^{\mathrm{T}} X B\right)^{-1} \tilde{A}_{\sigma(k)}$$

$$\leq (1 + \kappa^{-1})(A + \Delta A(k))^{\mathrm{T}} X B \left(B^{\mathrm{T}} X B\right)^{-1} B^{\mathrm{T}} X (A + \Delta A(k))$$

$$+ (1 + \kappa) C^{\mathrm{T}} K_{\sigma(k)}^{\mathrm{T}} \Phi_{\sigma(k)} B^{\mathrm{T}} X B \Phi_{\sigma(k)} K_{\sigma(k)} C,$$

which implies that $\bar{\Psi}_{\sigma(k),\sigma(k+1)}^{(1,1)} > \tilde{\Psi}_{\sigma(k),\sigma(k+1)}^{(1,1)}$. Therefore, we conclude that the conditions (3.37)–(3.39) in Theorem 3.2 are guaranteed by the inequalities (3.72)–(3.74).

Letting $\sigma(k) = s \in \mathcal{M}$ and noticing that

$$\begin{cases} \sigma(k+1) - \sigma(k) = 1, & \text{if } k \neq rm, \\ \sigma(k) = m \text{ and } \sigma(k+1) = 1, & \text{if } k = rm, \ r \in \mathbb{Z}^+ \end{cases} \tag{3.75}$$

one can obtain LMIs (3.57), (3.59)–(3.62), and (3.64)–(3.67) from the inequalities (3.69)–(3.73) directly by using the notations in (3.54) and the Schur complement.

Next, we consider the following inequality:

$$XB \left(B^{\mathrm{T}} X B\right)^{-1} B^{\mathrm{T}} X \leq \epsilon I, \tag{3.76}$$

which can be ensured by the condition (3.58) with $\bar{\epsilon} \triangleq \epsilon^{-1}$ since the relationships $-\epsilon \bar{X}^2 \leq -2\tau_1 \bar{X} + \tau_1^2 \bar{\epsilon} I$ and $-B^{\mathrm{T}} X B \leq -2\tau_2 B^{\mathrm{T}} B + \tau_2^2 B^{\mathrm{T}} \bar{X} B$ hold for $\tau_1 > 0$ and $\tau_2 > 0$ by Lemma 3.2. Moreover, taking the inequality (3.76) and the parameter uncertainty into account, it is easy to find that the inequality (3.74) is guaranteed by the LMIs (3.58), (3.63), and (3.68). The proof is complete.

Remark 3.8 *It is shown from Theorem 3.3 that the token-dependent output-feedback SMC law (3.11) can not only drive the state trajectories of the closed-loop system (3.13) into the sliding domain Ω around the specified sliding surface (3.8), but also ensure the asymptotic stability of the system state $x(k)$ and the control input signal $u(k-1)$, which shows that the effects from the periodic scheduler $\sigma(k)$ and the ZOHs have been handled effectively.*

Remark 3.9 *It is noted that the selection of the parameters ϑ, κ, τ_1, and τ_2 would largely affect the feasibility of the Theorem 3.3. For practicality, one can simply determine these parameters by checking the feasibility of the LMIs (3.57)–(3.68) through the usage of a four-dimensional search.*

From the above analysis, the design procedures of the static output-feedback SMC (3.11) under Round-Robin protocol (3.4)–(3.5) can be summarized in Algorithm 1.

Algorithm 1

▶ **Step 1.** Solve LMIs (3.57)–(3.68) by selecting adjustable scalars $\vartheta > 0$, $\kappa > 0$, $\tau_1 > 0$, and $\tau_2 > 0$, if at least a set of feasible solutions are found, go to Step 2; else, the design parameters must be modified.

▶ **Step 2.** Obtain the matrix $X = \vec{X}^{-1} > 0$, and determine the matrix F in sliding function (3.8) by solving the convex optimization problem (3.10) with a sufficiently small $\varrho > 0$.

▶ **Step 3.** Get the matrices $\vec{K}_s$, $\vec{L}_s$ and $\vec{P}_s > 0$ ($s \in \mathcal{M}$), then attain the gain matrix K_s in SMC law (3.11) by solving the convex optimization problem (3.56) with a sufficiently small $\tilde{\varrho} > 0$.

▶ **Step 4.** Produce the static output-feedback SMC law (3.11) and apply it to the networked system (3.1) under the Round-Robin protocol (3.4)–(3.5).

3.3 Example

Consider an uncertain discrete-time system (3.1)–(3.2) with the following parameters:

$$A = \begin{bmatrix} 0.9 & -0.1 \\ 0.02 & 1 \end{bmatrix}, \ B = \begin{bmatrix} 0.1 & 0.02 \\ 0.08 & 0.4 \end{bmatrix},$$

$$C = \begin{bmatrix} 1 & 0 \\ 0 & 1 \end{bmatrix}, \ M = \begin{bmatrix} -0.1 \\ 0.5 \end{bmatrix}, \ N = \begin{bmatrix} -0.2 & -0.3 \end{bmatrix},$$

$$Z(k) = \cos(k), \ g(x,y) = \begin{bmatrix} \sqrt{x_1^2 + y_2^2} \\ \sin\left(x_2^2 + y_1^2\right) \end{bmatrix}.$$

Clearly, the parameters α and β in (3.3) can be chosen as $\alpha = \beta = 1$. It is easy to verify that the uncontrolled uncertain system (when $u(k) = 0$) under the initial condition $x(0) = \begin{bmatrix} 0.5 & -1 \end{bmatrix}^{\mathrm{T}}$ is unstable as shown in Fig. 3.2.

Choose the scalars $\tau_1 = \tau_2 = 1$, $\kappa = 0.9$ in Theorem 3.3, $\vartheta = 0.01$ in the SMC law (3.11), and $\varrho = \tilde{\varrho} = 10^{-8}$ in Algorithm 1. To examine the influence of the Round-Robin protocol, we design the static output-feedback SMC law in the following two cases, that is, without and with Round-Robin protocol.

• **Case 1 (without Round-Robin protocol).** Set the scheduling signal to be $\sigma(k) = 1$ for all $k \in \mathbb{N}^+$, and the updating matrix to be $\Phi_1 = I_2$ in the Round-Robin protocol (3.6). Then, the underlying communication network reduces to the common type without Round-Robin protocol, and the actual control input (3.6) in plant side becomes $u(k) = \bar{u}(k)$. Applying Algorithm 1, we obtain the matrix F in sliding function (3.8) and the

Example

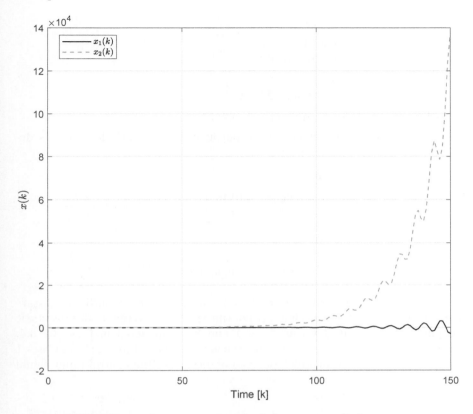

FIGURE 3.2: State trajectories of the uncontrolled system.

matrix K_1 in SMC law (3.11) as follows:

$$F = \begin{bmatrix} 0.0786 & 0.0556 \\ 0.0136 & 0.2807 \end{bmatrix}, \ K_1 = \begin{bmatrix} 8.2191 & -1.3662 \\ -1.5926 & 2.5254 \end{bmatrix},$$

with obtaining the parameter $\mu = 0.2560$.

Subsequently, the static output-feedback SMC law (3.11) is given as:

$$\bar{u}(k) = -\begin{bmatrix} 8.2191 & -1.3662 \\ -1.5926 & 2.5254 \end{bmatrix} y(k) - 0.01\|y(k)\|\frac{s(k)}{\|s(k)\|}. \quad (3.77)$$

- **Case 2 (with Round-Robin protocol).** In this case, we let the updating matrices be $\Phi_1 = \text{diag}\{1,0\}$ and $\Phi_2 = \text{diag}\{0,1\}$. By applying Algorithm 1, the matrix F in sliding function (3.8) and the matrix K_1 in SMC law

(3.11) are calculated as

$$F = \begin{bmatrix} 0.0673 & 0.0519 \\ 0.0113 & 0.2622 \end{bmatrix}, \quad K_1 = \begin{bmatrix} 5.0652 & 4.9148 \\ 0 & 0 \end{bmatrix},$$

$$K_2 = \begin{bmatrix} 0 & 0 \\ -0.1944 & 2.3526 \end{bmatrix}.$$

with obtaining the parameter $\mu = 0.2417$.

Now, the token-dependent static output-feedback SMC law $\bar{u}(k)$ is designed as follows:

— if $\sigma(k) = 1$, then

$$\bar{u}(k) = -\begin{bmatrix} 5.0652 & 4.9148 \\ 0 & 0 \end{bmatrix} y(k) - 0.01\|y(k)\| \frac{s(k)}{\|s(k)\|},$$

— if $\sigma(k) = 2$, then

$$\bar{u}(k) = -\begin{bmatrix} 0 & 0 \\ -0.1944 & 2.3526 \end{bmatrix} y(k) - 0.01\|y(k)\| \frac{s(k)}{\|s(k)\|}.$$

Under the initial condition $x(0) = \begin{bmatrix} -3 & 2 \end{bmatrix}^{\mathrm{T}}$, the evolutions of state trajectories and sliding variables $s(k)$ of the controlled system under the static output-feedback SMC laws in Cases 1 and 2 are depicted in Figs. 3.3–3.4, respectively. Figs. 3.5a and 3.5b show the actual control input $u(k)$ in Cases 1 and 2, respectively, from which one can observe that the control input signal

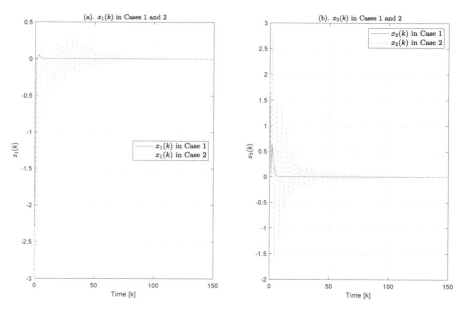

FIGURE 3.3: State trajectories of the closed-loop systems in Cases 1 & 2.

Example 59

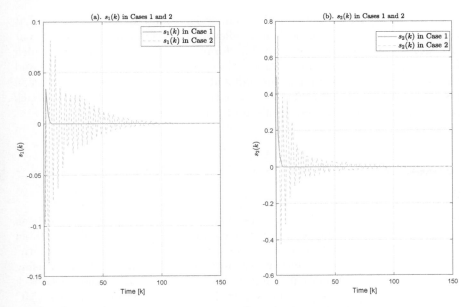

FIGURE 3.4: Sliding variables $s(k)$ of the closed-loop systems in Cases 1 & 2.

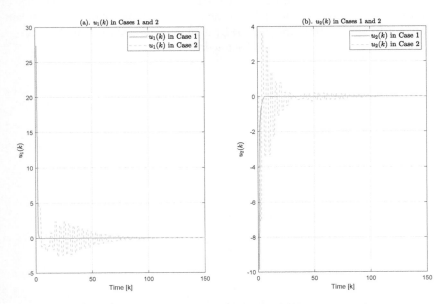

FIGURE 3.5: Actual control input $u(k)$ in Cases 1 & 2.

$u(k-1)$ converges to zero indeed. By comparing Cases 1 and 2, it is shown that the Round-Robin protocol mechanism alleviates the communication burden at the cost of sacrificing certain control performances of the SMC systems, which include the convergence of the state trajectories (see Figs. 3.3a and 3.3b) and the sliding variables (see Figs. 3.4a and 3.4b). Nevertheless, all the simulation results have confirmed the effectiveness of the proposed static output-feedback SMC approach under Round-Robin protocol scheduling.

3.4 Conclusion

In this chapter, the static output-feedback SMC design problem has been studied for uncertain NCSs under Round-Robin protocol. The Round-Robin protocol has been applied to determine which actuator gets access to the shared network between sliding mode controller and actuators. By only using the measurement information, a linear sliding surface and a output-feedback SMC law have been constructed properly. The proposed output-feedback SMC scheme exhibit the following distinct features: 1) to cope with the impacts from the introduction of the Round-Robin protocol scheduling and the ZOHs, the designed SMC law not only depends on the access token but also possesses the convergence property on the control signal $u(k-1)$; 2) the reachability of a sliding domain around the specified sliding surface is ensured by a token-dependent Lyapunov function; and 3) by exploiting a *new* separation strategy, the controller parameters can be designed easily.

4

Observer-Based Sliding Mode Control Under Weighted Try-Once-Discard Protocol

State-saturated systems are a typical kind of nonlinear systems whose states are constrained into a prescribed bounded area. The key motivation for studying state-saturated systems is that the state saturation occurs frequently in many real systems due to physical limitations of the devices or by protection equipment. Over the past three decades, much work has been done on dealing with the effects of saturation nonlinearities.

In this chapter, we endeavour to design the observer-based sliding mode controller for a class of state-saturated networked control systems (NCSs) under weighted Try-Once-Discard (WTOD) protocol. The key idea of WTOD protocol is to utilize dynamic quadratic functions to select most-needed-devices for accessing the resource which, in turn, seamlessly incorporates the corresponding scheduling behaviours into Lyapunov-theory-based designs. At each transmission instant, only one sensor obtains access to the communication network, which is determined by the WTOD scheduling scheme. By designing a state-saturated observer, the observer-based SMC law is constructed properly with aid of an updating rule on the *actually* received measurement. *The main contributions of this chapter are highlighted as follows: 1) The SMC problem is, for the first time, investigated for networked state-saturated systems under the WTOD protocol; 2) In order to cope with the impacts from the state saturation and the WTOD scheduling, a token-dependent state-saturated observer is constructed and then a suitable token-dependent sliding mode control (SMC) law is designed by using the state estimates and the buffer; and 3) A new algorithm is proposed to design the desired observer-based SMC scheme through solving a convex optimization problem by using standard software package.*

DOI: 10.1201/9781003309499-4

61

4.1 Problem Formulation

Consider the following class of state-saturated NCSs:

$$x(k+1) = \sigma\Big((A + \Delta A(k))\, x(k) + B\, (u(k) + f(x(k)))\Big), \qquad (4.1)$$

$$y(k) = Cx(k), \qquad (4.2)$$

where $x(k) \in \mathbb{R}^n$ is the state vector, $u(k) \in \mathbb{R}^m$ is the control input vector, $y(k) \in \mathbb{R}^p$ is the measured output vector. The time-varying uncertainty $\Delta A(k)$ is assumed to be norm-bounded, that is, $\Delta A(k) = M\Delta(k)N$ where M and N are the known constant matrices, and the unknown time-varying matrix $\Delta(k)$ satisfies $\Delta^{\mathrm{T}}(k)\Delta(k) \leq I$. $A \in \mathbb{R}^{n \times n}$, $B \in \mathbb{R}^{n \times m}$ and $C \in \mathbb{R}^{p \times n}$ are known constant matrices.

The uncertain NCS (4.1)–(4.2) satisfies the following assumptions.

Assumption 4.1 *The input matrix B has a full column rank, the output matrix C has a full row rank, and* rank $(CB) = m$.

Assumption 4.2 *(see Page 96 in [195]) The parameter variation $\Delta A(k)$ is unmatched, that is, $\Delta A(k) \notin$ range(B).*

Assumption 4.3 *The nonlinear function $f(x(k))$ with $f(0) = 0$ satisfies the following condition:*

$$\|f(x(k))\| \leq \nu \|x(k)\| \qquad (4.3)$$

where $\nu \geq 0$ is a known scalar.

The saturation function $\sigma(\cdot) : \mathbb{R}^n \to \mathbb{R}^n$ is expressed by

$$\sigma(\mu) \triangleq \begin{bmatrix} \sigma_1(\mu_1) & \sigma_2(\mu_2) & \cdots & \sigma_n(\mu_n) \end{bmatrix}^{\mathrm{T}}, \ \forall \mu \in \mathbb{R}^n, \qquad (4.4)$$

where the ith scalar-valued saturation function $\sigma_i(\mu_i)$ is defined as follows:

$$\sigma_i(\mu_i) \triangleq \mathrm{sgn}(\mu_i) \min \{\bar{s}_i, |\mu_i|\}, \qquad (4.5)$$

with μ_i presenting the ith element of the vector μ and $\bar{s}_i$ standing for the saturation level [128].

According to the definition of the saturation function in (4.5), it is easy to obtain the following technical lemmas.

Lemma 4.1 *[162] For the saturation function $\sigma_i(\cdot)$ in (4.5) and any $z_1, z_2 \in \mathbb{R}$, there exists a real number $\varepsilon_i \in [0, 1]$ such that*

$$\sigma_i(z_1) - \sigma_i(z_2) = \varepsilon_i(z_1 - z_2), \ i \in \{1, 2, \ldots, n\}.$$

Problem Formulation

Lemma 4.2 *For the saturation function $\sigma_i(\mu_i)$ in (4.5), there exists a real number $\varepsilon_i \in [0,1]$ such that*

$$\sigma_i(\mu_i) = \varepsilon_i \mu_i, \ i \in \{1, 2, \ldots, n\}.$$

Proof *Noting the fact that $\sigma_i(0) = 0$, the proof follows from Lemma 4.1 directly by letting $z_1 = \mu_i$ and $z_2 = 0$.*

In practice, the system states are not always available due to the limit of physical condition or the lack of expense to measure. In this case, it is important to estimate the system states by using the measured output. Unfortunately, in the context of SMC problems, the state estimation problem for our addressed networked state-saturated systems becomes much complicated than those investigated in existing works (e.g. [40, 83]) since the communication protocol has a major impact on the actually received measurement by the observer.

As shown in Fig. 4.1, the measured output $y(k)$ is transmitted to the controller and the observer via a shared communication channel subject to the following WTOD protocol: *at each transmission instant, only one sensor node, which has the largest difference between the latest transmitted values and the current values of the signals corresponding to the node, will be granted access to the network, and the other nodes keep the received values unchanged via the buffer [98, 99, 144] until the renewed data is arrived.*

Obviously, with the implementation of the WTOD protocol, the communication burden would be significantly reduced at each instant, thereby relieving the pressure on the communication channel.

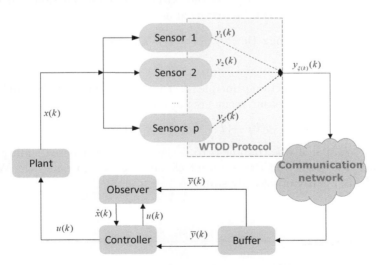

FIGURE 4.1: The control systems under the WTOD protocol scheduling.

Let the scheduling signal $\xi(k) \in \mathbf{T} \triangleq \{1, 2, \ldots, p\}$ denote the selected sensor node obtaining the *access* to the sensor-to-controller (S/C) network at the transmission instant k. For technical analysis, define

$$y(k) \triangleq \begin{bmatrix} y_1^{\mathrm{T}}(k) & y_2^{\mathrm{T}}(k) & \cdots & y_p^{\mathrm{T}}(k) \end{bmatrix}^{\mathrm{T}} \in \mathbb{R}^p,$$

$$\bar{y}(k) \triangleq \begin{bmatrix} \bar{y}_1^{\mathrm{T}}(k) & \bar{y}_2^{\mathrm{T}}(k) & \cdots & \bar{y}_p^{\mathrm{T}}(k) \end{bmatrix}^{\mathrm{T}} \in \mathbb{R}^p,$$

where $y_i(k)$ is the measurement of the ith sensor and $\bar{y}_i(k)$ is the ith measurement output after transmitted through S/C network.

The WTOD protocol is a dynamical protocol under which the value of the scheduling signal $\xi(k)$ is determined by the following selection principle:

$$\xi(k) \triangleq \arg \max_{1 \leq i \leq p} \left(y_i(k) - \bar{y}_i(k-1) \right) Q_i \left(y_i(k) - \bar{y}_i(k-1) \right), \qquad (4.6)$$

where $\bar{y}_i(k-1)$ represents the previously transmitted signal of the sensor node i, and $Q_i > 0$ ($i \in \mathbf{T}$) is the weighting matrix. Furthermore, we make the following assumptions for the selection principle (4.6):

- In case two or more nodes have the same minimal values, one of them is chosen arbitrarily;

- $\bar{y}(-1) = 0$.

Define $\bar{Q} \triangleq \mathrm{diag} \{Q_1, Q_2, \ldots, Q_p\}$ and the following updating matrix:

$$\Phi_i \triangleq \mathrm{diag} \left\{ \tilde{\delta}_i^1, \tilde{\delta}_i^2, \ldots, \tilde{\delta}_i^p \right\}, \ \forall i \in \{1, 2, \ldots, p\},$$

where $\tilde{\delta}_a^b \triangleq \delta(a-b)$ with $\delta(\cdot)$ as the Kronecker delta function. Then, the selection principle (4.6) can be rewritten as

$$\xi(k) \triangleq \arg \max_{1 \leq i \leq p} \left(y(k) - \bar{y}(k-1) \right)^{\mathrm{T}} \bar{Q}_i \left(y(k) - \bar{y}(k-1) \right), \qquad (4.7)$$

with $\bar{Q}_i = \bar{Q}\Phi_i$.

Based upon the selection principle (4.7) among the sensor nodes, we further express the received measurement $\bar{y}(k)$ at controller side as

$$\bar{y}(k) = \Phi_{\xi(k)} y(k) + \left(I - \Phi_{\xi(k)} \right) \bar{y}(k-1), \ k \in \mathbb{N}^+. \qquad (4.8)$$

Our objective of this chapter is to design an observer-based sliding mode controller $u(k)$ such that the resultant closed-loop system under the WTOD protocol (4.7) is asymptotically stable.

4.2 Main Results

In this section, by considering the WTOD protocol (4.7), a token-dependent state-saturated observer is first constructed to estimate the unmeasured states,

Main Results 65

and then an observer-based SMC law $u(k)$ is designed to asymptotically stabilize the state-saturated NCS (4.1)–(4.2).

4.2.1 Token-dependent state-saturated observer

Notice that, under the WTOD protocol scheduling, the measured output received by the observer actually becomes $\bar{y}(k)$. Therefore, to cope with the effects from the scheduling signal $\xi(k) \in \mathbf{T}$ and the state saturation function $\sigma(\cdot)$, we design the following token-dependent state-saturated observer for the NCS (4.1)–(4.2):

$$\hat{x}(k+1) = \sigma\Big(A\hat{x}(k) + Bu(k) + L_{\xi(k)}\left(\bar{y}(k) - C\hat{x}(k)\right)\Big), \qquad (4.9)$$

where $\hat{x}(k) \in \mathbb{R}^n$ is the state estimate of the NCS (4.1)–(4.2) and the matrices $L_{\xi(k)} \in \mathbb{R}^{n \times p}$ are the token-dependent observer gains to be designed.

Bearing Lemma 4.1 in mind, we obtain the estimation error dynamics from (4.1), (4.8), and (4.9) as follows:

$$\begin{aligned}
e(k+1) =& \sigma\Big(\left(A + \Delta A(k)\right)x(k) + B\left(u(k) + f(x(k))\right) \Big) \\
& - \sigma\Big(A\hat{x}(k) + Bu(k) + L_{\xi(k)}\left(\bar{y}(k) - C\hat{x}(k)\right) \Big) \\
=& \Psi\Big[\mathbb{L}_{\xi(k)}(k)x(k) + Bf(x(k)) + \left(A - L_{\xi(k)}C\right)e(k) \\
& - L_{\xi(k)}\left(I - \Phi_{\xi(k)}\right)\bar{y}(k-1) \Big].
\end{aligned} \qquad (4.10)$$

where $e(k) \triangleq x(k) - \hat{x}(k)$ denotes the estimation error and

$$\begin{aligned}
\mathbb{L}_{\xi(k)}(k) &\triangleq L_{\xi(k)}C - L_{\xi(k)}\Phi_{\xi(k)}C + \Delta A(k), \\
\Psi &\triangleq \text{diag}\left\{\varepsilon_1, \varepsilon_2, \cdots, \varepsilon_n\right\}
\end{aligned}$$

with the scalars $\varepsilon_j \in [0,1]$ $(j = 1, 2, \ldots, n)$.

4.2.2 Token-dependent sliding mode controller

Based on the state estimate $\hat{x}(k)$, we choose the following sliding function:

$$s(k) = B^{\mathrm{T}}X\hat{x}(k) \qquad (4.11)$$

where $X > 0$ is a given matrix. Clearly, the nonsingularity of the matrix $\mathbb{X} \triangleq B^{\mathrm{T}}XB$ can be readily guaranteed.

It follows from (4.9) and (4.11) that

$$s(k+1) = B^{\mathrm{T}}X\sigma\Big(A\hat{x}(k) + Bu(k) + L_{\xi(k)}\left(\bar{y}(k) - C\hat{x}(k)\right) \Big). \qquad (4.12)$$

66   *Observer-Based SMC under WTODP*

Now, we design a token-dependent SMC law as follows:

$$u(k) = K_{\xi(k)}\hat{x}(k) - \mathbb{X}^{-1}B^{\mathrm{T}}XL_{\xi(k)}\bar{y}(k-1) - \nu\|\hat{x}(k)\| \cdot \chi\left(s(k)\right), \qquad (4.13)$$

where the gain matrices $K_{\xi(k)} \in \mathbb{R}^{m \times n}$ ($\xi(k) \in \mathbf{T}$) are to be determined later, and the function $\chi\left(s(k)\right)$ is defined as follows:

$$\chi\left(s(k)\right) \triangleq \left\{ \begin{array}{ll} \frac{s(k)}{\|s(k)\|}, & s(k) \neq 0, \\ 0, & s(k) = 0. \end{array} \right.$$

Remark 4.1 *By using the state estimates $\hat{x}(k)$ and the measurement $\bar{y}(k-1)$, the token-dependent SMC law (4.13) is constructed appropriately, where the effect of dynamical scheduling from one node to another via the WTOD protocol (4.7) is just reflected in the gain matrices $K_{\xi(k)}$ and $L_{\xi(k)}$. Meanwhile, the term $-\mathbb{X}^{-1}B^{\mathrm{T}}XL_{\xi(k)}\bar{y}(k-1)$ is introduced to cope with the effect of the buffer.*

Substituting (4.13) into (4.1) gives

$$x(k+1) = \sigma\Big(\mathbb{A}_{\xi(k)}(k)x(k) - BK_{\xi(k)}e(k) - \tilde{\mathbb{X}}L_{\xi(k)}\bar{y}(k-1) + Bf_s(k)\Big), \tag{4.14}$$

where

$$\mathbb{A}_{\xi(k)}(k) \triangleq A + BK_{\xi(k)} + \Delta A(k), \ \tilde{\mathbb{X}} \triangleq B\mathbb{X}^{-1}B^{\mathrm{T}}X,$$
$$f_s(k) \triangleq f(x(k)) - \nu\|\hat{x}(k)\| \cdot \chi\left(s(k)\right).$$

It is clear from (4.3) that

$$\|f_s(k)\| \leq \nu\big(\|x(k)\| + \|\hat{x}(k)\|\big) \leq \nu\big(2\|x(k)\| + \|e(k)\|\big). \tag{4.15}$$

Furthermore, by resorting to Lemma 4.2, we rewrite the controlled system (4.14) as:

$$x(k+1) = \bar{\Psi}\Big[\mathbb{A}_{\xi(k)}(k)x(k) - BK_{\xi(k)}e(k) - \tilde{\mathbb{X}}L_{\xi(k)}\bar{y}(k-1) + Bf_s(k)\Big], \tag{4.16}$$

where $\bar{\Psi} \triangleq \mathrm{diag}\left\{\bar{\varepsilon}_1, \bar{\varepsilon}_2, \cdots, \bar{\varepsilon}_n\right\}$ with scalars $\bar{\varepsilon}_j \in [0,1]$ for any $j = 1, 2, \ldots, n$.

In the sequel, the asymptotic stability of the resulting closed-loop system in (4.10) and (4.16) will be analyzed, and the reachability of the specified sliding surface (4.11) will be discussed.

4.2.3   Analysis of the asymptotic stability

The following theorem presents a sufficient condition for guaranteeing the asymptotic stability of the closed-loop system (4.10) and (4.16) under the WTOD protocol (4.7).

Main Results 67

Theorem 4.1 *Consider the state-saturated NCS (4.1)–(4.2) with the WTOD protocol (4.7) and the token-dependent SMC law (4.13). Let the matrix X in sliding function (4.11) be given. For any $i \in \mathbf{T}$, assume that there exist matrices $K_i \in \mathbb{R}^{m \times n}$, $L_i \in \mathbb{R}^{n \times p}$, $P > 0$, $W > 0$, $Z > 0$, and scalars $\eta > 0$, $\gamma > 0$, $\vartheta_{ij} \geq 0$ $(i, j \in \mathbf{T})$ such that the following conditions hold:*

$$P \leq \eta I, \tag{4.17}$$

$$Z \leq \gamma I, \tag{4.18}$$

$$\sum_{j=1}^{p} \vartheta_{ij} = 1, \tag{4.19}$$

$$\Omega_i \triangleq \begin{bmatrix} \Omega_i^{(1,1)} & \Omega_i^{(1,2)} & \Omega_i^{(1,3)} \\ \star & \Omega_i^{(2,2)} & \Omega_i^{(2,3)} \\ \star & \star & \Omega_i^{(3,3)} \end{bmatrix} < 0, \tag{4.20}$$

where

$$\Omega_i^{(1,1)} \triangleq - P + 2\eta \mathbb{A}_i^{\mathrm{T}}(k)\mathbb{A}_i(k) + 8\nu^2 \lambda_{\max}(B^{\mathrm{T}}B)\eta I + C^{\mathrm{T}}\Phi_i W \Phi_i C$$
$$+ 2\gamma \mathbb{L}_i^{\mathrm{T}}(k)\mathbb{L}_i(k) + 2\nu^2 \lambda_{\max}(B^{\mathrm{T}}B)\gamma I + \sum_{j=1}^{p} \vartheta_{ij} C^{\mathrm{T}}\bar{Q}\left(\Phi_i - \Phi_j\right)C,$$

$$\Omega_i^{(1,2)} \triangleq - 2\eta \mathbb{A}_i^{\mathrm{T}}(k)\tilde{\mathbb{X}}L_i + C^{\mathrm{T}}\Phi_i W\left(I - \Phi_i\right) - 2\gamma \mathbb{L}_i^{\mathrm{T}}L_i\left(I - \Phi_i\right)$$
$$- \sum_{j=1}^{p} \vartheta_{ij} C^{\mathrm{T}}\bar{Q}\left(\Phi_i - \Phi_j\right),$$

$$\Omega_i^{(1,3)} \triangleq - 2\eta \mathbb{A}_i^{\mathrm{T}}(k)BK_i + 2\gamma \mathbb{L}_i^{\mathrm{T}}(k)\left(A - L_iC\right),$$

$$\Omega_i^{(2,2)} \triangleq - W + 2\eta L_i^{\mathrm{T}}\tilde{\mathbb{X}}^{\mathrm{T}}\tilde{\mathbb{X}}L_i + \left(I - \Phi_i\right)W\left(I - \Phi_i\right)$$
$$+ 2\gamma\left(I - \Phi_i\right)L_i^{\mathrm{T}}L_i\left(I - \Phi_i\right) + \sum_{j=1}^{p} \vartheta_{ij}\bar{Q}\left(\Phi_i - \Phi_j\right),$$

$$\Omega_i^{(2,3)} \triangleq 2\eta L_i^{\mathrm{T}}\tilde{\mathbb{X}}^{\mathrm{T}}BK_i - 2\gamma\left(I - \Phi_i\right)L_i^{\mathrm{T}}\left(A - L_iC\right),$$

$$\Omega_i^{(3,3)} \triangleq -Z + 2\eta K_i^{\mathrm{T}}B^{\mathrm{T}}BK_i + 4\nu^2 \lambda_{\max}(B^{\mathrm{T}}B)\eta I + 2\gamma\left(A - L_iC\right)^{\mathrm{T}}\left(A - L_iC\right).$$

Then, the closed-loop system (4.10) and (4.16) is asymptotically stable.

Proof *Define the following Lyapunov-like function:*

$$V\left(x(k), \bar{y}(k-1), e(k)\right)$$
$$\triangleq x^{\mathrm{T}}(k)Px(k) + \bar{y}^{\mathrm{T}}(k-1)W\bar{y}(k-1) + e^{\mathrm{T}}(k)Ze(k). \tag{4.21}$$

Firstly, it follows from (4.16) that

$$x^{\mathrm{T}}(k+1)Px(k+1)$$
$$=\Big[\mathbb{A}_{\xi(k)}(k)x(k) - BK_{\xi(k)}e(k) - \tilde{\mathbb{X}}L_{\xi(k)}\bar{y}(k-1) + Bf_s(k)\Big]^{\mathrm{T}}$$
$$\times \bar{\Psi}P\bar{\Psi}\Big[\mathbb{A}_{\xi(k)}(k)x(k) - BK_{\xi(k)}e(k) - \tilde{\mathbb{X}}L_{\xi(k)}\bar{y}(k-1) + Bf_s(k)\Big]$$
$$\leq \eta\|\bar{\Psi}\|^2\Big[\mathbb{A}_{\xi(k)}(k)x(k) - BK_{\xi(k)}e(k) - \tilde{\mathbb{X}}L_{\xi(k)}\bar{y}(k-1) + Bf_s(k)\Big]^{\mathrm{T}}$$
$$\times \Big[\mathbb{A}_{\xi(k)}(k)x(k) - BK_{\xi(k)}e(k) - \tilde{\mathbb{X}}L_{\xi(k)}\bar{y}(k-1) + Bf_s(k)\Big],$$

where the condition (4.17) has been utilized in deducing the above.

By exploiting the relationship (4.15) and the fact that $\|\bar{\Psi}\|^2 \leq 1$, we further have

$$x^{\mathrm{T}}(k+1)Px(k+1)$$
$$\leq \eta\Big[\mathbb{A}_{\xi(k)}(k)x(k) - BK_{\xi(k)}e(k) - \tilde{\mathbb{X}}L_{\xi(k)}\bar{y}(k-1) + Bf_s(k)\Big]^{\mathrm{T}}$$
$$\times \Big[\mathbb{A}_{\xi(k)}(k)x(k) - BK_{\xi(k)}e(k) - \tilde{\mathbb{X}}L_{\xi(k)}\bar{y}(k-1) + Bf_s(k)\Big]$$
$$\leq 2\eta\Big[\mathbb{A}_{\xi(k)}(k)x(k) - BK_{\xi(k)}e(k) - \tilde{\mathbb{X}}L_{\xi(k)}\bar{y}(k-1)\Big]^{\mathrm{T}}$$
$$\times \Big[\mathbb{A}_{\xi(k)}(k)x(k) - BK_{\xi(k)}e(k) - \tilde{\mathbb{X}}L_{\xi(k)}\bar{y}(k-1)\Big]$$
$$+ 2\eta\lambda_{\max}(B^{\mathrm{T}}B)f_s^{\mathrm{T}}(k)f_s(k)$$
$$\leq 2\eta\Big[\mathbb{A}_{\xi(k)}(k)x(k) - BK_{\xi(k)}e(k) - \tilde{\mathbb{X}}L_{\xi(k)}\bar{y}(k-1)\Big]^{\mathrm{T}}$$
$$\times \Big[\mathbb{A}_{\xi(k)}(k)x(k) - BK_{\xi(k)}e(k) - \tilde{\mathbb{X}}L_{\xi(k)}\bar{y}(k-1)\Big]$$
$$+ \eta\lambda_{\max}(B^{\mathrm{T}}B)\big[8\nu^2 x^{\mathrm{T}}(k)x(k) + 4\nu^2 e^{\mathrm{T}}(k)e(k)\big]. \qquad (4.22)$$

Second, it follows from the actually received measurement (4.8) that

$$\bar{y}^{\mathrm{T}}(k)W\bar{y}(k) = \big[\Phi_{\xi(k)}Cx(k) + \big(I - \Phi_{\xi(k)}\big)\bar{y}(k-1)\big]^{\mathrm{T}}W$$
$$\times \big[\Phi_{\xi(k)}Cx(k) + \big(I - \Phi_{\xi(k)}\big)\bar{y}(k-1)\big]. \qquad (4.23)$$

Next, we focus on the term $e^{\mathrm{T}}(k+1)Ze(k+1)$. It yields from (4.10) that

$$e^{\mathrm{T}}(k+1)Ze(k+1)$$
$$=\Big[\mathbb{L}_{\xi(k)}(k)x(k) + Bf(x(k)) - L_{\xi(k)}\big(I - \Phi_{\xi(k)}\big)\bar{y}(k-1)$$
$$+ \big(A - L_{\xi(k)}C\big)e(k)\Big]^{\mathrm{T}}\Psi Z\Psi\Big[\mathbb{L}_{\xi(k)}(k)x(k) + Bf(x(k))$$
$$- L_{\xi(k)}\big(I - \Phi_{\xi(k)}\big)\bar{y}(k-1) + \big(A - L_{\xi(k)}C\big)e(k)\Big]$$

Main Results

$$\leq\gamma\|\Psi\|^2\Big[\mathbb{L}_{\xi(k)}(k)x(k) + Bf(x(k)) - L_{\xi(k)}\left(I - \Phi_{\xi(k)}\right)\bar{y}(k-1)$$

$$+ \left(A - L_{\xi(k)}C\right)e(k)\Big]^{\mathrm{T}}\Big[\mathbb{L}_{\xi(k)}(k)x(k) + Bf(x(k))$$

$$- L_{\xi(k)}\left(I - \Phi_{\xi(k)}\right)\bar{y}(k-1) + \left(A - L_{\xi(k)}C\right)e(k)\Big],$$

where the condition (4.18) has been applied.

Now, it follows from Assumption 4.3 and the fact $\|\Psi\| \leq 1$ that

$$e^{\mathrm{T}}(k+1)Ze(k+1)$$

$$\leq 2\gamma\Big[\mathbb{L}_{\xi(k)}(k)x(k) - L_{\xi(k)}\left(I - \Phi_{\xi(k)}\right)\bar{y}(k-1) + \left(A - L_{\xi(k)}C\right)e(k)\Big]^{\mathrm{T}}$$

$$\times \Big[\mathbb{L}_{\xi(k)}(k)x(k) - L_{\xi(k)}\left(I - \Phi_{\xi(k)}\right)\bar{y}(k-1) + \left(A - L_{\xi(k)}C\right)e(k)\Big]$$

$$+ 2\gamma\lambda_{\max}(B^{\mathrm{T}}B)\nu^2 x^{\mathrm{T}}(k)x(k). \tag{4.24}$$

As per the selection principle (4.7) for the WTOD protocol, it is easy to see that the following inequality holds for any $j \in \mathbf{T}$:

$$\left(y(k) - \bar{y}(k-1)\right)^{\mathrm{T}}\bar{Q}\left(\Phi_{\xi(k)} - \Phi_j\right)\left(y(k) - \bar{y}(k-1)\right) \geq 0. \tag{4.25}$$

Combining (4.22)–(4.25) and the condition (4.19), i.e. $\sum_{j=1}^{p}\vartheta_{\xi(k)j} = 1$ for any $\xi(k) \in \mathbf{T}$, we obtain

$$\Delta V(x(k), \bar{y}(k-1), e(k))$$

$$= V(x(k+1), \bar{y}(k), e(k+1)) - V(x(k), \bar{y}(k-1), e(k))$$

$$\leq x^{\mathrm{T}}(k+1)Px(k+1) - x^{\mathrm{T}}(k)Px(k) + \bar{y}^{\mathrm{T}}(k)W\bar{y}(k) - \bar{y}^{\mathrm{T}}(k-1)W\bar{y}(k-1)$$

$$+ e^{\mathrm{T}}(k+1)Ze(k+1) - e^{\mathrm{T}}(k)Ze(k) + \left(y(k) - \bar{y}(k-1)\right)^{\mathrm{T}}\bar{Q}$$

$$\times \sum_{j=1}^{p}\vartheta_{\xi(k)j}\left(\Phi_{\xi(k)} - \Phi_j\right)\left(y(k) - \bar{y}(k-1)\right)$$

$$\leq \zeta^{\mathrm{T}}(k)\Omega_{\xi(k)}\zeta(k), \tag{4.26}$$

where $\zeta(k) \triangleq \begin{bmatrix} x^{\mathrm{T}}(k) & \bar{y}^{\mathrm{T}}(k-1) & e^{\mathrm{T}}(k) \end{bmatrix}^{\mathrm{T}}$.

Letting the scheduling signal $\xi(k) = i \in \mathbf{T}$, it is clear from (4.26) that the condition (4.20) guarantees

$$\Delta V(x(k), \bar{y}(k-1), e(k)) \leq \zeta^{\mathrm{T}}(k)\Omega_i\zeta(k) < 0, \tag{4.27}$$

which means that the asymptotic stability of the closed-loop system (4.10) and (4.16) is attained eventually, and this completes the proof.

4.2.4 Analysis of the reachability

This subsection analyzes the reachability of the specified sliding surface (4.11). More specifically, by exploiting a dedicated Lyapunov function, a sufficient

70 *Observer-Based SMC under WTODP*

condition will be derived to ensure that the trajectories of the observer dynamics (4.9) enter into a sliding domain $\mathbb{S}$ around the specified sliding surface (4.11).

Theorem 4.2 *Consider the state-saturated NCS (4.1)–(4.2) with the WTOD protocol (4.7) and the token-dependent SMC law (4.13). Let the matrix X in sliding function (4.11) be given. For any $i \in \mathbf{T}$, assume that there exist matrices $K_i \in \mathbb{R}^{m \times n}$, $L_i \in \mathbb{R}^{n \times p}$, $P > 0$, $W > 0$, $Z > 0$, $S > 0$, and scalars $\varrho > 0$, $\eta > 0$, $\gamma > 0$, $\vartheta_{ij} \geq 0$ $(i, j \in \mathbf{T})$ such that the conditions (4.17)–(4.19) and the following conditions hold:*

$$XBSB^{\mathrm{T}}X \leq \varrho I, \tag{4.28}$$

$$\hat{\Omega}_i < 0, \tag{4.29}$$

where

$$\hat{\Omega}_i \triangleq \Omega_i + 2\varrho \begin{bmatrix} \hat{\mathbb{A}}_i^{\mathrm{T}} \hat{\mathbb{A}}_i & \hat{\mathbb{A}}_i^{\mathrm{T}} \hat{\mathbb{L}}_i & -\hat{\mathbb{A}}_i^{\mathrm{T}} \hat{\mathbb{K}}_i \\ \star & \hat{\mathbb{L}}_i^{\mathrm{T}} \hat{\mathbb{L}}_i & -\hat{\mathbb{L}}_i^{\mathrm{T}} \hat{\mathbb{K}}_i \\ \star & \star & \hat{\mathbb{K}}_i^{\mathrm{T}} \hat{\mathbb{K}}_i \end{bmatrix},$$

$$\hat{\mathbb{A}}_i \triangleq A + BK_i + L_i \left(\Phi_i - I \right) C,$$

$$\hat{\mathbb{L}}_i \triangleq L_i \left(I - \Phi_i \right) - BX^{-1}B^{\mathrm{T}}XL_i,$$

$$\hat{\mathbb{K}}_i \triangleq A + BK_i - L_i C.$$

Then, the sliding variable $s(k)$ along the closed-loop system (4.10) and (4.16) is forced into the following domain $\mathbb{S}$ around the specified sliding surface $s(k) = 0$:

$$\mathbb{S} \triangleq \left\{ s(k) \;\middle|\; \|s(k)\| \leq \sqrt{\frac{2\varrho \lambda_{\max}(B^{\mathrm{T}}B)\nu^2}{\lambda_{\min}(S)}} \cdot \|\hat{x}(k)\| \right\}. \tag{4.30}$$

Proof *Let $\xi(k) = i \in \mathbf{T}$. Substituting the actually received measurement (4.8) and the designed SMC law (4.13) into (4.12), we have*

$$s(k+1) = B^{\mathrm{T}}X\sigma\left(\hat{\mathbb{A}}_i x(k) + \hat{\mathbb{L}}_i \bar{y}(k-1) - \hat{\mathbb{K}}_i e(k) - B\rho_s(k)\right)$$

where $\rho_s(k) \triangleq \nu\|\hat{x}(k)\| \cdot \chi\left(s(k)\right)$. Recalling Lemma 4.2, we rewrite the above dynamics as

$$s(k+1) = B^{\mathrm{T}}X\hat{\Psi}\left[\hat{\mathbb{A}}_i x(k) + \hat{\mathbb{L}}_i \bar{y}(k-1) - \hat{\mathbb{K}}_i e(k) - B\rho_s(k)\right] \tag{4.31}$$

where $\hat{\Psi} \triangleq \mathrm{diag}\left\{\hat{\varepsilon}_1, \hat{\varepsilon}_2, \cdots, \hat{\varepsilon}_n\right\}$ with scalars $\hat{\varepsilon}_j \in [0, 1]$ for $j = 1, 2, \ldots, n$.
Now, we choose the following Lyapunov-like function:

$$\hat{V}(x(k), \bar{y}(k-1), e(k)) \triangleq V(x(k), \bar{y}(k-1), e(k)) + s^{\mathrm{T}}(k)Ss(k). \tag{4.32}$$

Main Results 71

By employing the condition (4.28) and the fact that $\|\hat{\Psi}\|^2 \leq 1$, it follows from (4.31) that

$$s^{\mathrm{T}}(k+1)Ss(k+1)$$

$$=\Big[\hat{\mathbb{A}}_i x(k) + \hat{\mathbb{L}}_i \bar{y}(k-1) - \hat{\mathbb{K}}_i e(k) - B\rho_s(k)\Big]^{\mathrm{T}} \hat{\Psi} X B S B^{\mathrm{T}} X \hat{\Psi}$$

$$\times \Big[\hat{\mathbb{A}}_i x(k) + \hat{\mathbb{L}}_i \bar{y}(k-1) - \hat{\mathbb{K}}_i e(k) - B\rho_s(k)\Big]$$

$$\leq \varrho \|\hat{\Psi}\|^2 \Big[\hat{\mathbb{A}}_i x(k) + \hat{\mathbb{L}}_i \bar{y}(k-1) - \hat{\mathbb{K}}_i e(k) - B\rho_s(k)\Big]^{\mathrm{T}}$$

$$\times \Big[\hat{\mathbb{A}}_i x(k) + \hat{\mathbb{L}}_i \bar{y}(k-1) - \hat{\mathbb{K}}_i e(k) - B\rho_s(k)\Big]$$

$$\leq 2\varrho \Big[\hat{\mathbb{A}}_i x(k) + \hat{\mathbb{L}}_i \bar{y}(k-1) - \hat{\mathbb{K}}_i e(k)\Big]^{\mathrm{T}} \Big[\hat{\mathbb{A}}_i x(k) + \hat{\mathbb{L}}_i \bar{y}(k-1) - \hat{\mathbb{K}}_i e(k)\Big]$$

$$+ 2\varrho\lambda_{\max}(B^{\mathrm{T}}B)\nu^2\|\hat{x}(k)\|^2. \tag{4.33}$$

Bearing the conditions (4.17)–(4.19) and the WTOD protocol (4.7) in mind, it follows from (4.26) and (4.33) that

$$\Delta \hat{V}(x(k), \bar{y}(k-1), e(k))$$

$$=\hat{V}(x(k+1), \bar{y}(k), e(k+1)) - \hat{V}(x(k), \bar{y}(k-1), e(k))$$

$$\leq \zeta^{\mathrm{T}}(k)\hat{\Omega}_i \zeta(k) - \Big[\lambda_{\min}(S)\|s(k)\|^2 - 2\varrho\lambda_{\max}(B^{\mathrm{T}}B)\nu^2\|\hat{x}(k)\|^2\Big]. \tag{4.34}$$

If the sliding variable $s(k)$ escapes the domain $\mathbb{S}$, one will have $\|s(k)\| \geq \sqrt{\frac{2\varrho\lambda_{\max}(B^{\mathrm{T}}B)\nu^2}{\lambda_{\min}(S)}} \cdot \|\hat{x}(k)\|$. Then, it is concluded from the condition (4.29) that

$$\Delta \hat{V}(x(k), \bar{y}(k-1), e(k)) \leq \zeta^{\mathrm{T}}(k)\hat{\Omega}_i \zeta(k) < 0, \; \forall i \in \mathbf{T}, \tag{4.35}$$

which implies that the sliding variable $s(k)$ along the closed-loop system (4.10) and (4.16) is strictly deceasing outside the domain $\mathbb{S}$ defined in (4.30). The proof is now complete.

4.2.5 Solving algorithm

In order to guarantee both the reachability of the specified sliding surface (4.11) and the asymptotic stability of the closed-loop system (4.10) and (4.16), we establish a sufficient condition in the following theorem such that the conditions in Theorems 4.1 and 4.2 are ensured simultaneously.

Theorem 4.3 *Consider the state-saturated NCS (4.1)–(4.2) with the WTOD protocol (4.7) and the token-dependent SMC law (4.13). Let the matrix X in sliding function (4.11) be given. For any $i \in \mathbf{T}$, assume that there exist matrices $\mathcal{K}_i \in \mathbb{R}^{m \times n}$, $\mathcal{L}_i \in \mathbb{R}^{n \times p}$, $P > 0$, $W > 0$, $Z > 0$, $S > 0$, and scalars $\kappa > 0$, $\varrho > 0$, $\eta > 0$, $\gamma > 0$, $\vartheta_{ij} \geq 0$ $(i, j \in \mathbf{T})$ such that the following linear* *matrix inequalities (with an equality constraint) hold:*

$$-\eta I + P \le 0, \tag{4.36}$$

$$-\gamma I + Z \le 0, \tag{4.37}$$

$$-\varrho I + XBSB^{\mathrm{T}}X \le 0, \tag{4.38}$$

$$\bar{\Omega}_i \triangleq \begin{bmatrix} \bar{\Omega}_i^{(1,1)} & \bar{\Omega}_i^{(1,2)} & \bar{\Omega}_i^{(1,3)} \\ \star & \bar{\Omega}_i^{(2,2)} & \bar{\Omega}_i^{(2,3)} \\ \star & \star & \bar{\Omega}_i^{(3,3)} \end{bmatrix} < 0, \tag{4.39}$$

$$\sum_{j=1}^{p} \vartheta_{ij} = 1, \tag{4.40}$$

where

$$\bar{\Omega}_i^{(1,1)} \triangleq \begin{bmatrix} -P + \sum_{j=1}^{p} \vartheta_{ij} C^{\mathrm{T}} \bar{Q} \left(\Phi_i - \Phi_j \right) C & -\sum_{j=1}^{p} \vartheta_{ij} C^{\mathrm{T}} \bar{Q} \left(\Phi_i - \Phi_j \right) & 0 \\ \star & -W + \sum_{j=1}^{p} \vartheta_{ij} \bar{Q} \left(\Phi_i - \Phi_j \right) & 0 \\ \star & \star & -Z \end{bmatrix},$$

$$\bar{\Omega}_i^{(1,2)} \triangleq \begin{bmatrix} \Gamma_i^1 & \Gamma_i^2 \end{bmatrix}, \ \mathbb{B} = \sqrt{\lambda_{\max}(B^{\mathrm{T}}B)},$$

$$\Gamma_i^1 \triangleq \begin{bmatrix} -\sqrt{2}\bar{\mathbb{A}}_i^{\mathrm{T}} & 2\sqrt{2}\mathbb{B}\nu\eta I & C^{\mathrm{T}}\Phi_i W \\ \sqrt{2}\mathcal{L}_i^{\mathrm{T}}\breve{\mathbb{X}} & 0 & (I - \Phi_i)W \\ \sqrt{2}\mathcal{K}_i^{\mathrm{T}}B^{\mathrm{T}} & 0 & 0 \end{bmatrix},$$

$$\Gamma_i^2 \triangleq \begin{bmatrix} \sqrt{2}\bar{\mathbb{L}}_i^{\mathrm{T}} & \sqrt{2}\mathbb{B}\nu\gamma I & 0 & \sqrt{2}\breve{\mathbb{A}}_i^{\mathrm{T}} \\ -\sqrt{2}(I - \Phi_i)\mathcal{L}_i^{\mathrm{T}} & 0 & 0 & \sqrt{2}\breve{\mathbb{L}}_i^{\mathrm{T}} \\ \sqrt{2}(\eta A - \mathcal{L}_i C)^{\mathrm{T}} & 0 & 2\mathbb{B}\nu\eta I & -\sqrt{2}\breve{\mathbb{K}}_i^{\mathrm{T}} \end{bmatrix},$$

$$\bar{\mathbb{A}}_i \triangleq \eta A + B\mathcal{K}_i, \ \bar{\mathbb{L}}_i \triangleq \mathcal{L}_i C - \mathcal{L}_i \Phi_i C, \ \breve{\mathbb{A}}_i \triangleq \eta A + B\mathcal{K}_i + \mathcal{L}_i(\Phi_i - I)C,$$

$$\breve{\mathbb{L}}_i \triangleq \mathcal{L}_i(I - \Phi_i) - B\mathbb{X}^{-1}B^{\mathrm{T}}X\mathcal{L}_i, \ \breve{\mathbb{K}}_i \triangleq \eta A + B\mathcal{K}_i - \mathcal{L}_i C,$$

$$\bar{\Omega}_i^{(1,3)} \triangleq \begin{bmatrix} \sqrt{2}\eta N^{\mathrm{T}} & 0 \\ 0 & 0 \\ 0 & 0 \end{bmatrix},$$

$$\bar{\Omega}_i^{(2,2)} \triangleq -\mathrm{diag}\{\eta I, \eta I, W, (\gamma - 2\eta)I, \gamma I, \eta I, (\varrho - 2\eta)I\},$$

$$\bar{\Omega}_i^{(2,3)} \triangleq \begin{bmatrix} 0 & 0 & 0 & 0 & 0 & 0 & 0 \\ -\kappa M^{\mathrm{T}} & 0 & 0 & \kappa M^{\mathrm{T}} & 0 & 0 & 0 \end{bmatrix}^{\mathrm{T}}, \ \bar{\Omega}_i^{(3,3)} \triangleq -\mathrm{diag}\{\kappa I, \kappa I\}.$$

Then, the asymptotic stability of the closed-loop system (4.10) and (4.16) as well as the reachability of the sliding domain $\mathbb{S}$ defined in (4.30) can be guaranteed simultaneously by the token-dependent SMC law (4.13) with the matrices $K_i = \eta^{-1}\mathcal{K}_i$ and $L_i = \eta^{-1}\mathcal{L}_i$ ($i \in \mathbf{T}$).

Main Results

Proof *It is noted that* $\Omega_i < 0$ *is ensured by the condition* $\hat{\Omega}_i < 0$. *By resorting to the Schur Complement Lemma and the following relationships:*

$$0 \le (\eta - \gamma)\gamma^{-1}(\eta - \gamma) = \eta^2\gamma^{-1} - 2\eta + \gamma, \tag{4.41}$$

$$0 \le (\eta - \varrho)\varrho^{-1}(\eta - \varrho) = \eta^2\varrho^{-1} - 2\eta + \varrho, \tag{4.42}$$

one can see that the inequality $\hat{\Omega}_i < 0$ *is guaranteed by the condition* $\bar{\Omega}_i < 0$. *Thus, the proof is completed.*

To tackle the nonconvex problem of the equality constraint (4.40), we introduce the following inequality:

$$\left(-1 + \sum_{j=1}^{p} \vartheta_{ij}\right)^2 \le \alpha_i, \tag{4.43}$$

where $\alpha_i > 0$ is a given scalar. According to Schur complement, (4.43) holds if and only if the following linear matrix inequality (LMI) is true:

$$\begin{bmatrix} -\alpha_i & -1 + \sum_{j=1}^{p} \vartheta_{ij} \\ -1 + \sum_{j=1}^{p} \vartheta_{ij} & -1 \end{bmatrix} \le 0, \ i \in \mathbf{T}. \tag{4.44}$$

Obviously, when the scalar α_i is chosen to be sufficiently small, we could have $\sum_{j=1}^{p} \vartheta_{ij} \approx 1$. Such an approximation may bring certain conservatism to the obtained LMI-based sufficient conditions. Nevertheless, from the viewpoint of practical application, a set of sufficiently small scalars α_i would be definitely beneficial for the feasibility of the proposed algorithm.

It is noted that the selection of the matrix X in (4.11) would affect the feasibility of the inequalities in Theorem 4.3. For practicality, we can simply choose $X = \frac{\beta}{1-\beta}I$, where the scalar $\beta \in (0, 1)$ can be determined by checking the feasibility of the LMIs (4.36)–(4.39) and (4.44). As per the above discussion, we summarize the design procedure of the observer-based SMC law (4.13) as follows:

- **Step 1.** Choose a set of sufficiently small scalars $\{\alpha_i\}$, and a small fixed step size $d > 0$. Let $n = 1$.

- **Step 2.** Solve LMIs (4.36)–(4.39) and (4.44) with the matrix $X = \frac{nd}{1-nd}I$.

- **Step 3.** If these LMIs are feasible go to Step 4; else, if $n \le \lceil \frac{1}{d} \rceil - 2$, then set $n = n + 1$ and go to Step 2.

- **Step 4.** Get the scalar η, and the matrices $\mathcal{K}_i$, $\mathcal{L}_i$. Produce the observer (4.9) and the token-dependent SMC law (4.13) with matrices $K_i = \eta^{-1}\mathcal{K}_i$ and $L_i = \eta^{-1}\mathcal{L}_i$, and then apply it to the state-saturated NCS (4.1)–(4.2).

Remark 4.2 *It is shown from Theorem 4.3 that the token-dependent observer-based SMC law (4.13) can not only drive the trajectories of the observer dynamics (4.9) into the sliding domain $\mathbb{S}$ around the specified sliding surface (4.11), but also ensure the asymptotic stability of the closed-loop system (4.10) and (4.16). As such, the effects from the scheduling signal $\xi(k)$ and the state saturation have been effectively eliminated.*

4.3 Example

Consider a discrete-time state-saturated system in the form of (4.1)–(4.2) with

$$A = \begin{bmatrix} 1 & 0.1 \\ 0.1 & -0.3 \end{bmatrix}, B = \begin{bmatrix} 0.1 \\ 0.1 \end{bmatrix}, C = \begin{bmatrix} 1 & 0 \\ 0 & 1 \end{bmatrix}, M = \begin{bmatrix} 0.2 \\ -0.05 \end{bmatrix},$$
$$\Delta(k) = \cos(k),\ N = \begin{bmatrix} 0.1 & -0.3 \end{bmatrix},\ f(x(k)) = 0.1\sin\left(x_1^2 + x_2^2\right).$$

Clearly, the parameter ν in (4.3) can be chosen as $\nu = 0.1$. Under the saturation level $\bar{s}_1 = \bar{s}_2 = 1$, it is easy to verify that the uncontrolled state-saturated system (when $u(k) = 0$) under the initial condition $x(0) = \begin{bmatrix} -0.8 & -0.8 \end{bmatrix}^{\mathrm{T}}$ is unstable as shown in Fig. 4.2.

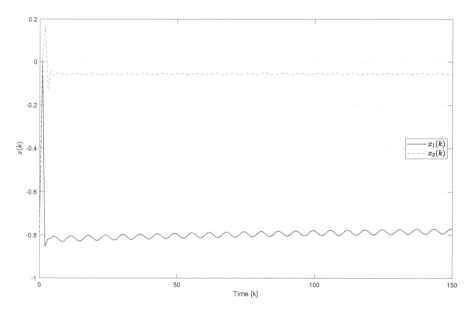

FIGURE 4.2: State trajectories of the uncontrolled state-saturated system.

Example

For the selection principle (4.7) with $\bar{Q} = \text{diag}\{0.5, 0.5\}$ and the sliding function (4.11) with $X = I$, we obtain the following controller gains K_i and observer gains L_i by using Theorem 4.3 with $\alpha_1 = \alpha_2 = 10^{-6}$:

$$K_1 = \begin{bmatrix} -5.4171 & 1.0997 \end{bmatrix}, K_2 = \begin{bmatrix} -5.3704 & 1.0351 \end{bmatrix},$$

$$L_1 = \begin{bmatrix} 0.1004 & -0.1009 \\ -0.1009 & 0.1008 \end{bmatrix}, L_2 = \begin{bmatrix} 0.0315 & -0.0327 \\ -0.0327 & 0.0325 \end{bmatrix}.$$

Then, the token-dependent observer-based SMC law $u(k)$ in (4.13) is synthesized as follows:

- if $\xi(k) = 1$, then

$$u(k) = K_1 \hat{x}(k) - \begin{bmatrix} -0.0025 & -0.0005 \end{bmatrix} \bar{y}(k-1) - 0.1 \|\hat{x}(k)\| \chi(s(k)),$$

- if $\xi(k) = 2$, then

$$u(k) = K_2 \hat{x}(k) - \begin{bmatrix} -0.0060 & -0.0010 \end{bmatrix} \bar{y}(k-1) - 0.1 \|\hat{x}(k)\| \chi(s(k)).$$

Under the initial conditions $x(0) = \begin{bmatrix} -0.8 & -0.8 \end{bmatrix}^{\text{T}}$ and $\hat{x}(0) = \begin{bmatrix} 0.5 & 0.5 \end{bmatrix}^{\text{T}}$, the simulation results are provided in Figs. 4.3–4.5. The selected sensor node $\xi(k)$ obtaining access to the communication network under the WTOD protocol is depicted in Fig. 4.3. The evolutions of the system states $x(k)$ and the observer states $\hat{x}(k)$ in the closed-loop case are

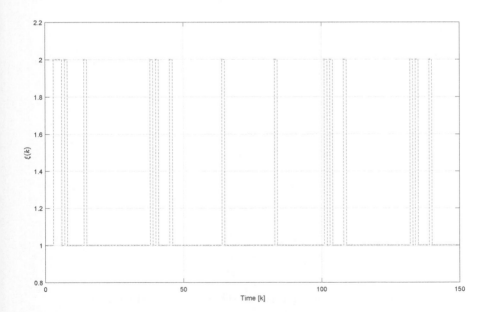

FIGURE 4.3: The selected sensor node $\xi(k)$ obtaining access to the network under the WTOD protocol.

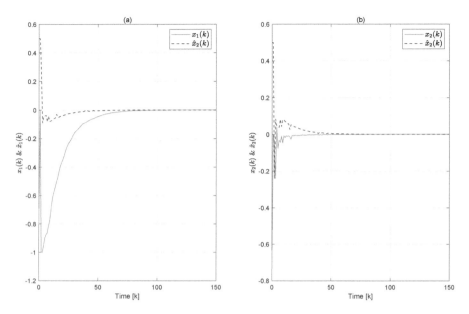

FIGURE 4.4: System state $x(k)$ and the observer state $\hat{x}(k)$ in the closed-loop case.

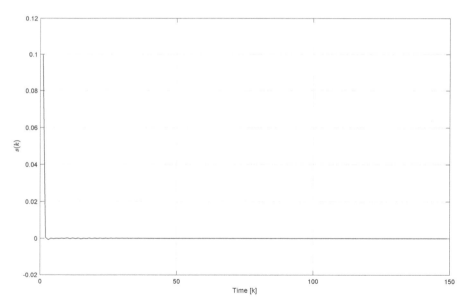

FIGURE 4.5: Sliding variable $s(k)$ in the closed-loop case.

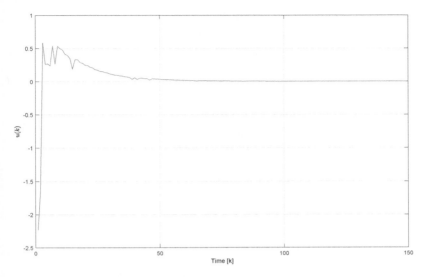

FIGURE 4.6: SMC input $u(k)$ in the closed-loop case.

shown in Fig. 4.4, from which it is observed that the observer states could track the system states well. Furthermore, the responses of the sliding variable $s(k)$ and the SMC input $u(k)$ in the closed-loop case are plotted in Figs. 4.5–4.6, respectively. All the simulation results have confirmed the effectiveness of the proposed observer-based SMC design scheme under the WTOD protocol scheduling and the state saturation.

4.4 Conclusion

This chapter has investigated the observer-based SMC problem for state-saturated NCSs under WTOD protocol. The WTOD protocol has been employed to determine which sensor node gets access to the shared network between controller and sensors. By exploiting the estimated state and the actual received measured output, a token-dependent observer-based SMC law has been designed properly for a linear sliding surface. The established observer-based SMC scheme exhibit the following distinct features: 1) to cope with the impacts from the WTOD protocol scheduling and the state-saturated constraint, a token-dependent state-saturated observer is constructed; 2) the designed token-dependent SMC law depends on the updating criterion for the actual received measurement; and 3) the reachability of a sliding domain around the specified sliding surface is ensured for the observer dynamics.

5

Asynchronous Sliding Mode Control Under Static Event-Triggered Protocol

Markovian jump systems (MJSs) are an important kind of hybrid systems in which the *mode signal*, responsible for controlling the switch among dynamic modes, is modelled by a time-homogeneous Markov chain. The key feature for this class of systems is that the Markov chain could model some abrupt changes in the dynamics of the system due to, for instance, environmental disturbances, component failures or repairs, changes in subsystems interconnections, etc. Examples of the MJSs can be found in economics, wind turbine and networked control. It is well-known that the nonlinearities are ubiquitous in practice and the control problem for nonlinear MJSs has been the hot topic in control field, especially, the so-called Markovian jump Lur'e systems composed of a linear part and a nonlinearity for each mode.

It should be noted that in networked control systems, the *mode signal* of the MJSs cannot be completely accessible because of communication delays and inevitable packet dropout, which may result in the asynchronization phenomenon between controller/filter modes and system modes. To date, the asynchronous control/filter of *linear* MJSs has gained some initial research interest and the corresponding research on *nonlinear* MJSs is still ongoing. In this chapter, we endeavour to study the event-triggered asynchronous SMC problem for the Markovian jump Lur'e systems under hidden mode detections. Based on a hidden Markov model (HMM), a detector is employed to observe the mode signal of the plant and emit an estimated mode to the controller. An event-triggered strategy is introduced to decrease the frequency of data transmission between the plant and the controller. *The main contributions of this chapter are highlighted as follows: 1) For the concerned Markovian jump Lur'e systems, a novel common sliding surface is developed and a novel event-triggered asynchronous SMC law is constructed appropriately by only using the detected modes; 2) The mean-square stability with H_∞ disturbance rejection performance and the reachability of the sliding mode dynamics are analyzed, respectively, by combining the stochastic Lur'e-type Lyapunov function and the HMM approach; and 3) The solving algorithm for the proposed event-triggered asynchronous SMC scheme is established in terms of a convex optimization problem, which can be computed directly offline via standard software package.*

DOI: 10.1201/9781003309499-5

79

5.1 Problem Formulation

This chapter discusses a networked Markovian jump Lur'e system under a remote sliding mode controller as shown in Fig. 5.1, in which an event detector is utilized to determine whether or not the current state should be transmitted to the controller and a mode detector is employed to emit an estimated mode signal to the controller.

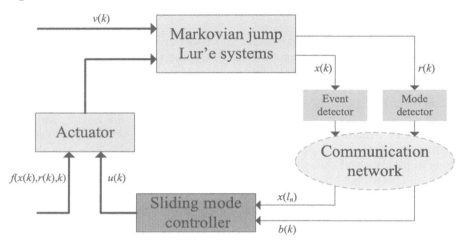

FIGURE 5.1: The event-triggered asynchronous SMC for the networked Markovian jump Lur'e systems.

Given the probability space $(\mathbb{U}, \mathcal{F}, \Pr\{\cdot\})$, we consider the following Markovian jump Lur'e systems:

$$\begin{cases} x(k+1) = [A(r(k)) + \Delta A(r(k), k)]x(k) + D(r(k))\varphi(r(k), H(r(k))x(k)) \\ \qquad\qquad + B(r(k))[u(k) + f(x(k), r(k), k)] + G(r(k))v(k), \\ z(k) = C(r(k))x(k), \end{cases} \tag{5.1}$$

where $x(k) \in \mathbb{R}^{n_x}$ is the system state, $u(k) \in \mathbb{R}^m$ is the control input, $z(k) \in \mathbb{R}^p$ is the controlled output, and $v(k) \in \mathbb{R}^{n_v}$ is the external disturbance belonging to $\ell_2[0, +\infty)$. $\varphi(r(k), H(r(k))x(k)) \in \mathbb{R}^q$ is a mode-dependent memoryless nonlinearity. The stochastic process $\{r(k) = i,\ k \geq 0\}$ is described via a homogeneous Markov chain which taking values on a finite set $\mathbb{S} \triangleq \{1, 2, \ldots, N\}$, and having the following transition probability from mode i at sample time k to mode j at sample time $k+1$:

$$\pi_{ij} = \Pr\{r(k+1) = j \mid r(k) = i\}, \forall i, j \in \mathbb{S}, \tag{5.2}$$

where $\pi_{ij} \in [0, 1]$, and $\sum_{j=1}^{N} \pi_{ij} = 1$. The transition probability matrix is defined as $\Pi = [\pi_{ij}]_{i,j \in \mathbb{S}}$.

Problem Formulation 81

For each possible $r(k) = i \in \mathbb{S}$, define the known constant matrices as $A_i \triangleq A(r(k))$, $B_i \triangleq B(r(k))$, $C_i \triangleq C(r(k))$, $D_i \triangleq D(r(k))$, $G_i \triangleq G(r(k))$, $H_i \triangleq H(r(k))$, and the known Lur'e nonlinearity as $\varphi_i(H_i x(k)) \triangleq \varphi(r(k), H(r(k))x(k))$. The input matrix B_i is full column rank, that is, $\text{rank}(B_i) = m$. The uncertainties $\Delta A_i(k) \triangleq \Delta A(r(k), k)$ satisfy $\Delta A_i(k) = M_i \Phi_i(k) E_i$, with M_i and E_i known constant matrices, and $\Phi_i(k)$ an unknown time-varying matrix satisfying $\Phi_i^{\mathrm{T}}(k)\Phi_i(k) \leq I$. Besides, for any $r(k) = i \in \mathbb{S}$, the external disturbance $f_i(x(k), k) \triangleq f(x(k), r(k), k)$ possesses

$$\|f_i(x(k), k)\| \leq \mu_i \|x(k)\|, \tag{5.3}$$

where $\mu_i \geq 0$ is a known scalar.

Here, we make the following assumption as in [55].

Assumption 5.1 *The nonlinear function $\varphi_i(\cdot) : \mathbb{R}^q \to \mathbb{R}^q$ verifies a cone-bounded sector condition and is decentralized being associated with each mode $r(k) = i \in \mathbb{S}$, which satisfies the conditions: 1) $\varphi(0) = 0$; and 2) there exists N diagonal positive definite matrices $\Omega_i \in \mathbb{R}^{q \times q}$, such that $\forall y \in \mathbb{R}^q$, $\forall h = 1, 2, \ldots, q$,*

$$\varphi_{i,(h)}(y) \left[\varphi_i(y) - \Omega_i y\right]_{(h)} \leq 0.$$

From Assumption 5.1, the following inequality holds, $\forall y \in \mathbb{R}^q$, $\forall i \in \mathbb{S}$,

$$\mathbf{SC}(i, y, \Lambda_i) \triangleq \varphi_i^{\mathrm{T}}(y)\Lambda_i \left[\varphi_i(y) - \Omega_i y\right] \leq 0, \tag{5.4}$$

where Λ_i are any *diagonal* positive semidefinite matrices. Clearly, the condition (5.4) implies the following relation for any $y \in \mathbb{R}^q$:

$$0 \leq \varphi_i^{\mathrm{T}}(y)\Lambda_i \varphi_i(y) \leq \varphi_i^{\mathrm{T}}(y)\Lambda_i \Omega_i y \leq y^{\mathrm{T}}\Omega_i \Lambda_i \Omega_i y. \tag{5.5}$$

In this chapter, the event-triggered strategy will be proposed to save computer resources. Denote by $\{l_0, l_1, \ldots\}$ the sequence of the time instants at which the current state of the plant (5.1) is transmitted to the controller. Each time instant l_n is defined according to the following event-triggered condition:

$$\varepsilon^{\mathrm{T}}(k)T\varepsilon(k) \geq \alpha x^{\mathrm{T}}(k)Tx(k), \tag{5.6}$$

where $\alpha \in [0, 1]$ is the prescribed event-triggering parameter, and $\varepsilon(k) \triangleq x(k) - x(l_n)$ is the error for $k \in [l_n, l_{n+1})$, $n = 0, 1, 2, \ldots, \infty$. $T > 0$ is the weighting matrix to be designed later.

In practical applications, it is not always possible to directly measure the information of system modes $r(k)$. Instead, a detector $b(k)$ may be utilized to obtain the estimation of $r(k)$ with some probability [27]. In this case, the emitted signal $b(k)$ from detector to controller may not synchronize with the

82　　　　　　　　　　　　　　　　　　　*Asynchronous SMC under SETP*

system mode $r(k)$. The HMM $(r(k), b(k))$ as in [27, 147] is introduced to characterize the above asynchronous phenomenon as follows:

$$\xi_{i\chi} = \Pr\{b(k) = \chi \mid r(k) = i\}, \quad i \in \mathbb{S}, \ \chi \in \mathbb{D}, \ \mathbb{D} \triangleq \{1, 2, \ldots, J\}, \quad (5.7)$$

where $\xi_{i\chi}$ is the mode detection probability belonging to $[0, 1]$. For any $i \in \mathbb{S}$, it has $\sum_{\chi=1}^{J} \xi_{i\chi} = 1$. The MDPM Θ is defined as $\Theta \triangleq [\xi_{i\chi}]$. It is clear that the HMM (5.7) covers the mode-dependent ($\mathbb{D} = \mathbb{S}$, $\xi_{i\chi} = 1$ for $\chi = i$) and mode-independent ($\mathbb{D} = \{1\}$) cases.

Definition 5.1 *[27] The system (5.1) with $v(k) = 0$ is said to be mean square stable (MSS) if the condition $\lim_{k\to\infty} \mathbb{E}\{\|x(k)\|^2\} \mid_{x(0), r(0)} = 0$ holds for any initial condition $\{x(0) \neq 0, \ r(0) \in \mathbb{S}\}$,*

Definition 5.2 *Given a scalar $\gamma > 0$, the system (5.1) is said to be MSS with an H_∞ disturbance attenuation level γ if the system is MSS, and under the zero initial condition $x(0) = 0$, the condition $\mathbb{E}\{\sum_{k=0}^{\infty} \|z(k)\|^2\} \leq \gamma^2 \sum_{k=0}^{\infty} \|v(k)\|^2$ holds for all $v(k) \in \ell_2[0, \infty)$.*

Now, our objective is to design an asynchronous SMC law $u(k)$, *depending on only the estimated mode $b(k)$*, such that the resultant closed-loop system is MJS with a prescribed H_∞ disturbance attention level under the event-triggered strategy (5.6) and the hidden mode detections (5.7).

5.2　Main Results

5.2.1　Designing of sliding surface and sliding mode controller

In this chapter, a novel *common* sliding function is constructed as follows:

$$s(k) = Zx(k), \quad (5.8)$$

where $Z \triangleq \sum_{i=1}^{N} \beta_i B_i^{\mathrm{T}}$, and scalars $\beta_i \in [0, 1]$ ($i \in \mathbb{S}$) will be chosen such that $X_i \triangleq ZB_i$ is nonsingular for any $i \in \mathbb{S}$. As discussed in [95], based on the assumption that $\mathrm{rank}(B_i) = m$ for any $i \in \mathbb{S}$, the above nonsingularity can be guaranteed easily by selecting the parameters β_i properly.

Remark 5.1 *In the recent years, the SMC problems of discrete-time MJSs have gained some particular research attenuations, see [61, 210] for example. It is noted that all of them employed the mode-dependent sliding surfaces, which render a shortcoming that the reachability of sliding surfaces may not always be attained due to the switching frequently from one mode to another. The above drawback may be overcome by the proposed common sliding surface (5.8).*

Main Results 83

Notice that the system mode $r(k)$ can be estimated via a detector $b(k)$. That means that the controller design for the system (5.1) can just utilize the mode information emitting from the detector. To this end, we design an appropriate event-triggered asynchronous SMC law as follows:

$$u(k) = K_\chi x(l_n) + L_\chi \varphi_\chi(H_\chi x(l_n)) - \varrho\|x(l_n)\| \cdot \mathrm{sgn}(s(l_n)), \qquad (5.9)$$

for $\chi \in \mathbb{D}$, $k \in [l_n, l_{n+1})$, $n = 0, 1, 2, \ldots, \infty$, where the matrices $K_\chi \in \mathbb{R}^{m \times n_x}$, $L_\chi \in \mathbb{R}^{m \times q}$ ($\chi \in \mathbb{D}$) will be determined later, and the scalar ϱ is given by

$$\varrho \triangleq \max_{i \in \mathcal{N}} \{\delta_i\}, \quad \delta_i \triangleq \|X_i^{-1} Z M_i\| \cdot \|E_i\| + \mu_i, \qquad (5.10)$$

with μ_i defined in (5.3).

Remark 5.2 *Actually, the effect of stochastic jumping from one detected mode to another is just reflected in the asynchronous SMC law (5.9) via matrices K_χ and L_χ, which are the solutions of some coupled matrix inequalities concerning the transition probabilities π_{ij} and the mode detection probabilities $\xi_{i\chi}$ (see Theorem 5.3 later). On the other hand, the asynchronous SMC law (5.9) takes the effects of the event-triggered mechanism (5.6) and the Lur'e-type nonlinearities into account appropriately. The introduction of the term $L_\chi \varphi_\chi(H_\chi x(l_n))$ in (5.9) is inspired by [55].*

By substituting (5.9) into (5.1), the closed-loop system is obtained as

$$\begin{aligned} x(k+1) = &\tilde{A}_{i\chi} x(k) - B_i K_\chi \varepsilon(k) + D_i \varphi_i(H_i x(k)) + B_i L_\chi \varphi_\chi(H_\chi x(l_n)) \\ &+ G_i v(k) + B_i \rho_i(k), \end{aligned} \qquad (5.11)$$

where $\tilde{A}_{i\chi} \triangleq A_i + \tilde{Z}_i \Delta A_i(k) + B_i K_\chi$, $\tilde{Z}_i \triangleq I - B_i X_i^{-1} Z$, $\rho_i(k) \triangleq U_i(k) - \varrho\|x(l_n)\|$
$\times \mathrm{sgn}(s(l_n))$, and $U_i(k) \triangleq X_i^{-1} Z \Delta A_i(k) x(k) + f_i(x(k), k)$.
Noticing that $\|\mathrm{sgn}(s(l_n))\| \le \sqrt{m}$, we have

$$\|\rho_i(k)\| \le (\delta_i + \varrho\sqrt{m}) \|x(k)\| + \varrho\sqrt{m}\|\varepsilon(k)\|. \qquad (5.12)$$

In the sequel, the mean square stability with a prescribed H_∞ disturbance rejection performance of the sliding mode dynamics (5.11) and the reachability for the specified sliding surface (5.8) will be analyzed, respectively.

5.2.2 Analysis of sliding mode dynamics

In the following theorem, by referring to the work [55], the stochastic Lur'e-type Lyapunov functional approach is utilized for reducing the conservatism in tackling the mode-dependent Lur'e-type nonlinearities $\varphi_i(H_i x(k))$, $i \in \mathbb{S}$.

Theorem 5.1 *Consider the Markovian jump Lur'e system (5.1) with given event-triggered parameter $\alpha \in [0, 1]$ and the event-triggered asynchronous SMC* *law (5.9). The closed-loop system (5.11) is MSS with a prescribed disturbance attenuation level $\gamma > 0$ if for any $i \in \mathbb{S}$ and $\chi \in \mathbb{D}$, there exist positive definite matrices P_i, $T \in \mathbb{R}^{n_x \times n_x}$, matrices $K_\chi \in \mathbb{R}^{m \times n_x}$, $L_\chi \in \mathbb{R}^{m \times q}$, diagonal positive semidefinite matrices $\Psi_{i\chi}$, $\Xi_{i\chi}$, $\Delta_i \in \mathbb{R}^{q \times q}$, and positive scalars ς_i, ω_i such that the following inequalities hold:*

$$B_i^{\mathrm{T}} \mathbb{P}_i B_i - \varsigma_i I \leq 0, \tag{5.13}$$

$$L_\chi^{\mathrm{T}} B_i^{\mathrm{T}} \mathbb{P}_i B_i L_\chi - \Psi_{i\chi} \leq 0, \tag{5.14}$$

$$B_i^{\mathrm{T}} \mathbb{H}_i B_i - \omega_i I \leq 0, \tag{5.15}$$

$$L_\chi^{\mathrm{T}} B_i^{\mathrm{T}} \mathbb{H}_i B_i L_\chi - \Xi_{i\chi} \leq 0, \tag{5.16}$$

$$\mathbb{X}_i + \sum_{\chi=1}^{J} \xi_{i\chi} \left[\mathbb{Y}_{i\chi}^{\mathrm{T}} \left(\mathbb{P}_i + \mathbb{H}_i \right) \mathbb{Y}_{i\chi} + \mathbb{I}_\chi^{\mathrm{T}} \left(\Psi_{i\chi} + \Xi_{i\chi} \right) \mathbb{I}_\chi \right] < 0, \tag{5.17}$$

where $\mathbb{P}_i \triangleq \sum_{j=1}^{N} \pi_{ij} P_j$, $\mathbb{H}_i \triangleq \sum_{j=1}^{N} \pi_{ij} H_j^{\mathrm{T}} \Omega_j \Delta_j \Omega_j H_j$, and

$$\mathbb{X}_i \triangleq \begin{bmatrix} \mathbb{C}_i & H_i^{\mathrm{T}} \Omega_i \Delta_i & 0 & 0 \\ \star & -3\Delta_i & 0 & 0 \\ \star & \star & \mathbb{T}_i & 0 \\ \star & \star & \star & -\gamma^2 I \end{bmatrix},$$

$$\mathbb{C}_i \triangleq -P_i + 6(\delta_i + \varrho\sqrt{m})^2(\varsigma_i + \omega_i)I + \alpha T + C_i^{\mathrm{T}} C_i,$$

$$\mathbb{T}_i \triangleq -T + 6m\varrho^2(\varsigma_i + \omega_i)I,$$

$$\mathbb{Y}_{i\chi} \triangleq \begin{bmatrix} \sqrt{3}\tilde{A}_{i\chi} & \sqrt{3}D_i & -\sqrt{3}B_i K_\chi & \sqrt{3}G_i \end{bmatrix},$$

$$\mathbb{I}_\chi \triangleq \begin{bmatrix} \sqrt{3}\Omega_\chi H_\chi & 0 & -\sqrt{3}\Omega_\chi H_\chi & 0 \end{bmatrix}.$$

Proof *For any system mode $r(k) = i \in \mathbb{S}$, we consider the following Lur'e-type Lyapunov functional candidate for sliding mode dynamics (5.11):*

$$V(x(k), i) \triangleq V_1(x(k), i) + V_2(x(k), i), \tag{5.18}$$

where $V_1(x(k), i) \triangleq x^{\mathrm{T}}(k) P_i x(k)$, and $V_2(x(k), i) \triangleq \varphi_i^{\mathrm{T}}(H_i x(k)) \Delta_i \Omega_i H_i x(k)$. Along the state trajectories of (5.11), one has

$$\mathbb{E}\left\{ \Delta V_1(x(k), i) \right\}$$
$$\triangleq \mathbb{E}\left\{ V_1(x(k+1), r(k+1)) \mid x(k), i \right\} - V_1(x(k), i)$$
$$\leq \sum_{\chi=1}^{J} \xi_{i\chi} \left\{ 3\left[\tilde{A}_{i\chi} x(k) - B_i K_\chi \varepsilon(k) + D_i \varphi_i(H_i x(k)) + G_i v(k) \right]^{\mathrm{T}} \mathbb{P}_i \right.$$
$$\times \left[\tilde{A}_{i\chi} x(k) - B_i K_\chi \varepsilon(k) + D_i \varphi_i(H_i x(k)) + G_i v(k) \right]$$
$$\left. + 3\varphi_\chi^{\mathrm{T}}(H_\chi x(l_n)) L_\chi^{\mathrm{T}} B_i^{\mathrm{T}} \mathbb{P}_i B_i L_\chi \varphi_\chi(H_\chi x(l_n)) \right\}$$
$$+ 3\rho_i^{\mathrm{T}}(k) B_i^{\mathrm{T}} \mathbb{P}_i B_i \rho_i(k) - x^{\mathrm{T}}(k) P_i x(k). \tag{5.19}$$

Main Results 85

Notice that from (5.12) and (5.13), we have

$$\rho_i^{\mathrm{T}}(k)B_i^{\mathrm{T}}\mathbb{P}_iB_i\rho_i(k) \leq 2\varsigma_i\left[\left(\delta_i + \varrho\sqrt{m}\right)^2 x^{\mathrm{T}}(k)x(k) + m\varrho^2\varepsilon^{\mathrm{T}}(k)\varepsilon(k)\right]. \quad (5.20)$$

Besides, by exploiting the condition (5.14) and the inequality (5.5), we obtain for any $\chi \in \mathbb{D}$,

$$\varphi_\chi^{\mathrm{T}}(H_\chi x(l_n))L_\chi^{\mathrm{T}}B_i^{\mathrm{T}}\mathbb{P}_iB_iL_\chi\varphi_\chi(H_\chi x(l_n))$$
$$\leq (x(k) - \varepsilon(k))^{\mathrm{T}} H_\chi^{\mathrm{T}}\Omega_\chi\Psi_{i\chi}\Omega_\chi H_\chi (x(k) - \varepsilon(k)). \quad (5.21)$$

Thus, substituting (5.20)–(5.21) into (5.19) yields

$$\mathbb{E}\{\Delta V_1(x(k), i)\}$$
$$\leq \sum_{\chi=1}^{J}\xi_{i\chi}\Bigg\{3\Big[\tilde{A}_{i\chi}x(k) - B_iK_\chi\varepsilon(k) + D_i\varphi_i(H_ix(k)) + G_iv(k)\Big]^{\mathrm{T}}\mathbb{P}_i$$
$$\times\Big[\tilde{A}_{i\chi}x(k) - B_iK_\chi\varepsilon(k) + D_i\varphi_i(H_ix(k)) + G_iv(k)\Big]$$
$$+ 3(x(k) - \varepsilon(k))^{\mathrm{T}} H_\chi^{\mathrm{T}}\Omega_\chi\Psi_{i\chi}\Omega_\chi H_\chi(x(k) - \varepsilon(k))\Bigg\}$$
$$+ 6\varsigma_i\left[\left(\delta_i + \varrho\sqrt{m}\right)^2 x^{\mathrm{T}}(k)x(k) + m\varrho^2\varepsilon^{\mathrm{T}}(k)\varepsilon(k)\right]$$
$$- x^{\mathrm{T}}(k)P_ix(k). \quad (5.22)$$

On the other hand, recalling the relationships in (5.5) gives

$$\mathbb{E}\{\Delta V_2(x(k))\}$$
$$\triangleq \mathbb{E}\{V_2(x(k+1), r(k+1)) \mid x(k), i\} - V_2(x(k), i)$$
$$\leq \sum_{\chi=1}^{J}\xi_{i\chi}\Bigg\{3\Big[\tilde{A}_{i\chi}x(k) - B_iK_\chi\varepsilon(k) + D_i\varphi_i(H_ix(k)) + G_iv(k)\Big]^{\mathrm{T}}$$
$$\times\mathbb{H}_i\Big[\tilde{A}_{i\chi}x(k) - B_iK_\chi\varepsilon(k) + D_i\varphi_i(H_ix(k)) + G_iv(k)\Big]$$
$$+ 3\varphi_\chi^{\mathrm{T}}(H_\chi x(l_n))L_\chi^{\mathrm{T}}B_i^{\mathrm{T}}\mathbb{H}_iB_iL_\chi\varphi_\chi(H_\chi x(l_n))\Bigg\}$$
$$+ 3\rho_i^{\mathrm{T}}(k)B_i^{\mathrm{T}}\mathbb{H}_iB_i\rho_i(k) - \varphi_i^{\mathrm{T}}(H_ix(k))\Delta_i\Omega_iH_ix(k). \quad (5.23)$$

By adopting the similar operations to (5.20)–(5.21), we have from the conditions (5.15)–(5.16)

$$\rho_i^{\mathrm{T}}(k)B_i^{\mathrm{T}}\mathbb{H}_iB_i\rho_i(k)$$
$$\leq 2\omega_i\left[\left(\delta_i + \varrho\sqrt{m}\right)^2 x^{\mathrm{T}}(k)x(k) + m\varrho^2\varepsilon^{\mathrm{T}}(k)\varepsilon(k)\right], \quad (5.24)$$
$$\varphi_\chi^{\mathrm{T}}(H_\chi x(l_n))L_\chi^{\mathrm{T}}B_i^{\mathrm{T}}\mathbb{H}_iB_iL_\chi\varphi_\chi(H_\chi x(l_n))$$
$$\leq \varphi_\chi^{\mathrm{T}}(H_\chi x(l_n))\Xi_{i\chi}\varphi_\chi(H_\chi x(l_n))$$
$$\leq (x(k) - \varepsilon(k))^{\mathrm{T}} H_\chi^{\mathrm{T}}\Omega_\chi\Xi_{i\chi}\Omega_\chi H_\chi(x(k) - \varepsilon(k)), \quad (5.25)$$

which implies that

$$\mathbb{E}\left\{\Delta V_2(x(k))\right\}$$

$$\leq \sum_{\chi=1}^{J} \xi_{i\chi} \left\{ 3\left[\tilde{A}_{i\chi}x(k) - B_iK_\chi\varepsilon(k) + D_i\varphi_i(H_ix(k)) + G_iv(k)\right]^{\mathrm{T}} \right.$$

$$\times \mathbb{H}_i \left[\tilde{A}_{i\chi}x(k) - B_iK_\chi\varepsilon(k) + D_i\varphi_i(H_ix(k)) + G_iv(k)\right]$$

$$+ 3\left(x(k) - \varepsilon(k)\right)^{\mathrm{T}} H_\chi^{\mathrm{T}}\Omega_\chi\Xi_{i\chi}\Omega_\chi H_\chi \left(x(k) - \varepsilon(k)\right) \Big\}$$

$$+ 6\omega_i \left[\left(\delta_i + \varrho\sqrt{m}\right)^2 x^{\mathrm{T}}(k)x(k) + m\varrho^2\varepsilon^{\mathrm{T}}(k)\varepsilon(k)\right]$$

$$- \varphi_i^{\mathrm{T}}(H_ix(k))\Delta_i\Omega_i H_ix(k). \tag{5.26}$$

Now, define an extended state vector as

$$\eta(k) \triangleq \left[\begin{array}{cccc} x^{\mathrm{T}}(k) & \varphi_i^{\mathrm{T}}(H_ix(k)) & \varepsilon^{\mathrm{T}}(k) & v^{\mathrm{T}}(k) \end{array}\right]^{\mathrm{T}}.$$

Then, combining the event-triggered scheme in (5.6) and the inequality in (5.4), we have

$$\mathbb{E}\{\Delta V(x(k), i)\}$$

$$\leq \mathbb{E}\{\Delta V_1(x(k), i)\} + \mathbb{E}\{\Delta V_2(x(k), i)\} + \alpha x^{\mathrm{T}}(k)Tx(k)$$

$$- \varepsilon^{\mathrm{T}}(k)T\varepsilon(k) - 3\mathbf{SC}(i, H_ix(k), \Delta_i). \tag{5.27}$$

When $v(k) = 0$, *it is easily shown from (5.22), (5.26), and (5.27) that the condition (5.17) guarantees* $\mathbb{E}\{\Delta V(x(k), i)\} < 0$ *for any* $i \in \mathbb{S}$, *which means that the sliding mode dynamics in (5.11) with* $v(k) = 0$ *is MSS.*

In what follows, we consider the H_∞ *disturbance attenuation performance of the sliding mode dynamics (5.11) under the zero initial condition. To this end, we exploit the following index for* $k \in [l_n, l_{n+1})$:

$$J_i(k) \triangleq \mathbb{E}\{\Delta V(x(k), i)\} + \mathbb{E}\left\{\|z(k)\|^2\right\} - \gamma^2\|v(k)\|^2.$$

Following a similar reasoning as above, it is shown that the condition (5.17) *implies* $J_i(k) \leq 0$. *Thus, it has* $\sum_{k=0}^{\infty} J_i(k) \leq 0$, *which renders* $\mathbb{E}\left\{\sum_{k=0}^{\infty}\|z(k)\|^2\right\} \leq \gamma^2\sum_{k=0}^{\infty}\|v(k)\|^2$ *under the zero initial condition.*

5.2.3 Analysis of reachability

This subsection carries out the reachability analysis by using an extended stochastic Lur'e-type Lyapunov functional. Around the specified sliding surface (5.8), i.e., $s(k) = 0$, we define a *time-varying* sliding region as follows:

$$\mathbf{O} \triangleq \left\{s(k) \mid \|s(k)\| \leq \tilde{\rho}(k)\right\}, \tag{5.28}$$

Main Results   87

where $\tilde{\rho}(k) \triangleq \max_{i \in \mathbb{S}} \left\{ \sqrt{\frac{\hat{\rho}_i(k)}{\lambda_{\min}(Q_i)}} \right\}$ with

$$\hat{\rho}_i(k) \triangleq 6 \big(\|G_i^{\mathrm{T}} \mathbb{P}_i G_i\| + \|G_i^{\mathrm{T}} \mathbb{H}_i G_i\| + \|G_i^{\mathrm{T}} Z^{\mathrm{T}} Q_i Z G_i\| \big) \cdot \|v(k)\|^2 + 12 \big(\|B_i^{\mathrm{T}} \mathbb{P} B_i\|$$
$$+ \|B_i^{\mathrm{T}} \mathbb{H}_i B_i\| + \|X_i^{\mathrm{T}} Q_i X_i\| \big) \left[\left(\delta_i + \varrho\sqrt{m} \right)^2 + m\varrho^2 \alpha \frac{\lambda_{\max}(T)}{\lambda_{\min}(T)} \right] \cdot \|x(k)\|^2,$$

and the matrices Q_i and T defined in Theorem 5.2.

Assuming that Theorem 5.1 hold and noticing the fact that $v(k) \in \ell_2[0, +\infty)$, it is easily shown that $\tilde{\rho}(k)$ is just a *vicinity* of the sliding surface $s(k) = 0$, that is, the quasi-sliding mode (QSM). The following theorem guarantees the quasi-sliding mode.

Theorem 5.2 *Consider the Markovian jump Lur'e system (5.1) with given event-triggered parameter $\alpha \geq 0$ and the event-triggered asynchronous SMC law (5.9). For any $i \in \mathbb{S}$ and $\chi \in \mathbb{D}$, there exist positive definite matrices P_i, $T \in \mathbb{R}^{n_x \times n_x}$, $Q_i \in \mathbb{R}^{m \times m}$, matrices $K_\chi \in \mathbb{R}^{m \times n_x}$, $L_\chi \in \mathbb{R}^{m \times q}$, and diagonal positive semidefinite matrices $\Psi_{i\chi}$, $\Xi_{i\chi}$, $\Gamma_{i\chi}$, $\Delta_i \in \mathbb{R}^{q \times q}$ satisfying the conditions (5.14), (5.16) and the following matrix inequalities:*

$$L_\chi^{\mathrm{T}} X_i^{\mathrm{T}} Q_i X_i L_\chi - \Gamma_{i\chi} \leq 0, \tag{5.29}$$

$$\bar{\mathbb{X}}_i + \sum_{\chi=1}^{J} \xi_{i\chi} \left[\bar{\mathbb{Y}}_{i\chi}^{\mathrm{T}} \left(\mathbb{P}_i + \mathbb{H}_i \right) \bar{\mathbb{Y}}_{i\chi} + \mathbb{Z}_{i\chi}^{\mathrm{T}} Q_i \mathbb{Z}_{i\chi} \bar{\mathbb{I}}_\chi^{\mathrm{T}} \left(\Psi_{i\chi} + \Xi_{i\chi} + \Gamma_{i\chi} \right) \bar{\mathbb{I}}_\chi \right] < 0,$$
$$\tag{5.30}$$

where

$$\bar{\mathbb{X}}_i \triangleq \begin{bmatrix} -P_i + \alpha T & H_i^{\mathrm{T}} \Omega_i \Delta_i & 0 \\ \star & -3\Delta_i & 0 \\ \star & \star & -T \end{bmatrix}, \quad \bar{\mathbb{Y}}_{i\chi} \triangleq \begin{bmatrix} \sqrt{3}\tilde{A}_{i\chi} & \sqrt{3}D_i & -\sqrt{3}B_i K_\chi \end{bmatrix},$$

$$\mathbb{Z}_{i\chi} \triangleq \begin{bmatrix} \sqrt{3}\bar{A}_{i\chi} & \sqrt{3}ZD_i & -\sqrt{3}X_i K_\chi \end{bmatrix}, \quad \bar{\mathbb{I}}_\chi \triangleq \begin{bmatrix} \sqrt{3}\Omega_\chi H_\chi & 0 & -\sqrt{3}\Omega_\chi H_\chi \end{bmatrix},$$

$$Q_i \triangleq \sum_{j=1}^{N} \pi_{ij} Q_j, \quad \bar{A}_{i\chi} \triangleq ZA_i + X_i K_\chi,$$

and other matrices are defined in Theorem 5.1, then the state trajectories of the closed-loop system (5.11) will be driven into the sliding region **O** *in mean square under the event-triggered asynchronous SMC law (5.9).*

Proof *Combining (5.8) and (5.11), we have*

$$s(k+1) = \bar{A}_{i\chi} x(k) - X_i K_\chi \varepsilon(k) + ZD_i \varphi_i(H_i x(k)) + X_i L_\chi \varphi_\chi(H_\chi x(l_n))$$
$$+ ZG_i v(k) + X_i \rho_i(k), \tag{5.31}$$

Now, we consider the following extended Lur'e-type Lyapunov functional:

$$\bar{V}(k, i) \triangleq V(x(k), i) + s^{\mathrm{T}}(k) Q_i s(k). \tag{5.32}$$

where the Lyapunov functional $V(x(k), i)$ is defined in (5.18). Along with the sliding function in (5.31), we obtain

$$\mathbb{E}\left\{s^{\mathrm{T}}(k+1)Q(r(k+1))s(k+1) \mid x(k), r(k) = i\right\}$$

$$\leq \sum_{\chi=1}^{J} \xi_{i\chi} \left\{ 3\left[\bar{A}_{i\chi}x(k) - X_i K_\chi \varepsilon(k) + ZD_i\varphi_i(H_i x(k))\right]^{\mathrm{T}}\right.$$

$$\times \mathbb{Q}_i \left[\bar{A}_{i\chi}x(k) - X_i K_\chi \varepsilon(k) + ZD_i\varphi_i(H_i x(k))\right]$$

$$\left. + 3\varphi_\chi^{\mathrm{T}}(H_\chi x(l_n))L_\chi^{\mathrm{T}} X_i^{\mathrm{T}} \mathbb{Q}_i X_i L_\chi \varphi_\chi(H_\chi x(l_n))\right\}$$

$$+ 3\left(ZG_i v(k) + X_i\rho_i(k)\right)^{\mathrm{T}} \mathbb{Q}_i \left(ZG_i v(k) + X_i\rho_i(k)\right).$$

$$(5.33)$$

By resorting to the condition (5.29), it follows from (5.5) that

$$\varphi_\chi^{\mathrm{T}}(H_\chi x(l_n))L_\chi^{\mathrm{T}} X_i^{\mathrm{T}} \mathbb{Q}_i X_i L_\chi \varphi_\chi(H_\chi x(l_n))$$

$$\leq \varphi_\chi^{\mathrm{T}}(H_\chi x(l_n))\Gamma_{i\chi}\varphi_\chi(H_\chi x(l_n))$$

$$\leq (x(k) - \varepsilon(k))^{\mathrm{T}} H_\chi^{\mathrm{T}}\Omega_\chi\Gamma_{i\chi}\Omega_\chi H_\chi (x(k) - \varepsilon(k)).$$

$$(5.34)$$

Substituting (5.34) into (5.33) renders

$$\mathbb{E}\left\{s^{\mathrm{T}}(k+1)Q(r(k+1))s(k+1) \mid x(k), r(k) = i\right\}$$

$$\leq \sum_{\chi=1}^{J} \xi_{i\chi} \left\{ 3\left[\bar{A}_{i\chi}x(k) - X_i K_\chi \varepsilon(k) + ZD_i\varphi_i(H_i x(k))\right]^{\mathrm{T}}\right.$$

$$\times \mathbb{Q}_i \left[\bar{A}_{i\chi}x(k) - X_i K_\chi \varepsilon(k) + ZD_i\varphi_i(H_i x(k))\right]$$

$$\left. + 3\left(x(k) - \varepsilon(k)\right)^{\mathrm{T}} H_\chi^{\mathrm{T}}\Omega_\chi\Gamma_{i\chi}\Omega_\chi H_\chi (x(k) - \varepsilon(k)) \right\}$$

$$+ 6\left(\|G_i^{\mathrm{T}} Z^{\mathrm{T}} \mathbb{Q}_i ZG_i\| \cdot \|v(k)\|^2 + \|X_i^{\mathrm{T}} \mathbb{Q}_i X_i\| \cdot \|\rho_i(k)\|^2\right).$$

$$(5.35)$$

Following a similar argument as above and bearing (5.4), (5.6), (5.19), and (5.23) in mind, we obtain from the conditions (5.14) and (5.16) that

$$\mathbb{E}\left\{\Delta\bar{V}(k, i)\right\}$$

$$\leq \sum_{\chi=1}^{J} \xi_{i\chi} \left\{ 3\left[\tilde{A}_{i\chi}x(k) - B_i K_\chi \varepsilon(k) + D_i\varphi_i(H_i x(k))\right]^{\mathrm{T}}\right.$$

$$\times \mathbb{P}_i \left[\tilde{A}_{i\chi}x(k) - B_i K_\chi \varepsilon(k) + D_i\varphi_i(H_i x(k))\right]$$

$$\left. + 3\left(x(k) - \varepsilon(k)\right)^{\mathrm{T}} H_\chi^{\mathrm{T}}\Omega_\chi\Psi_{i\chi}\Omega_\chi H_\chi (x(k) - \varepsilon(k)) \right\}$$

$$+ 6\left(\|G_i^{\mathrm{T}}\mathbb{P}_i G_i\| \cdot \|v(k)\|^2 + \|B_i^{\mathrm{T}}\mathbb{P}_i B_i\| \cdot \|\rho_i(k)\|^2\right)$$

$$
+ \sum_{\chi=1}^{J} \xi_{i\chi} \left\{ 3\Big[\tilde{A}_{i\chi} x(k) - B_i K_\chi \varepsilon(k) + D_i \varphi_i(H_i x(k)) \Big]^{\mathrm{T}} \right.
$$

$$
\times \mathbb{H}_i \Big[\tilde{A}_{i\chi} x(k) - B_i K_\chi \varepsilon(k) + D_i \varphi_i(H_i x(k)) \Big]
$$

$$
\left. + 3\left(x(k) - \varepsilon(k) \right)^{\mathrm{T}} H_\chi^{\mathrm{T}} \Omega_\chi \Xi_{i\chi} \Omega_\chi H_\chi \left(x(k) - \varepsilon(k) \right) \right\}
$$

$$
+ 6\big(\| G_i^{\mathrm{T}} \mathbb{H}_i G_i \| \cdot \| v(k) \|^2 + \| B_i^{\mathrm{T}} \mathbb{H}_i B_i \| \cdot \| \rho_i(k) \|^2 \big)
$$

$$
+ \sum_{\chi=1}^{J} \xi_{i\chi} \left\{ 3\Big[\bar{A}_{i\chi} x(k) - X_i K_\chi \varepsilon(k) + Z D_i \varphi_i(H_i x(k)) \Big]^{\mathrm{T}} \right.
$$

$$
\times \mathbb{Q}_i \Big[\bar{A}_{i\chi} x(k) - X_i K_\chi \varepsilon(k) + Z D_i \varphi_i(H_i x(k)) \Big]
$$

$$
\left. + 3\left(x(k) - \varepsilon(k) \right)^{\mathrm{T}} H_\chi^{\mathrm{T}} \Omega_\chi \Gamma_{i\chi} \Omega_\chi H_\chi \left(x(k) - \varepsilon(k) \right) \right\}
$$

$$
+ 6\big(\| G_i^{\mathrm{T}} Z^{\mathrm{T}} \mathbb{Q}_i Z G_i \| \cdot \| v(k) \|^2 + \| X_i^{\mathrm{T}} \mathbb{Q}_i X_i \| \cdot \| \rho_i(k) \|^2 \big)
$$

$$
- x^{\mathrm{T}}(k) P_i x(k) - \varphi_i^{\mathrm{T}}(H_i x(k)) \Delta_i \Omega_i H_i x(k) + \alpha x^{\mathrm{T}}(k) T x(k)
$$

$$
- \varepsilon^{\mathrm{T}}(k) T \varepsilon(k) - 3\mathbf{SC}(i, H_i x(k), \Delta_i) - s^{\mathrm{T}}(k) Q_i s(k)
$$

$$
\leq \bar{\eta}^{\mathrm{T}}(k) \left\{ \bar{\mathbb{X}}_i + \sum_{\chi=1}^{J} \xi_{i\chi} \Big[\bar{\mathbb{Y}}_{i\chi}^{\mathrm{T}} \left(\mathbb{P}_i + \mathbb{H}_i \right) \bar{\mathbb{Y}}_{i\chi} + \mathbb{Z}_{i\chi}^{\mathrm{T}} \mathbb{Q}_i \mathbb{Z}_{i\chi} \right.
$$

$$
\left. + \bar{\mathbb{I}}_\chi^{\mathrm{T}} \left(\Psi_{i\chi} + \Xi_{i\chi} + \Gamma_{i\chi} \right) \bar{\mathbb{I}}_\chi \Big] \right\} \bar{\eta}(k) - \lambda_{\min}(Q_i) \| s(k) \|^2 + \bar{\rho}_i(k), \qquad (5.36)
$$

where $\eta(k) \triangleq \begin{bmatrix} x^{\mathrm{T}}(k) & \varphi_i^{\mathrm{T}}(H_i x(k)) & \varepsilon^{\mathrm{T}}(k) \end{bmatrix}^{\mathrm{T}}$ *and*

$$
\bar{\rho}_i(k) \triangleq 6\big(\| G_i^{\mathrm{T}} \mathbb{P}_i G_i \| + \| G_i^{\mathrm{T}} \mathbb{H}_i G_i \| + \| G_i^{\mathrm{T}} Z^{\mathrm{T}} \mathbb{Q}_i Z G_i \| \big) \cdot \| v(k) \|^2 + 6\big(\| B_i^{\mathrm{T}} \mathbb{P} B_i \|
$$

$$
+ \| B_i^{\mathrm{T}} \mathbb{H}_i B_i \| + \| X_i^{\mathrm{T}} \mathbb{Q}_i X_i \| \big) \cdot \| \rho_i(k) \|^2.
$$

Notice that during any time interval $k \in [l_n, l_{n+1})$, *one has from (5.6) and (5.12) that*

$$
\| \rho_i(k) \|^2 \leq 2 \left[\left(\delta_i + \varrho \sqrt{m} \right)^2 + m \varrho^2 \alpha \frac{\lambda_{\max}(T)}{\lambda_{\min}(T)} \right] \| x(k) \|^2. \qquad (5.37)
$$

Thus, when the state trajectories escape from the region **O** *around the specified sliding surface (5.8), that is,*

$$
\| s(k) \| > \tilde{\rho}(k) \geq \sqrt{\frac{\hat{\rho}_i(k)}{\lambda_{\min}(Q_i)}} \geq \sqrt{\frac{\bar{\rho}_i(k)}{\lambda_{\min}(Q_i)}}, \forall i \in \mathbb{S},
$$

90 *Asynchronous SMC under SETP*

it yields from (5.30) and (5.36) that

$$\mathbb{E}\left\{\Delta\tilde{V}(k,i)\right\}$$

$$\leq \bar{\eta}^{\mathrm{T}}(k)\left\{\bar{\mathbb{X}}_i + \sum_{\chi=1}^{J}\xi_{i\chi}\left[\bar{\mathbb{Y}}_{i\chi}^{\mathrm{T}}\left(\mathbb{P}_i + \mathbb{H}_i\right)\bar{\mathbb{Y}}_{i\chi} + \mathbb{Z}_{i\chi}^{\mathrm{T}}\mathbb{Q}_i\mathbb{Z}_{i\chi}\right.\right.$$

$$\left.\left.+ \bar{\mathbb{I}}_{\chi}^{\mathrm{T}}\left(\Psi_{i\chi} + \Xi_{i\chi} + \Gamma_{i\chi}\right)\bar{\mathbb{I}}_{\chi}\right]\right\}\bar{\eta}(k) < 0. \tag{5.38}$$

*This implies that the state trajectories of the closed-loop system (5.11) are strictly deceasing (*with mean square*) outside the region* **O** *defined in (5.28).*

Remark 5.3 *Since the adopted event-triggered strategy schedules the system state $x(k)$ to the controller until the event-triggered condition (5.6) is attained, the communication burden will be greatly reduced. Nevertheless, it should be also noted that the introduction of event-triggered protocol may sacrifice certain SMC performance. Actually, it is seen from (5.28) that the sliding region* **O** *with event-triggered protocol is larger than the one without event-triggered protocol (i.e., $\alpha = 0$ in (5.6)) due to the additional term $m\varrho^2\alpha\frac{\lambda_{\max}(T)}{\lambda_{\min}(T)}\cdot\|x(k)\|^2$.*

5.2.4 Synthesis of SMC law

In order to achieve the mean square stability with a prescribed H_∞ performance of the sliding mode dynamics (5.11) and the reachability of the specified sliding function (5.8) simultaneously, the gain matrices K_χ and L_χ in the designed SMC scheme (5.9) should be determined by Theorems 5.1 and 5.2 simultaneously. In what follows, we derive some sufficient conditions to synthesize the event-triggered asynchronous SMC law (5.9).

Theorem 5.3 *Consider the following assertions:*

i) Given the parameters α and γ, there exist positive definite matrices P_i, $T \in \mathbb{R}^{n\times n}$, $Q_i \in \mathbb{R}^{m\times m}$, matrices $K_\chi \in \mathbb{R}^{m\times n}$, $L_\chi \in \mathbb{R}^{m\times q}$, diagonal positive semidefinite matrices $\Psi_{i\chi}$, $\Xi_{i\chi}$, $\Gamma_{i\chi}$, $\Delta_i \in \mathbb{R}^{q\times q}$, and positive scalars ς_i, ω_i such that the conditions (5.13)–(5.17) and (5.29)–(5.30) hold for any $i \in \mathbb{S}$ and $\chi \in \mathbb{D}$.

ii) Given the parameters α and γ, there exist positive definite matrices P_i, T, $X_{i\chi}$, $Z_{i\chi} \in \mathbb{R}^{n_x\times n_x}$, $Q_i \in \mathbb{R}^{m\times m}$, $Y_{i\chi} \in \mathbb{R}^{q\times q}$, $W_{i\chi} \in \mathbb{R}^{n_v\times n_v}$, matrices $K_\chi \in \mathbb{R}^{m\times n_x}$, $L_\chi \in \mathbb{R}^{m\times q}$, diagonal positive semidefinite matrices $\Psi_{i\chi}$, $\Xi_{i\chi}$, $\Gamma_{i\chi}$, $\Delta_i \in \mathbb{R}^{q\times q}$, and positive scalars ς_i, ω_i such that the conditions (5.13)–(5.16), (5.29) and the following matrix inequalities hold for any $i \in \mathbb{S}$ and $\chi \in \mathbb{D}$:

$$\mathbb{Y}_{i\chi}^{\mathrm{T}}\left(\mathbb{P}_i + \mathbb{H}_i\right)\mathbb{Y}_{i\chi} + \mathbb{I}_{\chi}^{\mathrm{T}}\left(\Psi_{i\chi} + \Xi_{i\chi} + \Gamma_{i\chi}\right)\mathbb{I}_{\chi} + \tilde{\mathbb{Z}}_{i\chi}^{\mathrm{T}}\mathbb{Q}_i\tilde{\mathbb{Z}}_{i\chi} \leq \mathbb{W}_{i\chi}, \tag{5.39}$$

$$\mathbb{X}_i + \sum_{\chi=1}^{J}\xi_{i\chi}\mathbb{W}_{i\chi} < 0, \tag{5.40}$$

Main Results 91

where $\tilde{\mathbb{Z}}_{i\chi} \triangleq [\sqrt{3}\bar{A}_{i\chi} \ \sqrt{3}ZD_i \ -\sqrt{3}X_iK_\chi \ 0]$, $\mathbb{W}_{i\chi} \triangleq \mathrm{diag}\{X_{i\chi}, Y_{i\chi}, Z_{i\chi}, W_{i\chi}\}$.

iii) *Given the parameters* α *and* γ, *there exist positive definite matrices* $\tilde{P}_i$, $\tilde{T}$, $\tilde{X}_{i\chi}$, $\tilde{Z}_{i\chi} \in \mathbb{R}^{n_x \times n_x}$, $\tilde{Q}_i \in \mathbb{R}^{m \times m}$, $\tilde{Y}_{i\chi} \in \mathbb{R}^{q \times q}$, $\tilde{W}_{i\chi} \in \mathbb{R}^{n_v \times n_v}$, *matrices* $\mathcal{K}_\chi \in \mathbb{R}^{m \times n_x}$, $\mathcal{L}_\chi \in \mathbb{R}^{m \times q}$, $\mathcal{J}_\chi \in \mathbb{R}^{n_x \times n_x}$, $\mathcal{V}_\chi \in \mathbb{R}^{q \times q}$, *diagonal positive semidefinite matrices* $\tilde{\Psi}_{i\chi}$, $\tilde{\Xi}_{i\chi}$, $\tilde{\Gamma}_{i\chi}$, $\tilde{\Delta}_i \in \mathbb{R}^{q \times q}$, *and positive scalars* $\tilde{\varsigma}_i$, $\tilde{\omega}_i$, ϑ_i *such that the following coupled LMIs hold for any* $i \in \mathbb{S}$ *and* $\chi \in \mathbb{D}$:

$$\begin{bmatrix} -\tilde{\varsigma}_i I & \mathbb{B}_i \\ \mathbb{B}_i^T & -\bar{\mathbb{P}} \end{bmatrix} \leq 0, \tag{5.41}$$

$$\begin{bmatrix} \tilde{\Psi}_{i\chi} - \mathcal{V}_\chi^T - \mathcal{V}_\chi & \mathbb{L}_{i\chi} \\ \mathbb{L}_{i\chi}^T & -\bar{\mathbb{P}} \end{bmatrix} \leq 0, \tag{5.42}$$

$$\begin{bmatrix} -\tilde{\omega}_i I & \tilde{\mathbb{B}}_i \\ \tilde{\mathbb{B}}_i & -\bar{\Delta} \end{bmatrix} \leq 0, \tag{5.43}$$

$$\begin{bmatrix} \tilde{\Xi}_{i\chi} - \mathcal{V}_\chi^T - \mathcal{V}_\chi & \tilde{\mathbb{L}}_{i\chi} \\ \tilde{\mathbb{L}}_{i\chi}^T & -\bar{\Delta} \end{bmatrix} \leq 0, \tag{5.44}$$

$$\begin{bmatrix} \tilde{\Gamma}_{i\chi} - \mathcal{V}_\chi^T - \mathcal{V}_\chi & \hat{\mathbb{L}}_{i\chi} \\ \hat{\mathbb{L}}_{i\chi}^T & -\bar{\mathbb{Q}} \end{bmatrix} \leq 0, \tag{5.45}$$

$$\begin{bmatrix} \tilde{\mathbb{W}}_{i\chi} & \bar{\mathbb{R}}_{i\chi} & \tilde{\mathbb{E}}_i \\ \bar{\mathbb{R}}_{i\chi}^T & \Sigma_{i\chi} & \tilde{\mathbb{M}}_i \\ \tilde{\mathbb{E}}_i^T & \tilde{\mathbb{M}}_i^T & -\mathrm{diag}\{\vartheta_i I, \vartheta_i I\} \end{bmatrix} \leq 0, \tag{5.46}$$

$$\begin{bmatrix} \tilde{\mathbb{X}}_i & \bar{\mathbb{T}}_i \\ \bar{\mathbb{T}}_i^T & \Upsilon_i \end{bmatrix} < 0, \tag{5.47}$$

where

$$\bar{\mathbb{P}} \triangleq \mathrm{diag}\left\{\tilde{P}_1, \ldots, \tilde{P}_N\right\}, \ \bar{\mathbb{Q}} \triangleq \mathrm{diag}\left\{\tilde{Q}_1, \ldots, \tilde{Q}_N\right\}, \ \bar{\Delta} \triangleq \mathrm{diag}\left\{\tilde{\Delta}_1, \ldots, \tilde{\Delta}_N\right\},$$

$$\mathbb{B}_i \triangleq \begin{bmatrix} \sqrt{\pi_{i1}}B_i^T\tilde{\varsigma}_i & \cdots & \sqrt{\pi_{iN}}B_i^T\tilde{\varsigma}_i \end{bmatrix},$$

$$\mathbb{L}_{i\chi} \triangleq \begin{bmatrix} \sqrt{\pi_{i1}}\mathcal{L}_\chi^T B_i^T & \cdots & \sqrt{\pi_{iN}}\mathcal{L}_\chi^T B_i^T \end{bmatrix},$$

$$\tilde{\mathbb{B}}_i \triangleq \begin{bmatrix} \sqrt{\pi_{i1}}B_i^T H_1^T \Omega_1 \tilde{\omega}_i & \cdots & \sqrt{\pi_{iN}}B_i^T H_N^T \Omega_N \tilde{\omega}_i \end{bmatrix},$$

$$\tilde{\mathbb{L}}_{i\chi} \triangleq \begin{bmatrix} \sqrt{\pi_{i1}}\mathcal{L}_\chi^T B_i^T H_1^T \Omega_1 & \cdots & \sqrt{\pi_{iN}}\mathcal{L}_\chi^T B_i^T H_N^T \Omega_N \end{bmatrix},$$

$$\hat{\mathbb{L}}_{i\chi} \triangleq \begin{bmatrix} \sqrt{\pi_{i1}}\mathcal{L}_\chi^T X_i^T & \cdots & \sqrt{\pi_{iN}}\mathcal{L}_\chi^T X_i^T \end{bmatrix},$$

$$\tilde{\mathbb{W}}_{i\chi} \triangleq \mathrm{diag}\left\{\tilde{X}_{i\chi} - \mathcal{J}_\chi^T - \mathcal{J}_\chi, -\tilde{Y}_{i\chi}, \tilde{Z}_{i\chi} - \mathcal{J}_\chi^T - \mathcal{J}_\chi, -\tilde{W}_{i\chi}\right\},$$

$$\Sigma_{i\chi} \triangleq -\mathrm{diag}\left\{\bar{\mathbb{P}}, \bar{\Delta}, \tilde{\Psi}_{i\chi}, \tilde{\Xi}_{i\chi}, \tilde{\Gamma}_{i\chi}, \bar{\mathbb{Q}}\right\},$$

$$\bar{\mathbb{R}}_{i\chi} \triangleq \begin{bmatrix} \tilde{\mathbb{A}}_{i\chi} & \hat{\mathbb{A}}_{i\chi} & \sqrt{3}\mathcal{J}_\chi^T H_\chi^T \Omega_\chi & \sqrt{3}\mathcal{J}_\chi^T H_\chi^T \Omega_\chi & \sqrt{3}\mathcal{J}_\chi^T H_\chi^T \Omega_\chi & \bar{\mathbb{A}}_{i\chi} \\ \tilde{\mathbb{D}}_{i\chi} & \hat{\mathbb{D}}_{i\chi} & 0 & 0 & 0 & \bar{\mathbb{D}}_{i\chi} \\ \tilde{\mathbb{K}}_{i\chi} & \hat{\mathbb{K}}_{i\chi} & -\sqrt{3}\mathcal{J}_\chi^T H_\chi^T \Omega_\chi & -\sqrt{3}\mathcal{J}_\chi^T H_\chi^T \Omega_\chi & -\sqrt{3}\mathcal{J}_\chi^T H_\chi^T \Omega_\chi & \bar{\mathbb{K}}_{i\chi} \\ \tilde{\mathbb{G}}_{i\chi} & \hat{\mathbb{G}}_{i\chi} & 0 & 0 & 0 & 0 \end{bmatrix},$$

$$\tilde{\mathbb{A}}_{i\chi} \triangleq \begin{bmatrix} \sqrt{3\pi_{i1}}\tilde{\mathbf{A}}_{i\chi}^T & \cdots & \sqrt{3\pi_{iN}}\tilde{\mathbf{A}}_{i\chi}^T \end{bmatrix},$$

$$\hat{\mathbb{A}}_{i\chi} \triangleq \left[\begin{array}{ccc} \sqrt{3\pi_{i1}}\tilde{\mathbf{A}}_{i\chi}^{\mathrm{T}}H_1^{\mathrm{T}}\Omega_1 & \cdots & \sqrt{3\pi_{iN}}\tilde{\mathbf{A}}_{i\chi}^{\mathrm{T}}H_N^{\mathrm{T}}\Omega_N \end{array}\right],$$

$$\bar{\mathbb{A}}_{i\chi} \triangleq \left[\begin{array}{ccc} \sqrt{3\pi_{i1}}\bar{\mathbf{A}}_{i\chi}^{\mathrm{T}} & \cdots & \sqrt{3\pi_{iN}}\bar{\mathbf{A}}_{i\chi}^{\mathrm{T}} \end{array}\right],$$

$$\tilde{\mathbb{D}}_{i\chi} \triangleq \left[\begin{array}{ccc} \sqrt{3\pi_{i1}}\tilde{Y}_{i\chi}D_i^{\mathrm{T}} & \cdots & \sqrt{3\pi_{iN}}\tilde{Y}_{i\chi}D_i^{\mathrm{T}} \end{array}\right],$$

$$\hat{\mathbb{D}}_{i\chi} \triangleq \left[\begin{array}{ccc} \sqrt{3\pi_{i1}}\tilde{Y}_{i\chi}D_i^{\mathrm{T}}H_1^{\mathrm{T}}\Omega_1 & \cdots & \sqrt{3\pi_{iN}}\tilde{Y}_{i\chi}D_i^{\mathrm{T}}H_N^{\mathrm{T}}\Omega_N \end{array}\right],$$

$$\bar{\mathbb{D}}_{i\chi} \triangleq \left[\begin{array}{ccc} \sqrt{3\pi_{i1}}\tilde{Y}_{i\chi}D_i^{\mathrm{T}}Z^{\mathrm{T}} & \cdots & \sqrt{3\pi_{iN}}\tilde{Y}_{i\chi}D_i^{\mathrm{T}}Z^{\mathrm{T}} \end{array}\right],$$

$$\tilde{\mathbb{K}}_{i\chi} \triangleq \left[\begin{array}{ccc} -\sqrt{3\pi_{i1}}\mathcal{K}_\chi^{\mathrm{T}}B_i^{\mathrm{T}} & \cdots & -\sqrt{3\pi_{iN}}\mathcal{K}_\chi^{\mathrm{T}}B_i^{\mathrm{T}} \end{array}\right],$$

$$\hat{\mathbb{K}}_{i\chi} \triangleq \left[\begin{array}{ccc} -\sqrt{3\pi_{i1}}\mathcal{K}_\chi^{\mathrm{T}}B_i^{\mathrm{T}}H_1^{\mathrm{T}}\Omega_1 & \cdots & -\sqrt{3\pi_{iN}}\mathcal{K}_\chi^{\mathrm{T}}B_i^{\mathrm{T}}H_N^{\mathrm{T}}\Omega_N \end{array}\right],$$

$$\bar{\mathbb{K}}_{i\chi} \triangleq \left[\begin{array}{ccc} -\sqrt{3\pi_{i1}}\mathcal{K}_\chi^{\mathrm{T}}X_i^{\mathrm{T}} & \cdots & -\sqrt{3\pi_{iN}}\mathcal{K}_\chi^{\mathrm{T}}X_i^{\mathrm{T}} \end{array}\right],$$

$$\tilde{\mathbb{G}}_{i\chi} \triangleq \left[\begin{array}{ccc} \sqrt{3\pi_{i1}}\tilde{W}_{i\chi}G_i^{\mathrm{T}} & \cdots & \sqrt{3\pi_{iN}}\tilde{W}_{i\chi}G_i^{\mathrm{T}} \end{array}\right],$$

$$\hat{\mathbb{G}}_{i\chi} \triangleq \left[\begin{array}{ccc} \sqrt{3\pi_{i1}}\tilde{W}_{i\chi}G_i^{\mathrm{T}}H_1^{\mathrm{T}}\Omega_1 & \cdots & \sqrt{3\pi_{iN}}\tilde{W}_{i\chi}G_i^{\mathrm{T}}H_N^{\mathrm{T}}\Omega_N \end{array}\right],$$

$$\tilde{\mathbf{A}}_{i\chi} \triangleq A_i\mathcal{J}_\chi + B_i\mathcal{K}_\chi,\ \bar{\mathbf{A}}_{i\chi} \triangleq ZA_i\mathcal{J}_\chi + X_i\mathcal{K}_\chi,$$

$$\tilde{\mathbb{E}}_i \triangleq \left[\begin{array}{cccc} E_i\mathcal{J}_\chi & 0 & 0 & 0 \\ 0 & 0 & 0 & 0 \end{array}\right]^{\mathrm{T}},$$

$$\tilde{\mathbb{M}}_i \triangleq \left[\begin{array}{cccccc} \underbrace{0\cdots 0}_{N} & \underbrace{0\cdots 0}_{N} & 0 & 0 & 0 & \underbrace{0\cdots 0}_{N} \\ \tilde{\mathbf{M}}_i & \hat{\mathbf{M}}_i & 0 & 0 & 0 & \underbrace{0\cdots 0}_{N} \end{array}\right]^{\mathrm{T}},$$

$$\tilde{\mathbf{M}}_i \triangleq \left[\begin{array}{ccc} \sqrt{3\pi_{i1}}M_i^{\mathrm{T}}\tilde{Z}_i^{\mathrm{T}}\vartheta_i & \cdots & \sqrt{3\pi_{iN}}M_i^{\mathrm{T}}\tilde{Z}_i^{\mathrm{T}}\vartheta_i \end{array}\right],$$

$$\hat{\mathbf{M}}_i \triangleq \left[\begin{array}{ccc} \sqrt{3\pi_{i1}}M_i^{\mathrm{T}}\tilde{Z}_i^{\mathrm{T}}H_1^{\mathrm{T}}\Omega_1\vartheta_i & \cdots & \sqrt{3\pi_{iN}}M_i^{\mathrm{T}}\tilde{Z}_i^{\mathrm{T}}H_N^{\mathrm{T}}\Omega_N\vartheta_i \end{array}\right],$$

$$\tilde{\mathbb{X}}_i \triangleq \left[\begin{array}{cccc} -\tilde{P}_i & \tilde{P}_iH_i^{\mathrm{T}}\Omega_i & 0 & 0 \\ \star & -3\tilde{\Delta}_i & 0 & 0 \\ \star & \star & -\tilde{T} & 0 \\ \star & \star & \star & -\gamma^2 I \end{array}\right],$$

$$\Upsilon_i \triangleq -\mathrm{diag}\{\tilde{\varsigma}_i I, \tilde{\omega}_i I, \tilde{T}, I, \tilde{\varsigma}_i I, \tilde{\omega}_i I, \mathbf{X}_i, \mathbf{Y}_i, \mathbf{Z}_i, \mathbf{W}_i\},$$

$$\mathbf{X}_i \triangleq \mathrm{diag}\left\{\tilde{X}_{i1}, \ldots, \tilde{X}_{iJ}\right\}, \mathbf{Y}_i \triangleq \mathrm{diag}\left\{\tilde{Y}_{i1}, \ldots, \tilde{Y}_{iJ}\right\},$$

$$\mathbf{Z}_i \triangleq \mathrm{diag}\left\{\tilde{Z}_{i1}, \ldots, \tilde{Z}_{iJ}\right\}, \mathbf{W}_i \triangleq \mathrm{diag}\left\{\tilde{W}_{i1}, \ldots, \tilde{W}_{iJ}\right\},$$

$$\bar{\mathbb{T}}_i \triangleq \left[\begin{array}{cccccccccc} \sqrt{6}\bar{\delta}_i\tilde{P}_i & \sqrt{6}\bar{\delta}_i\tilde{P}_i & \sqrt{\alpha}\tilde{P}_i & \tilde{P}_iC_i^{\mathrm{T}} & 0 & 0 & \hat{\mathbf{P}}_i & 0 & 0 & 0 \\ 0 & 0 & 0 & 0 & 0 & 0 & 0 & \hat{\Delta}_i & 0 & 0 \\ 0 & 0 & 0 & 0 & \sqrt{6m}\varrho\tilde{T} & \sqrt{6m}\varrho\tilde{T} & 0 & 0 & \hat{\mathbf{T}} & 0 \\ 0 & 0 & 0 & 0 & 0 & 0 & 0 & 0 & 0 & \hat{\mathbf{I}}_i \end{array}\right],$$

$$\bar{\delta}_i \triangleq \delta_i + \varrho\sqrt{m}, \hat{\mathbf{P}}_i \triangleq \left[\begin{array}{ccc} \sqrt{\xi_{i1}}\tilde{P}_i & \cdots & \sqrt{\xi_{iJ}}\tilde{P}_i \end{array}\right],$$

$$\hat{\Delta}_i \triangleq \left[\begin{array}{ccc} \sqrt{\xi_{i1}}\tilde{\Delta}_i & \cdots & \sqrt{\xi_{iJ}}\tilde{\Delta}_i \end{array}\right],$$

$$\hat{\mathbf{T}} \triangleq \left[\begin{array}{ccc} \sqrt{\xi_{i1}}\tilde{T} & \cdots & \sqrt{\xi_{iJ}}\tilde{T} \end{array}\right],$$

$$\hat{\mathbf{I}}_i \triangleq \left[\begin{array}{ccc} \sqrt{\xi_{i1}}I & \cdots & \sqrt{\xi_{iJ}}I \end{array}\right].$$

Main Results 93

We have that $iii) \Rightarrow ii) \Rightarrow i)$. *Moreover, if the coupled LMIs (5.41)–(5.47) hold, the mean square stability with a desired H_∞ disturbance attenuation level γ of the closed-loop system (5.11) and the reachability of the sliding region* **O** *defined in (5.28) around the sliding surface (5.8) can be guaranteed simultaneously by the event-triggered asynchronous SMC law (5.9) with $K_\chi = \mathcal{K}_\chi \mathcal{J}_\chi^{-1}$ and $L_\chi = \mathcal{L}_\chi \mathcal{V}_\chi^{-1}$.*

Proof $ii) \Rightarrow i)$. *Notice that $\Psi_{i\chi} + \Xi_{i\chi} + \Gamma_{i\chi} > \Psi_{i\chi} + \Xi_{i\chi}$. It is readily shown that the conditions (5.17) and (5.30) are ensured by the following matrix inequality:*

$$\mathbb{X}_i + \sum_{\chi=1}^{J} \xi_{i\chi} \left[\mathbb{Y}_{i\chi}^{\mathrm{T}} \left(\mathbb{P}_i + \mathbb{H}_i \right) \mathbb{Y}_{i\chi} + \mathbb{I}_\chi^{\mathrm{T}} \left(\Psi_{i\chi} + \Xi_{i\chi} + \Gamma_{i\chi} \right) \mathbb{I}_\chi + \tilde{\mathbb{Z}}_{i\chi}^{\mathrm{T}} \mathbb{Q}_i \tilde{\mathbb{Z}}_{i\chi} \right] < 0,$$

which is guaranteed by the conditions (5.39)–(5.40).

$iii) \Rightarrow ii)$. *Denote $\tilde{P}_i \triangleq P_i^{-1}$, $\tilde{Q}_i \triangleq Q_i^{-1}$, $\tilde{T} \triangleq T^{-1}$, $\tilde{X}_{i\chi} \triangleq X_{i\chi}^{-1}$, $\tilde{Y}_{i\chi} \triangleq Y_{i\chi}^{-1}$, $\tilde{Z}_{i\chi} \triangleq Z_{i\chi}^{-1}$, $\tilde{W}_{i\chi} \triangleq W_{i\chi}^{-1}$, $\tilde{\Phi}_{i\chi} \triangleq \Phi_{i\chi}^{-1}$, $\tilde{\Xi}_{i\chi} \triangleq \Xi_{i\chi}^{-1}$, $\tilde{\Gamma}_{i\chi} \triangleq \Gamma_{i\chi}^{-1}$, $\tilde{\Delta}_i \triangleq \Delta_i^{-1}$, $\tilde{\varsigma}_i \triangleq \varsigma_i^{-1}$, $\tilde{\omega}_i \triangleq \omega_i^{-1}$, $\mathcal{K}_\chi \triangleq K_\chi \mathcal{J}_\chi$, and $\mathcal{L}_i \triangleq L_\chi \mathcal{V}_\chi$. By resorting to the following matrix inequalities:*

$$\left(\Phi_{i\chi} - \mathcal{V}_\chi^{\mathrm{T}} \right) \Phi_{i\chi}^{-1} \left(\Phi_{i\chi} - \mathcal{V}_\chi \right) \geq 0, \quad \left(\Xi_{i\chi} - \mathcal{V}_\chi^{\mathrm{T}} \right) \Xi_{i\chi}^{-1} \left(\Xi_{i\chi} - \mathcal{V}_\chi \right) \geq 0,$$

$$\left(\Gamma_{i\chi} - \mathcal{V}_\chi^{\mathrm{T}} \right) \Gamma_{i\chi}^{-1} \left(\Gamma_{i\chi} - \mathcal{V}_\chi \right) \geq 0, \quad \left(X_{i\chi} - \mathcal{J}_\chi^{\mathrm{T}} \right) X_{i\chi}^{-1} \left(X_{i\chi} - \mathcal{J}_\chi \right) \geq 0,$$

$$\left(Z_{i\chi} - \mathcal{J}_\chi^{\mathrm{T}} \right) Z_{i\chi}^{-1} \left(Z_{i\chi} - \mathcal{J}_\chi \right) \geq 0.$$

it is concluded that the conditions (5.13)–(5.16), (5.29) are ensured by the LMIs (5.41)–(5.45) and the conditions (5.39)–(5.40) are guaranteed by the LMIs (5.46)–(5.47), respectively.

Remark 5.4 *In this chapter, the SMC problem is investigated for a class of Markovian jump Lur'e systems under the event-triggered strategy (5.6) and the hidden Markov mode detection model (5.7). The proposed event-triggered asynchronous SMC law exhibits the following distinct features: 1) the Lur'e nonlinearities are contained for enhancing the control performance; 2) an event-triggered protocol is cooperated for the energy-saving purpose; and 3) the designed SMC law just utilizes the detected mode signal but not the system mode signal.*

5.2.5 Solving algorithm

About the computational complexity of the conditions (5.41)–(5.47) in Theorem 5.3, it is required to solve $10N + 11NJ$ LMIs to get $6N + 4J + 7NJ + 1$ decision variables, or $(N + 2NJ + 1)\frac{n_x(n_x+1)}{2} + N \left(\frac{m(m+1)}{2} + J\frac{q(q+1)}{2} + J\frac{n_v(n_v+1)}{2} \right) + J(mn_x + mq + n_x^2 + q^2) + (3NJ + N)q + 3N$ scalar variables.

94   *Asynchronous SMC under SETP*

For the prescribed event-triggered parameter α in (5.6), we seek to find a minimum H_∞ performance level γ. As per Theorem 5.3, the design procedures of the SMC law with the minimum H_∞ performance can be summarized as follows:

- **Step 1.** Given the parameter $\alpha \in [0,1]$ in (5.6). Select the parameters β_i properly such that the matrix X_i is nonsingular for any $i \in \mathbb{S}$. Compute the scalars δ_i and ϱ in (5.10).

- **Step 2.** Get the matrices $\tilde{T}$, $\mathcal{K}_\chi$, $\mathcal{L}_\chi$, $\mathcal{J}_\chi$, and $\mathcal{V}_\chi$ by solving the following optimization problem:

$$\min_{\tilde{P}_i, \tilde{T}, \tilde{X}_{i\chi}, \tilde{Z}_{i\chi}, \tilde{Q}_i, \tilde{Y}_{i\chi}, \tilde{W}_{i\chi}, \tilde{\Psi}_{i\chi}, \tilde{\Xi}_{i\chi}, \tilde{\Gamma}_{i\chi}, \tilde{\Delta}_i, \tilde{\varsigma}_i, \tilde{\omega}_i, \vartheta_i} \gamma^*$$

$$\text{subject to : LMIs } (5.41) - (5.47) \text{ with } \gamma^* \triangleq \gamma^2. \tag{5.48}$$

- **Step 3.** Produce the event-triggered protocol (5.6) with $T = \tilde{T}^{-1}$, and the asynchronous SMC law (5.9) with matrices $K_\chi = \mathcal{K}_\chi \mathcal{J}_\chi^{-1}$ and $L_\chi = \mathcal{L}_\chi \mathcal{V}_\chi^{-1}$.

- **Step 4.** Apply the obtained asynchronous SMC scheme (5.9) to the Markovian jump Lur'e system (5.1) under the event-triggered protocol (5.6).

5.3   Example

In this section, a modified practical experiment from [113, 159] is employed to illustrate the proposed event-triggered asynchronous SMC scheme, where the angular velocity of a DC Motor device is controlled subject to abrupt failures. The Markov jump variable $r(k) = i$ characterize the random failures occurred on the power, which are generated by a computer. The power modes jump between three modes, that is, 0% of rotary (normal mode, $r(k) = 1$), +20% of rotary for improving the power (low mode, $r(k) = 2$), and -40% of rotary for decreasing the power (medium mode, $r(k) = 3$). Denote the system state as $x(k) = \begin{bmatrix} x_1(k) & x_2(k) & x_3(k) \end{bmatrix}^{\mathrm{T}}$, where $x_1(k)$, $x_2(k)$, and $x_3(k)$ stand for the angular velocity of the rotor, the electrical current consumed by the motor, and the integrative term written as a discrete sum, respectively.

Example 95

Now, the networked DC Motor device with Markov-driven power failures can be modelled as the discrete-time Markovian jump Lur'e system (5.1) with

$$A_i = \begin{bmatrix} a_{11}^{(i)} & a_{12}^{(i)} & 0 \\ a_{21}^{(i)} & a_{22}^{(i)} & 0 \\ a_{31}^{(i)} & 0 & a_{33}^{(i)} \end{bmatrix}, \quad B_i = \begin{bmatrix} b_1^{(i)} & b_2^{(i)} & 0 \end{bmatrix}^{\mathrm{T}},$$

$$G_i = 0.1I, \ C_i = \mathrm{diag}\{1, 1.2, 1.1\}, \ i \in \{1, 2, 3\},$$

$$M_1 = \begin{bmatrix} 0.9 & 0.5 & 0.12 \end{bmatrix}^{\mathrm{T}}, \ M_2 = \begin{bmatrix} 0.1 & 0.1 & 0.4 \end{bmatrix}^{\mathrm{T}},$$

$$M_3 = \begin{bmatrix} -0.9 & -0.1 & 0.1 \end{bmatrix}^{\mathrm{T}}, \ E_1 = \begin{bmatrix} -0.1 & 0.7 & 0.11 \end{bmatrix},$$

$$E_2 = \begin{bmatrix} 0.2 & 0.2 & 0.13 \end{bmatrix}, \ E_3 = \begin{bmatrix} 0.14 & 0.17 & 0.01 \end{bmatrix},$$

$$\Phi_1(k) = \frac{1}{1 + k^2}, \Phi_2(k) = \cos(0.03k), \Phi_3(k) = \sin(20k).$$

and the other parameters chosen as in Table 5.1. The transition probability matrix for the Markov jump failures is set as [113]:

$$\Pi = \begin{bmatrix} 0.95 & 0.05 & 0 \\ 0.36 & 0.6 & 0.04 \\ 0.1 & 0.1 & 0.8 \end{bmatrix}.$$

The following nonlinearity functions $\varphi_i(H_i x(k))$ as in [55, 216] are considered for the Markovian jump Lur'e system (5.1):

$$D_1 = \begin{bmatrix} 0.6 & 0.8 & 1 \end{bmatrix}^{\mathrm{T}}, \ D_2 = \begin{bmatrix} 0.4 & 0.7 & 0.1 \end{bmatrix}^{\mathrm{T}},$$

$$D_3 = \begin{bmatrix} 0.4 & 0.5 & -0.3 \end{bmatrix}^{\mathrm{T}}, \ H_1 = \begin{bmatrix} 0.5 & 0.3 & 0.1 \end{bmatrix},$$

$$H_2 = \begin{bmatrix} 0.4 & -0.5 & 0.3 \end{bmatrix}, H_3 = \begin{bmatrix} 1.5 & -0.9 & 0.4 \end{bmatrix},$$

$$\Omega_1 = 0.8, \ \Omega_2 = 1.5, \ \Omega_3 = 0.7,$$

$$\varphi_1(H_1 x(k)) = 0.5\Omega_1 H_1 x(k) \left(1 + \cos\left(H_1 x(k)\right)\right),$$

TABLE 5.1: Parameters in the networked DC Motor Device.

Parameters	$i = 1$	$i = 2$	$i = 3$
$a_{11}^{(i)}$	-0.4799	-1.6026	0.6346
$a_{12}^{(i)}$	0.51546	0.91632	0.9178
$a_{21}^{(i)}$	-0.38162	-0.5918	-0.5056
$a_{22}^{(i)}$	1.44723	0.30317	0.24811
$a_{31}^{(i)}$	0.1399	0.0740	0.3865
$a_{33}^{(i)}$	-0.9255	-0.4338	0.0982
$b_1^{(i)}$	0.58705	1.02851	0.7874
$b_2^{(i)}$	1.55010	0.22282	1.5302

$$\varphi_2(H_2x(k)) = 0.5\Omega_2 H_2x(k)\left(1 - e^{-0.1(H_2x(k))^2}\right),$$

$$\varphi_3(H_3x(k)) = 0.5\Omega_3 H_3x(k)\left(1 - \sin\left(H_3x(k)\right)\right).$$

In this example, we suppose that the system state $x(k)$ is transmitted to the controller under the event-triggered protocol (5.6) with the design parameter $\alpha = 0.4$, and the information of the power failure mode is emitted to the controller via a mode detector with the following MDPM:

$$\Theta = \begin{bmatrix} 0.6 & 0.2 & 0.2 \\ 0.1 & 0.7 & 0.2 \\ 0.3 & 0.3 & 0.4 \end{bmatrix}.$$

Our purpose now is to synthesize an event-triggered asynchronous SMC law (5.9) to stochastically stabilize the networked DC motor device subject to the actuator disturbances $f_i(x(k), k) = 0.3\sqrt{x_1^2 + x_2^2 + x_3^2}$. Letting $\beta_1 = \beta_2 = \beta_3 = \frac{1}{3}$ in (5.8), we have the sliding matrix $Z = \begin{bmatrix} 0.8010 & 1.1010 & 0 \end{bmatrix}$. It is easily verified that the nonsingularity of the matrix $X_i = ZB_i$ can be ensured for every mode i. By solving optimization problem (5.48), we obtain the minimum H_∞ disturbance attenuation performance $\gamma = 4.2650$ with the following event-triggered matrix and asynchronous SMC law for $k \in [l_n, l_{n+1})$:

$$T = \begin{bmatrix} 156.2101 & 0.5267 & -3.6297 \\ 0.5267 & 153.5401 & 0.4952 \\ -3.6297 & 0.4952 & 194.4801 \end{bmatrix}, \tag{5.49}$$

$$u(k) = \begin{cases} \begin{bmatrix} 0.2856 & -0.4651 & 0.0729 \end{bmatrix} x(l_n) - 0.0079\varphi_1\left(H_1x(l_n)\right) \\ \quad -0.7179\|x(l_n)\| \cdot \mathrm{sgn}(s(l_n)), & \chi = 1; \\ \begin{bmatrix} 1.8481 & -1.0199 & 0.0676 \end{bmatrix} x(k) - 0.0098\varphi_2\left(H_2x(l_n)\right) \\ \quad -0.7179\|x(l_n)\| \cdot \mathrm{sgn}(s(l_n)), & \chi = 2; \\ \begin{bmatrix} 0.1896 & -0.4268 & 0.0339 \end{bmatrix} x(k) - 0.0069\varphi_3\left(H_3x(l_n)\right) \\ \quad -0.7179\|x(l_n)\| \cdot \mathrm{sgn}(s(l_n)), & \chi = 3. \end{cases} \tag{5.50}$$

The simulation results are shown in Figs. 5.2–5.7. Under $v(k) = 0$, the initial condition $x(0) = \begin{bmatrix} 0.3 & -0.1 & 0.5 \end{bmatrix}^{\mathrm{T}}$ and a possible sequence of system and controller modes as depicted in Fig. 5.2, it is easily verified that the DC Motor device in open-loop case is unstable subject to the Markov-driven power failures. However, under the same scenarios, the MSS of the networked DC Motor device can be attained by the event-triggered asynchronous SMC law (5.50) as shown in Fig. 5.3. The evolutions of the release instants and release interval, the sliding variable $s(k)$ and the event-triggered asynchronous SMC input $u(k)$ are depicted in Figs. 5.4–5.6, respectively.

In order to illustrate the H_∞ disturbance attenuation performance of the obtained event-triggered asynchronous SMC law (5.50), we define an auxiliary function as $\gamma_D(k) = \sqrt{\frac{\sum_{j=0}^{k} \|z(j)\|^2}{\sum_{j=0}^{k} \|v(j)\|^2}}$. Under the zero initial condition and

Example

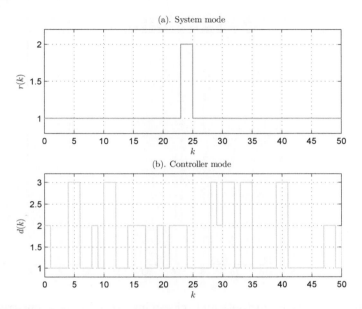

FIGURE 5.2: A possible sequence of system and controller modes.

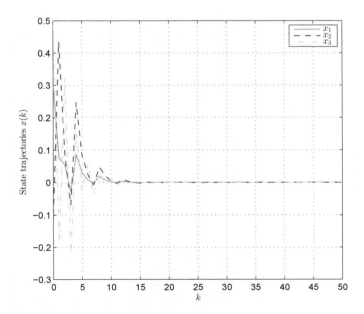

FIGURE 5.3: State trajectories $x(k)$ in closed-loop case.

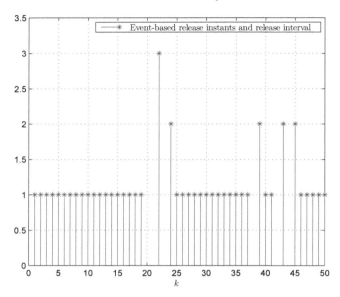

FIGURE 5.4: Release instants and release intervals for the event-triggered condition (5.6).

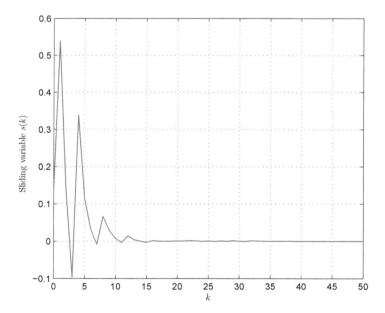

FIGURE 5.5: Sliding variable $s(k)$.

Example

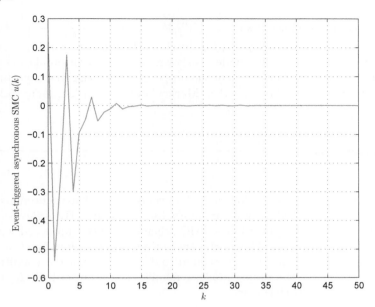

FIGURE 5.6: Event-triggered asynchronous SMC input $u(k)$.

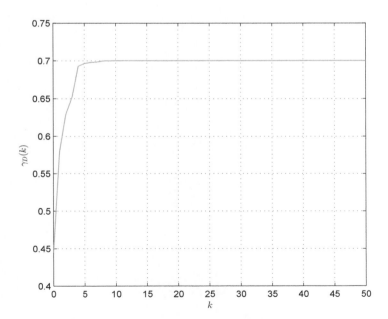

FIGURE 5.7: Time response of the function $\gamma_D(k)$.

the external disturbance $v(k) = \begin{bmatrix} \frac{\sin(100k)}{(k+1)^2} & -\frac{0.9}{(k+1)^2} & \frac{\cos(0.1k^2)}{(k+1)^3} \end{bmatrix}^{\mathrm{T}}$, Fig. 5.7 shows the time response of $\gamma_D(k)$ for the closed-loop DC Motor device subject to the system and controller modes as shown in Fig. 5.2. All simulation results have confirmed the effectiveness of the proposed event-triggered asynchronous H_∞ SMC scheme for the networked Markovian jump Lur'e system (5.1).

5.4 Conclusion

In this chapter, we have addressed the event-triggered asynchronous SMC problem for networked Markovian jump Lur'e systems. By *only* using the detected modes, an asynchronous SMC scheme has been proposed with the event-triggered protocol. The stochastic Lur'e-type Lyapunov functional and the HMM approach have been exploited to derive the sufficient conditions to ensure the MMS of the sliding mode dynamics with a prescribed H_∞ disturbance attenuation performance.

6

Sliding Mode Control Under Dynamic Event-Triggered Protocol

Multiple time-scale phenomena frequently occurs in many dynamical systems, such as the electrical circuits, engineering biomolecular systems, and advanced heavy water reactor. A traditional approach to dealing with a dynamical system with multiple time-scale characteristic is to model it as a so-called singularly perturbed system (SPS), where the degree of separation between the "slow" and "fast" modes of the systems are multiplied via a small positive parameter, i.e. the singular perturbed parameter ε. Over the past few decades, a large amount of attention has been paid to the analysis and synthesis of *continuous-time* SPSs. On the other hand, in the network-based communication or computer simulation, it is necessary to discretize a continuous-time SPSs to its discrete-time counterpart. According to the different sampling rates, there are two representative discrete-time SPS models, one is the fast-sampling SPSs and another is the slow-sampling ones.

In this chapter, we endeavour to solve the *dynamic* event-triggered SMC problem for the slow sampling SPSs. Between the plant and the sliding mode controller, an dynamic event-triggered mechanism is utilized to decrease the network transmission burden via introducing an additional internal dynamical variable. *The main contributions of this chapter are highlighted as follows: 1) A novel ε-dependent sliding surface is developed for the slow sampling SPSs by considering its specific structural characteristics; 2) A systematic approach is put forward to analyze the stability of the ideal sliding motion and the reachability of the specified sliding surface under the introduced dynamic event-triggering mechanism; and 3) The effect from the dynamic even-triggering parameter to the convergence of the quasi-sliding motion and the estimated ε-bound is examined within a convex optimization framework.*

6.1 Problem Formulation

This chapter concerns the following slow-sampling discrete-time SPS:

$$x_1(k+1) = [A_{11}+\Delta A_{11}(k)]\, x(k)+\varepsilon\, [A_{12}+\Delta A_{12}(k)]\, x_2(k)+B_1\, [u(k)+\xi(k)]\,,$$
$$(6.1)$$

DOI: 10.1201/9781003309499-6

101

$$x_2(k+1) = [A_{21} + \Delta A_{21}(k)]\, x(k) + \varepsilon\, [A_{22} + \Delta A_{22}(k)]\, x_2(k) + B_2\, [u(k) + \xi(k)]\,,$$
$$(6.2)$$

where $x_1(k) \in \mathbb{R}^{n_s}$ and $x_2(k) \in \mathbb{R}^{n_f}$ $(n_s + n_f = n)$ are the system slow and fast state vectors, respectively; $u(k) \in \mathbb{R}^m$ is the control input; $\varepsilon > 0$ is a known singularly perturbed parameter. $\xi(k) \in \mathbb{R}^m$ is a unknown external disturbance. The matrices A_{11}, A_{12}, A_{21}, A_{22}, B_1, and B_2 are known real matrices.

Let $x(k) \triangleq \begin{bmatrix} x_1^{\mathrm{T}}(k) & x_2^{\mathrm{T}}(k) \end{bmatrix}^{\mathrm{T}}$, the system (6.1)–(6.2) is rewritten as

$$x(k + 1) = [A + \Delta A(k)]\, E_\varepsilon x(k) + B\, [u(k) + \xi(k)]\,, \qquad (6.3)$$

where $E_\varepsilon \triangleq \begin{bmatrix} I_{n_s} & 0 \\ 0 & \varepsilon I_{n_f} \end{bmatrix}$, $A \triangleq \begin{bmatrix} A_{11} & A_{12} \\ A_{21} & A_{22} \end{bmatrix}$, $B \triangleq \begin{bmatrix} B_1 \\ B_2 \end{bmatrix}$, and $\Delta A(k) \triangleq \begin{bmatrix} \Delta A_{11}(k) & \Delta A_{12}(k) \\ \Delta A_{21}(k) & \Delta A_{22}(k) \end{bmatrix}$.

Remark 6.1 *Depending on the different sampling rate, it is well-recognized that discretization of the continuous-time SPS [49] will lead to a different discrete-time SPS model. One is the fast-sampling model, which is obtained by choosing the sampling rate appropriately to the fast subsystem as discussed in [33,96,177]. Another is the slow-sampling model, which is obtained by choosing the sampling rate appropriately to the slow subsystem as shown in [164]. Based on the slow- or fast-sampling models, control can be implemented with single-rate sample-data. This chapter focuses on studying the SMC problem of the slow-sampling SPS (6.1)–(6.2), which is still an open research issue.*

It is assumed that the system (6.3) satisfies the following standard conditions:

- The input matrix B is full column rank, that is, $\mathrm{rank}(B) = m$;

- The external disturbance $\xi(k)$ possesses $\|\xi(k)\| \leq \varrho$, where $\varrho \geq 0$ is a known scalar;

- The parameter uncertainty $\Delta A(k)$ has the form of $\Delta A(k) = MF(k)N$, where M, N are the known matrices and the time-varying matrix $F(k)$ satisfying $\|F(k)\| \leq 1$.

The interesting of this chapter is to design SMC law $u(k)$ such that the controlled system (6.3) is asymptotically stable under dynamic event-triggering protocol.

6.2 Main Results

6.2.1 A novel sliding surface

In this chapter, we employ the following novel sliding function:

$$s(k) = Gx(k) - G\left(A + BK_\varepsilon\right)E_\varepsilon x(k-1), \tag{6.4}$$

where $K_\varepsilon \in \mathbb{R}^{m \times n}$ is a ε-dependent gain matrix to be determined later and $G \in \mathbb{R}^{m \times n}$ is a real matrix to be chosen such that GB is nonsingular. Without loss of generality, we define $x(-1) = 0$.

Remark 6.2 *Notice that in the novel sliding function (6.4), the singular perturbed matrix E_ε is introduced skillfully by considering the specific singular perturbed structure of discrete-time slow sampling SPSs (6.3). By doing so, it enables to avoid the possible ill-conditioned numerical problems in the stability analysis of the sliding mode dynamics subsequently.*

6.2.2 Dynamic event-triggered SMC law

In this chapter, dynamic event-triggered strategy will be proposed to save computer resources. Denote by $\{l_0, l_1, \ldots\}$ the sequence of the instants at which the current state of the plant (6.3) is transmitted to the controller. Each instant l_n is defined according to the following event-triggered condition:

$$l_{n+1} = \min_{k \in \mathbb{N}_0} \left\{ k \Big| k > l_n, \frac{1}{\theta}\eta(k) + \sigma x^{\mathrm{T}}(k)x(k) - e^{\mathrm{T}}(k)e(k) \leq 0 \right\} \tag{6.5}$$

with $l_0 = 0$, where σ and θ are given positive scalars, the error $e(k)$ is defined by $e(k) = x(k) - x(l_n)$, and $\eta(k)$ is an internal dynamical variable satisfying

$$\eta(k+1) = \lambda\eta(k) + \sigma x^{\mathrm{T}}(k)x(k) - e^{\mathrm{T}}(k)e(k), \tag{6.6}$$

with $\lambda \in (0,1)$ being a given constant and $\eta(0) = \eta_0 \geq 0$ being the initial condition.

Remark 6.3 *It is easily seen from the event-triggering condition (6.5) that if the event-triggering parameter θ is large enough, the condition (6.5) will reduce to be $e^{\mathrm{T}}(k)e(k) \geq \sigma x^{\mathrm{T}}(k)x(k)$, which is just the so-called static event-triggering condition in [88, 137, 172]. Furthermore, one can observe that if the interval dynamical variable $\eta(k) \geq 0$ in (6.5), then the signal transmission between the plant and the sliding mode controller is occurred if and only if the error $e(k)$ satisfies $e^{\mathrm{T}}(k)e(k) \geq \frac{1}{\theta}\eta(k) + \sigma x^{\mathrm{T}}(k)x(k)$. Thus, when the dynamic event-triggering condition (6.5) is satisfied, then the static event-triggering condition $e^{\mathrm{T}}(k)e(k) \geq \sigma x^{\mathrm{T}}(k)x(k)$ must be also satisfied; but not vice versa. This relationship implies that the dynamic event-triggering mechanism could*

104   *SMC Under DETP*

render to fewer transmission times than the static one. The dynamic event-triggering mechanism (6.5)–(6.6) in this chapter is a modified version of the one in [77].

In what follows, we will carry out the reachability analysis for the specified sliding surface with the aid of designing a proper SMC law under the dynamic event-triggering mechanism (6.5)–(6.6). First, from the system (6.3) and the sliding function (6.4), it yields

$$s(k+1) = GBu(k) - GBK_\varepsilon E_\varepsilon x(k) + H(x(k), k), \qquad (6.7)$$

where $H(x(k), k) \triangleq G\Delta A(k)E_\varepsilon x(k) + GB\xi(k)$. Obviously, one has

$$\|H(x(k), k)\| \leq \chi(x(k)) \triangleq \kappa\|x(k)\| + \varrho\|GB\|, \qquad (6.8)$$

where $\kappa \triangleq \|GM\| \cdot \|NE_\varepsilon\|$.

Thus, under the dynamic event-triggered mechanism (6.5), we construct the following SMC law:

$$u(k) = (GB)^{-1}\left[GBK_\varepsilon E_\varepsilon x(l_n) - \chi(x(l_n))\mathrm{sgn}(s(l_n))\right], \qquad (6.9)$$

for $k \in [l_n, l_{n+1})$.

In what follows, we will analyze the reachability of the specified sliding surface (6.4) and the stability of the resultant sliding mode dynamics, respectively. To this end, the following lemmas are introduced to deal with the singular perturbation parameter.

Lemma 6.1 *[186] For a positive scalar $\bar{\varepsilon}$ and symmetric matrices S_1, S_2, and S_3 with appropriate dimensions, if $S_1 \geq 0$, $S_1 + \bar{\varepsilon}S_2 > 0$, $S_1 + \bar{\varepsilon}S_2 + \bar{\varepsilon}^2 S_3 > 0$ hold, then $S_1 + \varepsilon S_2 + \varepsilon^2 S_3 > 0$, $\forall \varepsilon \in (0, \bar{\varepsilon}]$.*

Lemma 6.2 *[186] If there exist matrices Q_i $(i = 1, 2, \ldots, 5)$ with $Q_i = Q_i^\mathrm{T}$ $(i = 1, 2, 3, 4)$ satisfying $Q_1 > 0$, $\begin{bmatrix} Q_1 + \bar{\varepsilon}Q_3 & \bar{\varepsilon}Q_5^\mathrm{T} \\ \bar{\varepsilon}Q_5 & \bar{\varepsilon}Q_2 \end{bmatrix} > 0$ and $\begin{bmatrix} Q_1 + \bar{\varepsilon}Q_3 & \bar{\varepsilon}Q_5^\mathrm{T} \\ \bar{\varepsilon}Q_5 & \bar{\varepsilon}Q_2 + \bar{\varepsilon}^2 Q_4 \end{bmatrix} > 0$, then $E_\varepsilon Q_\varepsilon = Q_\varepsilon^\mathrm{T} E_\varepsilon > 0$, $\forall \varepsilon \in (0, \bar{\varepsilon}]$, where $Q_\varepsilon = \begin{bmatrix} Q_1 + \varepsilon Q_3 & \varepsilon Q_5^\mathrm{T} \\ Q_5 & Q_2 + \varepsilon Q_4 \end{bmatrix}$.*

6.2.3   The reachability of sliding surface

This subsection will shows that the event-triggered SMC law (6.9) could derive the sliding variable $s(k)$ into a *vicinity* of the sliding surface $s(k) = 0$ in a finite time, that is, the *quasi-sliding motion*. By substituting the SMC law (6.9) into (6.7), it is obtained that during the event-triggering interval $k \in [l_n, l_{n+1})$, the sliding variable $s(k)$ possesses

$$s(k+1) = -GBK_\varepsilon E_\varepsilon e(k) + \tilde{H}(k), \qquad (6.10)$$

Main Results 105

where $\tilde{H}(k) \triangleq H(x(k), k) - \chi(x(l_n))\mathrm{sgn}(s(l_n))$. It is clear that

$$\|\tilde{H}(k)\| \leq \left\|G\Delta A(k)E_\varepsilon x(k) - \kappa\|x(l_n)\|\mathrm{sgn}(s(l_n))\right\|$$
$$+ \left\|GB\xi(k) - \varrho\|GB\|\mathrm{sgn}(s(l_n))\right\|$$
$$\leq (\kappa + \kappa\sqrt{m})\|x(k)\| + \kappa\sqrt{m}\|e(k)\| + 2\varrho\sqrt{m}\|GB\|. \tag{6.11}$$

The following theorem establishes a sufficient condition on the reachability of the specified sliding surface (6.4).

Theorem 6.1 *Consider the slow sampling SPS (6.3) with the dynamic event-triggering mechanism (6.5)–(6.6) and the event-triggered SMC law (6.9). Given a scalar $\bar{\varepsilon}$, if there exist symmetric matrices $\bar{W} \triangleq W^{-1} > 0$, $Q_1 > 0$, Q_2, Q_3 and Q_4, matrices Q_5 and $\bar{K} \in \mathbb{R}^{m\times n}$, and scalars $\bar{\beta} > 0$ and $\bar{\varpi} > 0$ satisfying the following LMIs:*

$$\begin{bmatrix} Q_1 + \bar{\varepsilon}Q_3 & \bar{\varepsilon}Q_5^{\mathrm{T}} \\ \bar{\varepsilon}Q_5 & \bar{\varepsilon}Q_2 \end{bmatrix} > 0, \tag{6.12}$$

$$\begin{bmatrix} Q_1 + \bar{\varepsilon}Q_3 & \bar{\varepsilon}Q_5^{\mathrm{T}} \\ \bar{\varepsilon}Q_5 & \bar{\varepsilon}Q_2 + \bar{\varepsilon}^2 Q_4 \end{bmatrix} > 0, \tag{6.13}$$

$$\begin{bmatrix} -\bar{\varpi}I & \bar{\varpi}I \\ \star & -\bar{W} \end{bmatrix} \leq 0, \tag{6.14}$$

$$\begin{bmatrix} \bar{\beta}I - (J^{\mathrm{T}} + J) & \sqrt{2}E_0\bar{K}^{\mathrm{T}}B^{\mathrm{T}}G^{\mathrm{T}} & \kappa\sqrt{6m}J^{\mathrm{T}} \\ \star & -\bar{W} & 0 \\ \star & \star & -\bar{\varpi}I \end{bmatrix} \leq 0, \tag{6.15}$$

$$\begin{bmatrix} \bar{\beta}I - (Q_{\bar{\varepsilon}}^{\mathrm{T}} + Q_{\bar{\varepsilon}}) & \sqrt{2}E_{\bar{\varepsilon}}\bar{K}^{\mathrm{T}}B^{\mathrm{T}}G^{\mathrm{T}} & \kappa\sqrt{6m}Q_{\bar{\varepsilon}}^{\mathrm{T}} \\ \star & -\bar{W} & 0 \\ \star & \star & -\bar{\varpi}I \end{bmatrix} \leq 0, \tag{6.16}$$

where $Q_\varepsilon \triangleq J + \varepsilon R$, $J \triangleq \begin{bmatrix} Q_1 & 0 \\ Q_5 & Q_2 \end{bmatrix}$, $R \triangleq \begin{bmatrix} Q_3 & Q_5^{\mathrm{T}} \\ 0 & Q_4 \end{bmatrix}$, $E_0 \triangleq \begin{bmatrix} I_{n_s} & 0 \\ 0 & 0 \end{bmatrix}$,

$E_{\bar{\varepsilon}} \triangleq \begin{bmatrix} I_{n_s} & 0 \\ 0 & \bar{\varepsilon}I_{n_f} \end{bmatrix}$ and $Q_{\bar{\varepsilon}} \triangleq J + \bar{\varepsilon}R$, then for the singularly perturbed parameter $\varepsilon \in (0, \bar{\varepsilon}]$ and the dynamic event-triggering mechanism (6.5)–(6.6), the sliding variable $s(k)$ will be driven into the following sliding region $\mathbf{O}$ by the SMC law (6.9) with $K_\varepsilon = \bar{K}Q_\varepsilon^{-\mathrm{T}}$:

$$\mathbf{O} \triangleq \left\{ s(k) \mid \|s(k)\| \leq \tilde{\gamma}(k) \right\}, \tag{6.17}$$

where $\tilde{\gamma}(k) \triangleq \sqrt{\frac{\gamma(k)}{\lambda_{\min}(W)}}$ with $\gamma(k) \triangleq \left[6\varpi(\kappa + \kappa\sqrt{m})^2 + \sigma\beta\right]\|x(k)\|^2 + \frac{\beta}{\theta}\eta(k) + 24\varpi m\varrho^2\|GB\|^2$.

Proof *Consider the Lyapunov function $\tilde{V}(k) \triangleq s^{\mathrm{T}}(k)Ws(k)$. Along the solution of (6.10) and by using the condition (6.14), we evaluate the difference of $\tilde{V}(k)$ as follows:*

$$
\begin{aligned}
\Delta\tilde{V}(k) =& s^{\mathrm{T}}(k+1)Ws(k+1) - s^{\mathrm{T}}(k)Ws(k)\\
\leq& 2e^{\mathrm{T}}(k)E_\varepsilon K^{\mathrm{T}}B^{\mathrm{T}}G^{\mathrm{T}}WGBKE_\varepsilon e(k) + 2\varpi\|\tilde{H}(k)\|^2 - s^{\mathrm{T}}(k)Ws(k)\\
\leq& 2e^{\mathrm{T}}(k)E_\varepsilon K^{\mathrm{T}}B^{\mathrm{T}}G^{\mathrm{T}}WGBKE_\varepsilon e(k) + 2\varpi\Big[3(\kappa + \kappa\sqrt{m})^2\|x(k)\|^2\\
&+ 3m\kappa^2\|e(k)\|^2 + 12m\varrho^2\|GB\|^2\Big] - s^{\mathrm{T}}(k)Ws(k).
\end{aligned}
\tag{6.18}
$$

Furthermore, as per the dynamic event-triggering condition (6.5), for any scalar $\beta > 0$, one has

$$
\begin{aligned}
\Delta\tilde{V}(k) \leq& \Delta\tilde{V}(k) + \beta\Big[\frac{1}{\theta}\eta(k) + \sigma x^{\mathrm{T}}(k)x(k) - e^{\mathrm{T}}(k)e(k)\Big]\\
\leq& e^{\mathrm{T}}(k)\big[2E_\varepsilon K^{\mathrm{T}}B^{\mathrm{T}}G^{\mathrm{T}}WGBKE_\varepsilon + 6m\kappa^2\varpi I - \beta I\big]e(k) + \gamma(k)\\
&- \lambda_{\min}(W)\|s(k)\|^2.
\end{aligned}
\tag{6.19}
$$

Denote $\bar{K} \triangleq K_\varepsilon Q_\varepsilon^{\mathrm{T}}$, $\bar{\beta} \triangleq \beta^{-1}$ and $\bar{\varpi} \triangleq \varpi^{-1}$. Notice that the conditions (6.12)–(6.13) imply $Q_\varepsilon^{\mathrm{T}}E_\varepsilon = E_\varepsilon Q_\varepsilon > 0$ for any $\varepsilon \in (0, \bar{\varepsilon}]$. Thus, by resorting to Lemma 6.1 and the relationship $(Q_\varepsilon^{\mathrm{T}} - \bar{\beta}I)\beta(Q_\varepsilon - \bar{\beta}I) \geq 0$, it is shown that the conditions (6.15)–(6.16) guarantee

$$
Q_\varepsilon^{\mathrm{T}}\left(2E_\varepsilon K^{\mathrm{T}}B^{\mathrm{T}}G^{\mathrm{T}}WGBKE_\varepsilon + 6m\kappa^2\varpi I - \beta I\right)Q_\varepsilon < 0,
\tag{6.20}
$$

for any $\varepsilon \in (0, \bar{\varepsilon}]$.

Obviously, when the state trajectories escape from the region $\mathbf{O}$ around the specified sliding surface (6.4), that is, $\|s(k)\| \geq \sqrt{\frac{\gamma(k)}{\lambda_{\min}(W)}}$, from (6.19)–(6.20), the conditions (6.15)–(6.16) render $\Delta\tilde{V}(k) < 0$ for any $\varepsilon \in (0, \bar{\varepsilon}]$, which means that the sliding variable $s(k)$ is strictly deceasing outside the region $\mathbf{O}$.

6.2.4 The stability of sliding mode dynamics

It is known that when the ideal quasi-sliding mode occurs, the sliding variable satisfies $s(k+1) = s(k) = 0$. Thus, the equivalent control law can be obtained from (6.3) and (6.4) that

$$
u_{eq}(k) = -(GB)^{-1}G\Delta A(k)E_\varepsilon x(k) + K_\varepsilon E_\varepsilon x(k) - \xi(k).
\tag{6.21}
$$

Under the dynamic event-triggered condition (6.5)–(6.6), the event-triggered equivalent control for the sliding motion is as follows:

$$
u_{eq}(k) = -(GB)^{-1}G\Delta A(k)E_\varepsilon x(l_n) + K_\varepsilon E_\varepsilon x(l_n) - \xi(k),
\tag{6.22}
$$

during the time interval $k \in [l_n, l_{n+1})$.

Main Results 107

Submitting (6.22) into (6.3), we have the following sliding mode dynamics:

$$x(k+1) = \bar{A}(k)E_\varepsilon x(k) + \bar{A}_e(k)E_\varepsilon e(k), \qquad (6.23)$$

where $k \in [l_n, l_{n+1})$, $\bar{A}(k) \triangleq A + BK_\varepsilon + \tilde{G}\Delta A(k)$, $\bar{A}_e(k) \triangleq B(GB)^{-1}G\Delta A(k) - BK_\varepsilon$, and $\tilde{G} \triangleq I - B(GB)^{-1}G$.

In the following theorem, we analyze the asymptotic stability of the sliding mode dynamics (6.23).

Theorem 6.2 *Consider the slow sampling SPS (6.3) with the dynamic event-triggering mechanism (6.5)–(6.6). For given scalars $\bar{\varepsilon} > 0$ and $\mu > 0$, if the prescribed dynamic event-triggering parameters in (6.5)–(6.6) satisfy*

$$\lambda\theta \geq 1, \qquad (6.24)$$

and there exist symmetric matrices $\bar{P} > 0$, $Q_1 > 0$, Q_2, Q_3 and Q_4, matrices Q_5 and $\bar{K} \in \mathbb{R}^{m\times n}$, and scalars $\gamma_1 > 0$, $\gamma_2 > 0$ and $\bar{\alpha} > 0$ such that the conditions (6.12)–(6.13) and the following LMIs hold:

$$\begin{bmatrix} \hat{\Omega} & \hat{\Sigma} \\ \star & -\Lambda \end{bmatrix} < 0, \qquad (6.25)$$

$$\begin{bmatrix} \check{\Omega} & \check{\Sigma} \\ \star & -\Lambda \end{bmatrix} < 0, \qquad (6.26)$$

$$\begin{bmatrix} \acute{\Omega} & \acute{\Sigma} \\ \star & -\Lambda \end{bmatrix} < 0, \qquad (6.27)$$

where

$$\hat{\Omega} \triangleq \begin{bmatrix} \hat{\Gamma} & 0 & 0 & TA^{\mathrm{T}} + E_0\bar{K}^{\mathrm{T}}B^{\mathrm{T}} \\ \star & \hat{\Psi} & 0 & -E_0\bar{K}^{\mathrm{T}}B^{\mathrm{T}} \\ \star & \star & \frac{\lambda-1}{\theta} & 0 \\ \star & \star & \star & -\bar{P} \end{bmatrix},$$

$$\check{\Omega} \triangleq \begin{bmatrix} \check{\Gamma} & 0 & 0 & (T+\bar{\varepsilon}U)A^{\mathrm{T}} + E_{\bar{\varepsilon}}\bar{K}^{\mathrm{T}}B^{\mathrm{T}} \\ \star & \check{\Psi} & 0 & -E_{\bar{\varepsilon}}\bar{K}^{\mathrm{T}}B^{\mathrm{T}} \\ \star & \star & \frac{\lambda-1}{\theta} & 0 \\ \star & \star & \star & -\bar{P} \end{bmatrix},$$

$$\acute{\Omega} \triangleq \begin{bmatrix} \check{\Gamma} & 0 & 0 & (T+\bar{\varepsilon}U+\bar{\varepsilon}^2 Z)A^{\mathrm{T}} + E_{\bar{\varepsilon}}\bar{K}^{\mathrm{T}}B^{\mathrm{T}} \\ \star & \check{\Psi} & 0 & -E_{\bar{\varepsilon}}\bar{K}^{\mathrm{T}}B^{\mathrm{T}} \\ \star & \star & \frac{\lambda-1}{\theta} & 0 \\ \star & \star & \star & -\bar{P} \end{bmatrix},$$

$$\hat{\Sigma} \triangleq \begin{bmatrix} TN^{\mathrm{T}} & 0 & 0 & 0 & J^{\mathrm{T}} & J^{\mathrm{T}} & 0 \\ 0 & TN^{\mathrm{T}} & 0 & 0 & 0 & 0 & 0 \\ 0 & 0 & 0 & 0 & 0 & 0 & 1 \\ 0 & 0 & \gamma_1\tilde{G}M & \gamma_2 B(GB)^{-1}G & 0 & 0 & 0 \end{bmatrix},$$

$$\check{\Sigma} \triangleq \begin{bmatrix} (T+\bar{\varepsilon}U)N^{\mathrm{T}} & 0 & 0 & 0 & Q_{\bar{\varepsilon}}^{\mathrm{T}} & Q_{\bar{\varepsilon}}^{\mathrm{T}} & 0 \\ 0 & (T+\bar{\varepsilon}U)N^{\mathrm{T}} & 0 & 0 & 0 & 0 & 0 \\ 0 & 0 & 0 & 0 & 0 & 0 & 1 \\ 0 & 0 & \gamma_1\tilde{G}M & \gamma_2 B(GB)^{-1}G & 0 & 0 & 0 \end{bmatrix},$$

$$\acute{\Sigma} \triangleq \begin{bmatrix} (T+\bar{\varepsilon}U+\bar{\varepsilon}^2 Z)N^{\mathrm{T}} & 0 & 0 & 0 & Q_{\bar{\varepsilon}}^{\mathrm{T}} & Q_{\bar{\varepsilon}}^{\mathrm{T}} & 0 \\ 0 & (T+\bar{\varepsilon}U+\bar{\varepsilon}^2 Z)N^{\mathrm{T}} & 0 & 0 & 0 & 0 & 0 \\ 0 & 0 & 0 & 0 & 0 & 0 & 1 \\ 0 & 0 & \gamma_1\tilde{G}M & \gamma_2 B(GB)^{-1}G & 0 & 0 & 0 \end{bmatrix},$$

$$\Lambda \triangleq \mathrm{diag}\left\{\gamma_1 I, \gamma_2 I, \gamma_1 I, \gamma_2 I, \frac{\theta}{\sigma}I, \frac{\bar{\alpha}}{\sigma}I, \theta\bar{\alpha}\right\},$$

$$\hat{\Gamma} \triangleq \bar{P} - (J^{\mathrm{T}} + J), \hat{\Psi} \triangleq (\mu^2\theta + \bar{\alpha})I - (1+\mu)(J^{\mathrm{T}} + J),$$

$$\check{\Gamma} \triangleq \bar{P} - (Q_{\bar{\varepsilon}}^{\mathrm{T}} + Q_{\varepsilon}), \check{\Psi} \triangleq (\mu^2\theta + \bar{\alpha})I - (1+\mu)(Q_{\bar{\varepsilon}}^{\mathrm{T}} + Q_{\bar{\varepsilon}}),$$

and $T \triangleq \begin{bmatrix} Q_1 & 0 \\ 0 & 0 \end{bmatrix}$, $U \triangleq \begin{bmatrix} Q_3 & Q_5^{\mathrm{T}} \\ Q_5 & Q_2 \end{bmatrix}$, $Z \triangleq \begin{bmatrix} 0 & 0 \\ 0 & Q_4 \end{bmatrix}$ *and the matrices* Q_ε, J, R, E_0, $E_{\bar{\varepsilon}}$, $Q_{\bar{\varepsilon}}$ *are defined in Theorem 6.1, then the sliding mode dynamics (6.23) is asymptotically stable for any* $\varepsilon \in (0, \bar{\varepsilon}]$. *In this case, the gain matrix* K_ε *is given as:* $K_\varepsilon = \bar{K} \cdot (J^{\mathrm{T}} + \varepsilon R^{\mathrm{T}})^{-1}$.

Proof *Notice that the dynamic event-triggered condition (6.5) implies that for any* $k \in [l_n, l_{n+1})$, *one has*

$$\frac{1}{\theta}\eta(k) + \sigma x^{\mathrm{T}}(k)x(k) - e^{\mathrm{T}}(k)e(k) \geq 0. \tag{6.28}$$

Furthermore, by using the dynamical equation (6.6), we have

$$\eta(k+1) \geq \left(\lambda - \frac{1}{\theta}\right)\eta(k) \geq \cdots \geq \left(\lambda - \frac{1}{\theta}\right)^{k+1}\eta(0). \tag{6.29}$$

Obviously, under the initial condition $\eta(0) \geq 0$, *the condition (6.24) ensures* $\eta(k) \geq 0$ *for any* $k \geq 0$.

Now, we consider the following Lyapunov function:

$$V(k) \triangleq x^{\mathrm{T}}(k)Px(k) + \frac{1}{\theta}\eta(k). \tag{6.30}$$

Along the sliding mode dynamics (6.23) and the dynamical equation (6.6), we compute the difference of $V(k)$ *as follows:*

$$\begin{aligned} \Delta V(k) =& x^{\mathrm{T}}(k+1)Px(k+1) - x^{\mathrm{T}}(k)Px(k) + \frac{1}{\theta}\eta(k+1) - \frac{1}{\theta}\eta(k) \\ =& x^{\mathrm{T}}(k)\left[E_\varepsilon\bar{A}^{\mathrm{T}}(k)P\bar{A}(k)E_\varepsilon - P + \frac{\sigma}{\theta}I\right]x(k) \\ &+ e^{\mathrm{T}}(k)\left[E_\varepsilon\bar{A}_e^{\mathrm{T}}(k)P\bar{A}_e(k)E_\varepsilon - \frac{1}{\theta}I\right]e(k) \\ &+ 2x^{\mathrm{T}}(k)E_\varepsilon\bar{A}^{\mathrm{T}}(k)P\bar{A}_e(k)E_\varepsilon e(k) + \frac{\lambda-1}{\theta}\eta(k). \end{aligned} \tag{6.31}$$

Main Results 109

Thus, for any scalar $\alpha > 0$, it follows from (6.28) and (6.31) that

$$\Delta V(k) \leq \Delta V(k) + \alpha \left[\frac{1}{\theta} \eta(k) + \sigma x^{\mathrm{T}}(k) x(k) - e^{\mathrm{T}}(k) e(k) \right]$$
$$= \zeta^{\mathrm{T}}(k) \Omega \zeta(k), \qquad (6.32)$$

where $\zeta(k) \triangleq \left[x^{\mathrm{T}}(k) \; e^{\mathrm{T}}(k) \; \sqrt{\eta(k)} \right]^{\mathrm{T}}$ *and* $\Omega \triangleq \begin{bmatrix} \Gamma & E_\varepsilon \bar{A}^{\mathrm{T}}(k) P \bar{A}_e(k) E_\varepsilon & 0 \\ \star & \Upsilon & 0 \\ \star & \star & \frac{\lambda - 1 + \alpha}{\theta} \end{bmatrix}$

with $\Gamma \triangleq E_\varepsilon \bar{A}^{\mathrm{T}}(k) P \bar{A}(k) E_\varepsilon - P + \frac{\sigma}{\theta} I + \alpha \sigma I$ *and* $\Upsilon \triangleq E_\varepsilon \bar{A}_e^{\mathrm{T}}(k) P \bar{A}_e(k) E_\varepsilon - \frac{1}{\theta} I - \alpha I$.

By resorting to the Schur complement, it is easily shown that $\Omega < 0$ is equivalent to $\tilde{\Omega} < 0$ where

$$\tilde{\Omega} \triangleq \bar{\Omega} + \gamma_1^{-1} \mathcal{N}_1 \mathcal{N}_1^{\mathrm{T}} + \gamma_1 \mathcal{M}_1 \mathcal{M}_1^{\mathrm{T}} + \gamma_2^{-1} \mathcal{N}_2 \mathcal{N}_2^{\mathrm{T}} + \gamma_2 \mathcal{M}_2 \mathcal{M}_2^{\mathrm{T}}, \qquad (6.33)$$

with the scalars $\gamma_1 > 0$, $\gamma_2 > 0$, and

$$\bar{\Omega} \triangleq \begin{bmatrix} \bar{\Gamma} & 0 & 0 & E_\varepsilon (A + BK)^{\mathrm{T}} \\ \star & -\frac{1}{\theta} I - \alpha I & 0 & -E_\varepsilon K^{\mathrm{T}} B^{\mathrm{T}} \\ \star & \star & \frac{\lambda - 1 + \alpha}{\theta} & 0 \\ \star & \star & \star & -P^{-1} \end{bmatrix},$$

$$\bar{\Gamma} \triangleq -P + \frac{\sigma}{\theta} I + \alpha \sigma I, \; \mathcal{N}_1 \triangleq \begin{bmatrix} NE_\varepsilon & 0 & 0 & 0 \end{bmatrix}^{\mathrm{T}}, \mathcal{N}_2 \triangleq \begin{bmatrix} 0 & NE_\varepsilon & 0 & 0 \end{bmatrix}^{\mathrm{T}},$$

$$\mathcal{M}_1 \triangleq \begin{bmatrix} 0 & 0 & 0 & M^{\mathrm{T}} \tilde{G}^{\mathrm{T}} \end{bmatrix}^{\mathrm{T}}, \mathcal{M}_2 \triangleq \begin{bmatrix} 0 & 0 & 0 & M^{\mathrm{T}} G^{\mathrm{T}} (GB)^{-\mathrm{T}} B^{\mathrm{T}} \end{bmatrix}^{\mathrm{T}}.$$

By pre- and post-multiplying matrix $\tilde{\Omega}$ by $\mathrm{diag}\{Q_\varepsilon^{\mathrm{T}}, Q_\varepsilon^{\mathrm{T}}, I, I\}$ and its transposed matrix, respectively, then one can find from Lemma 6.2 that the conditions (6.12)–(6.13) implies $Q_\varepsilon^{\mathrm{T}} E_\varepsilon = E_\varepsilon Q_\varepsilon > 0$ for any $\varepsilon \in (0, \bar{\varepsilon}]$. Now, we denote $\bar{\alpha} \triangleq \alpha^{-1}$, $\bar{P} \triangleq P^{-1}$, and $\bar{K} \triangleq K_\varepsilon Q_\varepsilon^{\mathrm{T}}$, and further employ the following inequalities:

$$\left(Q_\varepsilon^{\mathrm{T}} - \bar{P} \right) P \left(Q_\varepsilon - \bar{P} \right) \geq 0, \; \left(Q_\varepsilon^{\mathrm{T}} - \mu \theta I \right) \frac{1}{\theta} \left(Q_\varepsilon - \mu \theta I \right) \geq 0,$$
$$\left(Q_\varepsilon^{\mathrm{T}} - \bar{\alpha} I \right) \alpha \left(Q_\varepsilon - \bar{\alpha} I \right) \geq 0,$$

where $\mu > 0$ is an adjustable parameter, then, it renders from Lemma 6.1 that the conditions (6.25)–(6.27) ensure $\tilde{\Omega} < 0$ for any $\varepsilon \in (0, \bar{\varepsilon}]$ due to the fact of $E_\varepsilon Q_\varepsilon = T + \varepsilon U + \varepsilon^2 Z$. Therefore, when $\zeta(k) \neq 0$, the conditions (6.12)–(6.27) guarantees $\Delta V(k) \leq \zeta^{\mathrm{T}}(k) \Omega \zeta(k) < 0$, which means that the sliding mode dynamics (6.23) is asymptotically stable for any $\varepsilon \in (0, \bar{\varepsilon}]$.

6.2.5 Further discussions

6.2.5.1 Special case: Static event-triggered SMC of slow-sampling SPSs

As shown in the dynamic event-triggering condition (6.5), when the parameter $\theta \to +\infty$, the condition (6.5) will reduce to the following *static*

110   *SMC Under DETP*

event-triggering condition:

$$l_{n+1} = \min\left\{ k \in \mathbb{N}_0 \,\middle|\, k > l_n, e^{\mathrm{T}}(k)e(k) \geq \sigma x^{\mathrm{T}}(k)x(k) \right\}, \tag{6.34}$$

where $\sigma > 0$ is a given parameter. And then, the results in Theorems 6.1 and 6.2 can be extended to the case under *static* event-triggering mechanism (6.34).

Corollary 6.1 *Consider the slow sampling SPS (6.3) with the static event-triggering mechanism (6.34) and the event-triggered SMC law (6.9). Given a scalar $\bar{\varepsilon}$, if there exist symmetric matrices $\bar{W} > 0$, $Q_1 > 0$, Q_2, Q_3 and Q_4, matrices Q_5 and $\bar{K} \in \mathbb{R}^{m \times n}$, and scalars $\bar{\beta} > 0$ and $\bar{\varpi} > 0$ satisfying the LMIs (6.12)–(6.16), then for the singularly perturbed parameter $\varepsilon \in (0, \bar{\varepsilon}]$, the sliding variable $s(k)$ will be driven into the following sliding region $\mathbf{O}$ by the SMC law (6.9) with $K_\varepsilon = \bar{K} Q_\varepsilon^{-\mathrm{T}}$:*

$$\mathbf{O} \triangleq \left\{ s(k) \,\middle|\, \|s(k)\| \leq \tilde{\gamma}(k) \right\}, \tag{6.35}$$

where $\tilde{\gamma}(k) \triangleq \sqrt{\frac{\gamma(k)}{\lambda_{\min}(W)}}$ with $\gamma(k) \triangleq \left[6\varpi(\kappa + \kappa\sqrt{m})^2 + \sigma\beta \right] \|x(k)\|^2 + 24\varpi m \varrho^2 \times \|GB\|^2$.

The proof can be obtained easily by letting the parameter $\theta \to +\infty$ in the proof of Theorem 6.1.

The following corollary further states the stability condition under the *static* event-triggering mechanism (6.34).

Corollary 6.2 *Consider the slow sampling SPS (6.3) with the static event-triggering mechanism (6.34). For a given scalar $\bar{\varepsilon} > 0$, if there exist symmetric matrices $\bar{P} > 0$, $Q_1 > 0$, Q_2, Q_3, and Q_4, matrices Q_5 and $\bar{K} \in \mathbb{R}^{m \times n}$, and scalars $\gamma_1 > 0$, $\gamma_2 > 0$, and $\bar{\alpha} > 0$ satisfying the conditions (6.12)–(6.13) and the following LMIs:*

$$\begin{bmatrix} \hat{\Omega} & \hat{\Sigma} \\ \star & -\Lambda \end{bmatrix} < 0, \tag{6.36}$$

$$\begin{bmatrix} \check{\Omega} & \check{\Sigma} \\ \star & -\Lambda \end{bmatrix} < 0, \tag{6.37}$$

$$\begin{bmatrix} \acute{\Omega} & \acute{\Sigma} \\ \star & -\Lambda \end{bmatrix} < 0, \tag{6.38}$$

Main Results

where

$$\hat{\Omega} \triangleq \begin{bmatrix} \hat{\Gamma} & 0 & TA^{\mathrm{T}} + E_0\bar{K}^{\mathrm{T}}B^{\mathrm{T}} \\ \star & \hat{\Psi} & -E_0\bar{K}^{\mathrm{T}}B^{\mathrm{T}} \\ \star & \star & -\bar{P} \end{bmatrix},$$

$$\check{\Omega} \triangleq \begin{bmatrix} \check{\Gamma} & 0 & (T+\bar{\varepsilon}U)A^{\mathrm{T}} + E_{\bar{\varepsilon}}\bar{K}^{\mathrm{T}}B^{\mathrm{T}} \\ \star & \check{\Psi} & -E_{\bar{\varepsilon}}\bar{K}^{\mathrm{T}}B^{\mathrm{T}} \\ \star & \star & -\bar{P} \end{bmatrix},$$

$$\acute{\Omega} \triangleq \begin{bmatrix} \check{\Gamma} & 0 & (T+\bar{\varepsilon}U+\bar{\varepsilon}^2Z)A^{\mathrm{T}} + E_{\bar{\varepsilon}}\bar{K}^{\mathrm{T}}B^{\mathrm{T}} \\ \star & \check{\Psi} & -E_{\bar{\varepsilon}}\bar{K}^{\mathrm{T}}B^{\mathrm{T}} \\ \star & \star & -\bar{P} \end{bmatrix},$$

$$\hat{\Sigma} \triangleq \begin{bmatrix} TN^{\mathrm{T}} & 0 & 0 & 0 & J^{\mathrm{T}} \\ 0 & TN^{\mathrm{T}} & 0 & 0 & 0 \\ 0 & 0 & \gamma_1\tilde{G}M & \gamma_2B(GB)^{-1}G & 0 \end{bmatrix},$$

$$\check{\Sigma} \triangleq \begin{bmatrix} (T+\bar{\varepsilon}U)N^{\mathrm{T}} & 0 & 0 & 0 & Q_{\bar{\varepsilon}}^{\mathrm{T}} \\ 0 & (T+\bar{\varepsilon}U)N^{\mathrm{T}} & 0 & 0 & 0 \\ 0 & 0 & \gamma_1\tilde{G}M & \gamma_2B(GB)^{-1}G & 0 \end{bmatrix},$$

$$\acute{\Sigma} \triangleq \begin{bmatrix} (T+\bar{\varepsilon}U+\bar{\varepsilon}^2Z)N^{\mathrm{T}} & 0 & 0 & 0 & Q_{\bar{\varepsilon}}^{\mathrm{T}} \\ 0 & (T+\bar{\varepsilon}U+\bar{\varepsilon}^2Z)N^{\mathrm{T}} & 0 & 0 & 0 \\ 0 & 0 & \gamma_1\tilde{G}M & \gamma_2B(GB)^{-1}G & 0 \end{bmatrix},$$

$$\Lambda \triangleq \mathrm{diag}\left\{\gamma_1I, \gamma_2I, \gamma_1I, \gamma_2I, \frac{\bar{\alpha}}{\sigma}I\right\}, \hat{\Gamma} \triangleq \bar{P} - (J^{\mathrm{T}}+J), \hat{\Psi} \triangleq \bar{\alpha}I - (J^{\mathrm{T}}+J),$$

$$\check{\Gamma} \triangleq \bar{P} - (Q_{\bar{\varepsilon}}^{\mathrm{T}}+Q_{\bar{\varepsilon}}), \check{\Psi} \triangleq \bar{\alpha}I - (Q_{\bar{\varepsilon}}^{\mathrm{T}}+Q_{\bar{\varepsilon}}),$$

and the matrices Q_ε, J, R, T, U, Z, E_0, $E_{\bar{\varepsilon}}$, and $Q_{\bar{\varepsilon}}$ are defined in Theorems 6.1–6.2, then the sliding mode dynamics (6.23) is asymptotically stable for any $\varepsilon \in (0, \bar{\varepsilon}]$. In this case, the gain matrix K_ε is given as: $K_\varepsilon = \bar{K}\cdot\left(J^{\mathrm{T}} + \varepsilon R^{\mathrm{T}}\right)^{-1}$.

Proof *Taking the parameter $\theta \to \infty$ in the proof of Theorem 6.2, then the matrix inequality (6.33) will become*

$$\tilde{\Omega} \triangleq \bar{\Omega} + \gamma_1^{-1}\mathcal{N}_1\mathcal{N}_1^{\mathrm{T}} + \gamma_1\mathcal{M}_1\mathcal{M}_1^{\mathrm{T}} + \gamma_2^{-1}\mathcal{N}_2\mathcal{N}_2^{\mathrm{T}} + \gamma_2\mathcal{M}_2\mathcal{M}_2^{\mathrm{T}} < 0, \qquad (6.39)$$

with the scalars $\gamma_1 > 0$, $\gamma_2 > 0$, and

$$\bar{\Omega} \triangleq \begin{bmatrix} -P+\alpha\sigma I & 0 & E_\varepsilon(A+BK)^{\mathrm{T}} \\ \star & -\alpha I & -E_\varepsilon K^{\mathrm{T}}B^{\mathrm{T}} \\ \star & \star & -P^{-1} \end{bmatrix},$$

$$\mathcal{N}_1 \triangleq \begin{bmatrix} NE_\varepsilon & 0 & 0 \end{bmatrix}^{\mathrm{T}}, \mathcal{N}_2 \triangleq \begin{bmatrix} 0 & NE_\varepsilon & 0 \end{bmatrix}^{\mathrm{T}},$$

$$\mathcal{M}_1 \triangleq \begin{bmatrix} 0 & 0 & M^{\mathrm{T}}\tilde{G}^{\mathrm{T}} \end{bmatrix}^{\mathrm{T}}, \mathcal{M}_2 \triangleq \begin{bmatrix} 0 & 0 & M^{\mathrm{T}}G^{\mathrm{T}}(GB)^{-\mathrm{T}}B^{\mathrm{T}} \end{bmatrix}^{\mathrm{T}}.$$

We pre- and post-multiply matrix $\tilde{\Omega}$ by $\mathrm{diag}\{Q_\varepsilon^{\mathrm{T}}, Q_\varepsilon^{\mathrm{T}}, I\}$ and its transposed matrix, respectively. Furthermore, the following inequalities are recalled:

$$\left(Q_\varepsilon^{\mathrm{T}} - \bar{P}\right) P \left(Q_\varepsilon - \bar{P}\right) \geq 0, \left(Q_\varepsilon^{\mathrm{T}} - \bar{\alpha}I\right) \alpha \left(Q_\varepsilon - \bar{\alpha}I\right) \geq 0,$$

112   SMC Under DETP

then the rest of the proof will be attained readily by similar lines to the ones in the proof of Theorem 6.2.

6.2.5.2 Convergence of the quasi-sliding motion

It is shown in Theorem 6.1 that under the designed SMC law (6.9) with the dynamic event-triggering mechanism (6.5)–(6.6), the bound $\tilde{\gamma}(k)$ of the resultant quasi-sliding motion depends on the system state $x(k)$, the internal dynamical variable $\eta(k)$ and the bound ϱ of the external disturbance $\xi(k)$ explicitly. The following corollary further analyzes the convergence of the time-varying bound $\tilde{\gamma}(k)$.

Corollary 6.3 *If the conditions in Theorems 6.1 and 6.2 hold simultaneously for any $\varepsilon \in (0, \bar{\varepsilon}]$, then the time-varying bound $\tilde{\gamma}(k)$ satisfies:*

$$\lim_{k \to \infty} \tilde{\gamma}(k) = \tilde{\varrho}, \tag{6.40}$$

where $\tilde{\varrho} \triangleq \sqrt{\frac{24\varpi m \varrho^2 \|GB\|^2}{\lambda_{\min}(W)}}$.

Proof *It is obtained from Theorem 6.2 that both of $x(k)$ and $\eta(k)$ converge to zero as $k \to \infty$. Thus, from (6.17), the equation (6.40) is followed directly.*

Remark 6.4 *Although the dynamic event-triggering protocol alleviates more communication burden than the static one, it is seen from (6.17) that the sliding region $\mathbf{O}$ with dynamic event-triggered protocol is larger than the one with the static event-triggered protocol due to the additional term $\frac{\beta}{\theta}\eta(k)$. That is means that, the dynamic event-triggering mechanism reduces more communication burden at the cost of certain SMC performances, such as the convergence bound $\tilde{\varrho}$ of the quasi-sliding motion (see numerical example later).*

6.2.6 Solving algorithm

In order to achieve the asymptotic stability of the sliding mode dynamics (6.23) and the reachability of the specified sliding surface (6.4) simultaneously, the gain matrix K_ε in the designed SMC scheme (6.9) should be determined by Theorems 6.1 and 6.2 simultaneously. Meanwhile, it is quite significant practically to determine the upper bound $\bar{\varepsilon}$ for the singular perturbation parameter. To this end, we give a synthesis algorithm to solve the dynamic event-triggered SMC law (6.9) as follows.

- ***Step 1.*** Select dynamic event-triggering parameters $\sigma > 0$, $\theta > 0$ and $0 < \lambda < 1$ with satisfying the condition (6.24). Choose a real matrix G in the sliding function (6.4) such that the matrix GB is nonsingular.

- ***Step 2.*** Choose a sufficiently small scalar $\Delta\varepsilon$, and set the scalars $\bar{\varepsilon} = \Delta\varepsilon$ and $t = 1$.

Example 113

- **Step 3.** Check the feasibility of the LMIs in Theorems 6.1 and 6.2 simultaneously for a prescribed adjustable scalar $\mu > 0$.

- **Step 4.** If the conditions have feasible solutions and $t \leq L$ (L denotes the given maximum iterative steps), set $\bar{\varepsilon} = \bar{\varepsilon} + \Delta\varepsilon$ and $t = t + 1$, go to Step 3; otherwise, modify the scalar μ until no feasible solution can be found, go to Step 5.

- **Step 5.** Output the maximum feasible bound $\bar{\varepsilon}$, and get the matrices $\bar{K}$, J and R. Produce the dynamic event-triggered SMC law (6.9) with matrices $K_\varepsilon = \bar{K} \cdot (J^{\mathrm{T}} + \varepsilon R^{\mathrm{T}})^{-1}$ for any known singularly perturbed parameter $\varepsilon \in (0, \bar{\varepsilon}]$.

Remark 6.5 *For the first attempt, this chapter investigates the SMC problem for the slow-sampling SPS (6.3) under the dynamic event-triggered mechanism (6.5)–(6.6). The proposed dynamic event-triggered SMC law exhibits the following distinct features: 1) by considering the structural characteristics of the slow-sampling SPS, a novel ε-dependent sliding surface is constructed properly for benefiting the stability analysis; 2) compared with the existing works on SMC under the static event-triggering protocol, the proposed SMC law is designed under the dynamic event-triggering condition, which achieves more energy-saving at the expense of certain SMC performance; and 3) the ε-bound estimation is cooperated with the design of sliding mode controller.*

6.3 Example

Consider the slow-sampling SPS (6.1)–(6.2) with the following parameters:

$$A_{11} = \begin{bmatrix} 0.7761 & 0.0736 \\ 0.7214 & 0.7343 \end{bmatrix}, \ A_{12} = \begin{bmatrix} 0.0560 \\ 0.8580 \end{bmatrix}, A_{21} = \begin{bmatrix} 0.5540 \\ 0.8588 \end{bmatrix}^{\mathrm{T}},$$

$$A_{22} = 2.1660, B_1 = \begin{bmatrix} 0.0022 \\ 0.0565 \end{bmatrix}, \ B_2 = 0.8377,$$

$$M = \begin{bmatrix} 0.1 \\ 0.1 \\ 0.4 \end{bmatrix}, \ N = \begin{bmatrix} 0.14 \\ 0.17 \\ 0.01 \end{bmatrix}^{\mathrm{T}}, \ F(k) = \frac{\cos(0.03k)}{1 + k^2},$$

and the matched external disturbance is $\xi(k) = 0.1\sin(20k)$ with the scalar $\varrho = 0.1$. It is noted that θ is a key parameter in the utilized dynamic event-triggering mechanism (6.5)–(6.6). In what follows, we will analyze the relation between the parameters θ and $\tilde{\varrho}$ (representing the SMC performance) or $\bar{\varepsilon}$ (representing the ε-bound estimation). To this end, we set the other parameters in the dynamic event-triggering mechanism (6.5)–(6.6) as $\lambda = 0.5$,

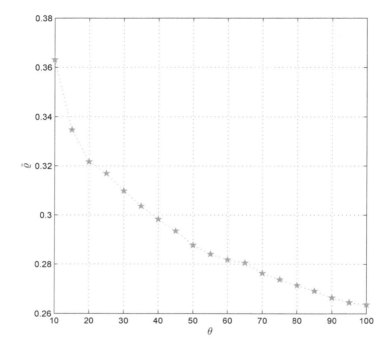

FIGURE 6.1: θ vs. $\tilde{\varrho}$.

$\sigma = 0.8$, $\eta_0 = 1$, the sliding matrix $G = B^{\mathrm{T}}$, and the initial condition as $x(0) = \begin{bmatrix} 3 & -1 & -5 \end{bmatrix}^{\mathrm{T}}$.

- θ vs. $\tilde{\varrho}$.

Choose the scalar $\mu = 1$ in Theorem 6.2 and the singularly perturbed parameter $\bar{\varepsilon} = \varepsilon = 0.21$. Fig. 6.1 shows the evolution of the converged bound $\tilde{\varrho}$ of the quasi-sliding motion versus the dynamic event-triggering parameter θ, from which one can observe that the scalar $\tilde{\varrho}$ decreases monotonically as the parameter θ increases. That is, the SMC performance meliorates as the parameter θ increases. In fact, this is true since a larger θ means less event-triggered times; therefore, the better SMC performance can be attained.

Furthermore, we consider two special cases: $\theta = 40$ and $\theta = 100$. By solving Theorems 6.1 and 6.2 simultaneously, we obtain $K_\varepsilon = \begin{bmatrix} -0.5376 & -1.0123 & -3.0040 \end{bmatrix}$ for $\theta = 40$ with $\tilde{\varrho} = 0.2982$; and $K_\varepsilon = \begin{bmatrix} -0.4764 & -1.0532 & -3.2075 \end{bmatrix}$ for $\theta = 100$ with $\tilde{\varrho} = 0.2636$. For $\theta = 40$, the evolutions of the internal dynamical variable $\eta(k)$, the event-based release interval, the state trajectories $x(k)$ and the sliding variable $s(k)$ are depicted in Fig. 6.2, respectively. In this case, the triggering time is 71

Example 115

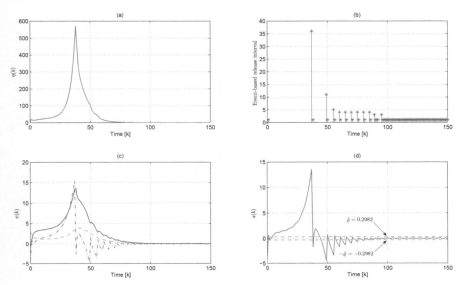

FIGURE 6.2: The simulation results under $\theta = 40$.

corresponding to triggering rate 47.02%. Fig. 6.3 shows the case under $\theta = 100$, where the triggering time 29 corresponding to triggering rate 19.21%. Obviously, the bigger θ achieves faster convergence of the internal dynamical variable $\eta(k)$, the system states $x(k)$ and the sliding variable $s(k)$. Besides, one can observe from Figs. 6.2(d)–6.3(d) that the quasi-sliding motion is bounded by $\tilde{\varrho}$ indeed. All simulation results illustrate the aforementioned statements well.

- θ vs. $\bar{\varepsilon}$.

In this part, we explore the relation between the dynamic event-triggering parameter θ and the upper bound $\bar{\varepsilon}$ of the singularly perturbed parameter ε. Table 6.1 shows that the available ε-bound decreases as the parameter θ increases. Actually, one can see from the expression of $\tilde{\Omega}$ in (6.33) that as the parameter θ increases, the term, $\frac{1}{\theta}I + \alpha I$, may be decreased, which implies that it is more difficult to satisfy the condition $\tilde{\Omega} < 0$. Therefore, it is reasonable to expect that as the parameter θ increases, the ε-bound decreases accordingly. In practice, one may determine the dynamic event-triggering parameter θ through comprehensive consideration of the SMC performance, the communication burden and the singularly perturbed parameter.

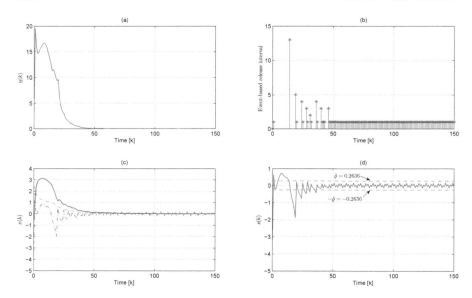

FIGURE 6.3: The simulation results under $\theta = 100$.

TABLE 6.1: θ vs. $\bar{\varepsilon}$.

θ	20	40	60	80	100
$\bar{\varepsilon}$	0.3649	0.3639	0.3080	0.3070	0.2565

6.4 Conclusion

This chapter has addressed the SMC problem of the slow-sampling singularly perturbed systems under the dynamic event-triggering mechanism. A novel sliding function has been introduced by considering the structure characteristics of the slow-sampling singularly perturbed systems. By resorting to some appropriate Lyapunov functionals, the asymptotic stability of the sliding mode dynamics and the reachability of the specified sliding surface have been analyzed, respectively, with the estimation of the upper bound for the singularly perturbed parameter. Besides, the convergence of the quasi-sliding motion and the relations between the dynamic event-triggering parameter and the SMC performance or the ε-bound have been detailedly analyzed.

7

Reliable Sliding Mode Control Under Redundant Channel Transmission Protocol

Singularly perturbed system (SPS) has been widely employed to model the dynamic system with multiple time-scale phenomena, which frequently occur in, for example, the electrical circuits, engineering biomolecular systems, and advanced heavy water reactor. In a SPS, a key feature is that the degree of separation between the "slow" and "fast" modes of the systems are multiplied via a small positive parameter, i.e. the singular perturbed parameter ε. Over the past few decades, a considerable research attention has been focused on the stability analysis and the control/filter synthesis issues for the SPSs.

In this chapter, we endeavour to solve the output-feedback sliding mode control (SMC) problem for the fast-sampling SPSs. In order to improve the communication reliability, the RCTP is applied between the sensors and the controller. The key idea in redundant channel transmission protocol (RCTP) is that if the primary channel suffers certain communication failure, which can be detected by means of software or hardware devices, other channels (i.e. redundant channels) will be automatically activated to protect the key data and thus improve the reliability of the communication network in a great sense. *The main contributions of this chapter are highlighted as follows: 1) To cope with the impact of RCTP, a measured output-based sliding function is developed for the fast-sampling SPSs; 2) By utilizing some stochastic analysis techniques, sufficient conditions for ensuring the stability of the sliding mode dynamics and the reachability of the specified sliding surface are established; 3) A convex optimization problem is formulated to design the SMC strategy with estimating the upper bound of the singular perturbation parameter; and 4) The impacts from the RCTP to the SMC performance and the estimation of ε-bound are explored via an operational amplifier circuit.*

7.1 Problem Formulation

Consider the following fast-sampling discrete-time SPS:

$$x(k+1) = A_\varepsilon x(k) + B_\varepsilon \left(u(k) + w(k) \right), \tag{7.1}$$

DOI: 10.1201/9781003309499-7

117

with $x(k) \triangleq \begin{bmatrix} x_1^{\mathrm{T}}(k) & x_2^{\mathrm{T}}(k) \end{bmatrix}^{\mathrm{T}}$ and

$$A_\varepsilon \triangleq \begin{bmatrix} I + \varepsilon A_{11} & \varepsilon A_{12} \\ A_{21} & A_{22} \end{bmatrix}, \quad B_\varepsilon \triangleq \begin{bmatrix} \varepsilon B_1 \\ B_2 \end{bmatrix},$$

where $\varepsilon > 0$ is a singularly perturbed parameter; $x_1(k) \in \mathbb{R}^{n_s}$ and $x_2(k) \in \mathbb{R}^{n_f}$ $(n_s + n_f = n)$ are the slow and fast state vectors, respectively; $u(k) \in \mathbb{R}^m$ is the control input; $w(k) \in \mathbb{R}^m$ is an unknown external disturbance. The matrices A_{11}, A_{12}, A_{21}, A_{22}, B_1, and B_2 are known real matrices.

Besides, the system (7.1) satisfies the following assumptions:

- The input matrix B_ε is full column rank, that is, $\mathrm{rank}(B_\varepsilon) = m$;

- The external disturbance $w(k)$ possesses $\|w(k)\| \leq \varpi$, where $\varpi \geq 0$ is a known scalar.

Remark 7.1 *In many electrical circuits, the continuous-time SPS has widely employed to model the multiple time-scale phenomena, which are often occurred due to some small "parasitic" circuit elements, such as the "parasitic" capacitance in the van der Pol oscillator circuit [1, 185] and the modified Chua's circuit with mixed-mode oscillations [100]. With the development of the network-based communication, it seems naturally to discretize a continuous-time SPSs to its discrete-time counterpart for realizing the digital control/filtering. It is worth mentioning that different sampling rate will lead to different discrete-time SPS model [70]. As shown in [70], the fast-sampling SPS (7.1) is obtained from the practical continuous-time SPS by selecting the fast sampling rate as $T_f = \varepsilon$ and neglecting $o(\varepsilon)$ errors. One is the fast-sampling model [33, 96] and another is the slow-sampling model [164]. It is worth stressing that the stability of a circuit system is a key problem in practical applications. Therefore, it is valuable to study the control problem of the slow- or fast-sampling circuit models under a single-rate sample-data. This chapter focuses on investigating the SMC problem of the circuit systems modelled by the fast-sampling SPS (7.1), which is still an open research issue.*

In networked systems, it is quite common that the network communication through a single channel is unreliable due mainly to the network-induced phenomena, such as packet dropouts. In order to improve the reliability of data transmission services, the RCTP is utilized between the sensors and the controller as shown in Fig. 7.1, where the channel 1 is the primary channel and the channels $i \in \{2, 3, \ldots, N\}$ are called redundant channels [22,146]. The packet dropouts over these channels are governed by the stochastic variables $\gamma_i(k)$ $(i = 1, 2, \ldots, N)$, which are mutually independent Bernoulli distributed white sequences taking values of 0 or 1 as follows:

$$\Pr\{\gamma_i(k) = 1\} = \bar{\gamma}_i, \text{ and } \Pr\{\gamma_i(k) = 0\} = 1 - \bar{\gamma}_i, \tag{7.2}$$

where $\bar{\gamma}_i \in [0, 1]$ are known constants.

Problem Formulation

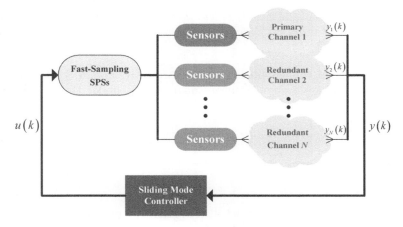

FIGURE 7.1: Reliable SMC of fast-sampling SPSs under RCTP.

Under the above RCTP, the actual received output $y(k)$ at the controller side can be formulated as follows:

$$y(k) = \gamma_1(k)y_1(k) + \sum_{i=2}^{N}\left\{\prod_{j=1}^{i-1}(1-\gamma_j(k))\gamma_i(k)y_i(k)\right\} \quad (7.3)$$

where $y_i(k) \triangleq C_i x(k)$ denotes the measurement output of the ith transmission channel with the known output matrices $C_i \in \mathbb{R}^{p \times n}$. In fact, the RCTP is executed under the assumption that all the measured outputs $\{y_1(k), y_2(k), \ldots, y_N(k)\}$ are transmitted to controller simultaneously without communication time-delays. In the case that more than one measured outputs are received by the controller, the actual employed output will be determined by (7.3).

Remark 7.2 *It is observed from the actual measurement model (7.3) that if no packet dropout occurs at the primary channel, the measured output will be $y(k) = y_1(k)$, which implies that the other redundant channels ($i = 2, 3, \ldots, N$) will not be activated. When the channels from 1 to $i - 1$ suffer from packet dropouts and the ith channel transmits successfully, the measured output will be $y(k) = y_i(k)$. That is, no information would be received by the controller (i.e., $y(k) = 0$ in this case) if and only if all channels transmit unsuccessfully. Although the introduction of RCTP to the control systems will increase the cost of equipment/energy, the probability of packet dropouts is significantly reduced from $1 - \bar{\gamma}_1$ to $\prod_{i=1}^{N}(1 - \bar{\gamma}_i)$ and thus the network reliability is much improved via RCTP mechanism. In practical applications, the number of the redundant channels will be determined as per different engineering requirements and capacities. For example, in some cases with high requirement for the reliability, the number of the redundant channels can be determined so that*

120   *Reliable SMC under RCTP*

the probability of packet dropouts is less than $\prod_{i=1}^{N}(1 - \bar{\gamma}_i)$. However, in some cases with limited resources, one may only employ the redundant channels as more as possible.

The interest of this chapter is to design a SMC law $u(k)$ such that the resultant closed-loop system is mean-square stable under the RCTP (7.2)–(7.3).

7.2 Main Results

7.2.1 Sliding function and sliding mode controller

It is worth stressing that in order to ensure the engineering feasibility, the sliding function and the sliding mode controller are designed just by utilizing the actual measured output signal $y(k)$ in (7.3), which is composed of $\{y_1(k), y_2(k), \ldots, y_N(k)\}$ under RCTP. To cope with the effect of RCTP, we introduce the following mathematical notations first:

$$\Upsilon(k) \triangleq \gamma_1(k)C_1 + \sum_{i=2}^{N}\left\{\prod_{j=1}^{i-1}(1 - \gamma_j(k))\gamma_i(k)C_i\right\}, \tag{7.4}$$

$$\bar{\Upsilon} \triangleq \mathbb{E}\{\Upsilon(k)\} = \bar{\gamma}_1 C_1 + \sum_{i=2}^{N}\left\{\prod_{j=1}^{i-1}(1 - \bar{\gamma}_j)\bar{\gamma}_i C_i\right\}. \tag{7.5}$$

Clearly, with the aid of above notations, the actual measured output under RCTP (7.3) becomes $y(k) = \Upsilon(k)x(k)$.

Now, based on the measurement outputs $\{y_1(k), y_2(k), \ldots, y_N(k)\}$ of the N channels, we introduce the following output-feedback sliding function:

$$\begin{aligned}
s(k) &= G_\varepsilon\left\{\bar{\gamma}_1 y_1(k) + \sum_{i=2}^{N}\left\{\prod_{j=1}^{i-1}(1 - \bar{\gamma}_j)\bar{\gamma}_i y_i(k)\right\}\right\} \\
&= G_\varepsilon \bar{\Upsilon} x(k),
\end{aligned} \tag{7.6}$$

where $G_\varepsilon \in \mathbb{R}^{m \times p}$ should be satisfied:

$$G_\varepsilon \bar{\Upsilon} = B_\varepsilon^{\mathrm{T}} P_\varepsilon, \tag{7.7}$$

with the ε-dependent matrix $P_\varepsilon > 0$ to be determined later. Clearly, it yields that $G_\varepsilon \bar{\Upsilon} B_\varepsilon = B_\varepsilon^{\mathrm{T}} P_\varepsilon B_\varepsilon > 0$ is nonsingular.

Remark 7.3 *To the authors' best knowledge, this chapter represents the first attempt to investigate the SMC problem under the RCTP (7.2)–(7.3). A key*

Main Results 121

challenge here is how to design the sliding function by just utilizing the measurement output signals $\{y_1(k), y_2(k), \ldots, y_N(k)\}$. *To this end, the sliding function (7.6) is constructed skillfully by considering the specific communication structure of RCTP (7.3), which will be helpful in the stability analysis of the sliding mode dynamics subsequently. Similar to [62], if the matrix* P_ε *is known in the equality condition (7.7), the gain matrix* G_ε *can be obtained readily by solving the following convex optimization problem:*

$$\begin{bmatrix} -\kappa I & G_\varepsilon \bar{\Upsilon} - B_\varepsilon^{\mathrm{T}} P_\varepsilon \\ \star & -I \end{bmatrix} < 0, \tag{7.8}$$

where $\kappa > 0$ *is a specified sufficiently small scalar.*

From the system (7.1) and the sliding function (7.6), we have

$$s(k+1) = G_\varepsilon \bar{\Upsilon} A_\varepsilon x(k) + G_\varepsilon \bar{\Upsilon} B_\varepsilon u(k) + D_\varepsilon(k), \tag{7.9}$$

where $D_\varepsilon(k) \triangleq G_\varepsilon \bar{\Upsilon} B_\varepsilon w(k)$. According to the assumption on the external disturbance $w(k)$, there exist bounds $\underline{d}_i$, $\overline{d}_i$ ($i = 1, 2, \ldots, m$) satisfying $\underline{d}_i \leq d_i(k) \leq \overline{d}_i$, where $d_i(k)$ are the ith element of $D_\varepsilon(k)$. Define

$$d_{io} = \frac{\overline{d}_i + \underline{d}_i}{2}, \quad D_o = \begin{bmatrix} d_{1o} & d_{2o} & \cdots & d_{mo} \end{bmatrix}^{\mathrm{T}}, \tag{7.10}$$

$$d_{is} = \frac{\overline{d}_i - \underline{d}_i}{2}, \quad D_s = \mathrm{diag}\{d_{1s}, d_{2s}, \ldots, d_{ms}\}. \tag{7.11}$$

By just utilizing the measurement outputs $\{y_i(k)\}$, we construct the following output-feedback SMC scheme:

$$u(k) = -\left(G_\varepsilon \bar{\Upsilon} B_\varepsilon\right)^{-1} [F_\varepsilon y(k) + D_o + D_s \mathrm{sgn}(s(k))], \tag{7.12}$$

where the ε-dependent gain matrix $F_\varepsilon \in \mathbb{R}^{m \times p}$ will be determined later. Here, we replace the sliding variable $s(k)$ by $\bar{s}(k)$, that is, replacing $\bar{\Gamma} x(k)$ by $y(k)$ in (6) since only the actual received output $y(k)$ can be utilized for the controller. Substituting (7.12) into (7.1), the following closed-loop system is yielded:

$$x(k+1) = \left[A_\varepsilon - B_\varepsilon \left(G_\varepsilon \bar{\Upsilon} B_\varepsilon\right)^{-1} F_\varepsilon \Upsilon(k)\right] x(k)$$
$$+ B_\varepsilon \left(G_\varepsilon \bar{\Upsilon} B_\varepsilon\right)^{-1} [D_\varepsilon(k) - D_o - D_s \mathrm{sgn}(s(k))]. \tag{7.13}$$

In the sequel, we will analyze the stability of the closed-loop system (7.13) and the reachability of the sliding surface (7.6). To this end, the following definition and lemmas are introduced.

Definition 7.1 *The closed-loop system (7.13) is said to be mean-square exponentially ultimately bounded (MSEUB) if there exist constants* $0 < \beta < 1$, $\alpha > 0$ *and* $\bar{\chi} \geq 0$ *such that*

$$\mathbb{E}\left\{\|x(k)\|^2\right\} \leq \alpha \beta^k \|x(0)\|^2 + \chi(k) \quad and \quad \lim_{k \to \infty} \chi(k) = \bar{\chi}.$$

Lemma 7.1 *[186] For a positive scalar $\bar{\varepsilon}$ and symmetric matrices S_1, S_2 and S_3 with appropriate dimensions, if $S_1 \geq 0$, $S_1 + \bar{\varepsilon}S_2 > 0$, $S_1 + \bar{\varepsilon}S_2 + \bar{\varepsilon}^2 S_3 > 0$ hold, then $S_1 + \varepsilon S_2 + \varepsilon^2 S_3 > 0$, $\forall \varepsilon \in (0, \bar{\varepsilon}]$ can be achieved.*

Lemma 7.2 *Defining the stochastic varying matrix $\tilde{\Upsilon}(k) \triangleq \Upsilon(k) - \bar{\Upsilon}$, then for a positive-definite matrix P and a real matrix M, it has:*

$$\mathbb{E}\left\{ M\tilde{\Upsilon}(k) \right\} = 0; \tag{7.14}$$

$$\mathbb{E}\left\{ (M\tilde{\Upsilon}(k))^{\mathrm{T}} P(M\tilde{\Upsilon}(k)) \right\}$$

$$= -\bar{\Upsilon}^{\mathrm{T}} M^{\mathrm{T}} PM\bar{\Upsilon} + \bar{\gamma}_1 C_1^{\mathrm{T}} M^{\mathrm{T}} PMC_1 + \sum_{i=2}^{N}\left\{ \prod_{j=1}^{i-1}(1 - \bar{\gamma}_j)\bar{\gamma}_i C_i^{\mathrm{T}} M^{\mathrm{T}} PMC_i \right\}. \tag{7.15}$$

Proof *The relationship in (7.14) can be obtained directly. Now, we prove the relationship (7.15). According the notations in (7.4)–(7.5), it yields*

$$\mathbb{E}\left\{ (M\tilde{\Upsilon}(k))^{\mathrm{T}} P(M\tilde{\Upsilon}(k)) \right\}$$

$$= \mathbb{E}\left\{ \Upsilon^{\mathrm{T}}(k)M^{\mathrm{T}} PM\Upsilon(k)) \right\} - 2\mathbb{E}\left\{ \Upsilon^{\mathrm{T}}(k)M^{\mathrm{T}} PM\bar{\Upsilon} \right\} + \bar{\Upsilon}^{\mathrm{T}} M^{\mathrm{T}} PM\bar{\Upsilon}$$

$$= -\bar{\Upsilon}^{\mathrm{T}} M^{\mathrm{T}} PM\bar{\Upsilon} + \bar{\gamma}_1 C_1^{\mathrm{T}} M^{\mathrm{T}} PMC_1 + \sum_{i=2}^{N}\left\{ \prod_{j=1}^{i-1}(1 - \bar{\gamma}_j)\bar{\gamma}_i C_i^{\mathrm{T}} M^{\mathrm{T}} PMC_i \right\}.$$

The proof is completed.

Lemma 7.3 *According to (7.10)–(7.11), it has*

$$\|D_\varepsilon(k) - D_o - D_s\mathrm{sgn}(s(k))\| \leq 2\|D_s\|. \tag{7.16}$$

Proof *For the ith element of $D_\varepsilon(k) - D_o - D_s\mathrm{sgn}(s(k))$, we consider the following three cases:*

- *Case 1: $s_i(k) = 0$. Then, it has*

$$| d_i(k) - d_{io} - d_{is}\mathrm{sgn}(s_i(k)) |$$

$$= \left| \frac{d_i(k) - \bar{d}_i}{2} + \frac{d_i(k) - \underline{d}_i}{2} \right| \leq \left| \frac{d_i(k) - \bar{d}_i}{2} \right| + \left| \frac{d_i(k) - \underline{d}_i}{2} \right| \leq 2d_{is}.$$

- *Case 2: $s_i(k) > 0$. Now, it gets*

$$| d_i(k) - d_{io} - d_{is}\mathrm{sgn}(s_i(k)) | = | d_i(k) - \underline{d}_i \leq 2d_{is}.$$

- *Case 3: $s_i(k) < 0$. One has*

$$| d_i(k) - d_{io} - d_{is}\mathrm{sgn}(s_i(k)) | = | d_i(k) - \bar{d}_i \leq 2d_{is}.$$

Main Results 123

To sum up the above three cases, it always has $\mid d_i(k) - d_{io} - d_{is}\mathrm{sgn}(s_i(k)) \mid \leq 2d_{is}$. *Next, by resorting to the definition of Euclidean norm, we obtain*

$$\|D_\varepsilon(k) - D_o - D_s\mathrm{sgn}(s(k))\|$$
$$= \sqrt{\sum_{i=1}^{m} |d_i(k) - d_{io} - d_{is}\mathrm{sgn}(s_i(k))|^2} \leq 2\sqrt{\sum_{i=1}^{m} d_{is}^2} = 2\|D_s\|.$$

This completes the proof.

7.2.2 MSEUB of closed-loop system

The following theorem proposes a sufficient condition to guarantee the MSEUB of the closed-loop system (7.13) under the output-feedback SMC law (7.12) with the RCTP (7.2)–(7.3).

Theorem 7.1 *Consider the fast sampling SPS (7.1) and the output-feedback SMC law (7.12) with the RCTP (7.2)–(7.3). For prescribed scalars $\delta > 0$, $\mu > 0$, and $\bar{\varepsilon} > 0$, if there exist symmetric matrices $P_{11} > 0$, $P_{22} > 0$, $Q > 0$, matrices P_{12} and $F_\varepsilon \in \mathbb{R}^{m \times p}$, and a scalar $\xi > 0$ such that the following LMIs hold:*

$$\begin{bmatrix} -P_0 & \tilde{\Xi} \\ \star & -\tilde{\Lambda} \end{bmatrix} < 0, \tag{7.17}$$

$$\begin{bmatrix} -P_{\bar{\varepsilon}} & \hat{\Xi} \\ \star & -\hat{\Lambda} \end{bmatrix} < 0, \tag{7.18}$$

$$\begin{bmatrix} -P_{\bar{\varepsilon}} & \check{\Xi} \\ \star & -\check{\Lambda} \end{bmatrix} < 0, \tag{7.19}$$

$$\begin{bmatrix} -\xi I & \varsigma_3 I \\ \star & -B_2^{\mathrm{T}} P_{22} B_2 \end{bmatrix} < 0, \tag{7.20}$$

$$\begin{bmatrix} -\xi I & \varsigma_3 I \\ \star & -H \end{bmatrix} < 0, \tag{7.21}$$

$$\begin{bmatrix} -\xi I & \varsigma_3 I \\ \star & -\check{H} \end{bmatrix} < 0, \tag{7.22}$$

where $P_0 \triangleq \begin{bmatrix} P_{11} & P_{12} \\ P_{12}^{\mathrm{T}} & P_{22} \end{bmatrix}$, $P_{\bar{\varepsilon}} \triangleq \begin{bmatrix} P_{11} & P_{12} \\ P_{12}^{\mathrm{T}} & P_{22} + \bar{\varepsilon}Q \end{bmatrix}$, $\varsigma_1 \triangleq \sqrt{(1+\delta)(1+\mu)}$, $\varsigma_3 \triangleq \sqrt{1 + \delta^{-1}}$, $\varsigma_2 \triangleq \sqrt{(1+\delta)(1+\mu^{-1}) - 1}$, *and*

$$U \triangleq \begin{bmatrix} P_{11} + A_{21}^{\mathrm{T}} P_{12}^{\mathrm{T}} & P_{12} + A_{21}^{\mathrm{T}} P_{22} \\ A_{12}^{\mathrm{T}} P_{11} + A_{22}^{\mathrm{T}} P_{12}^{\mathrm{T}} & A_{12}^{\mathrm{T}} P_{12} + A_{22}^{\mathrm{T}} P_{22} \end{bmatrix},$$

$$T \triangleq \begin{bmatrix} A_{11}^{\mathrm{T}} P_{11} & A_{11}^{\mathrm{T}} P_{12} + A_{21}^{\mathrm{T}} Q \\ 0 & A_{22}^{\mathrm{T}} Q \end{bmatrix},$$

$$\tilde{\Xi} \triangleq \begin{bmatrix} \varsigma_1 U^{\mathrm{T}} \\ \varsigma_2 F_\varepsilon \bar{\Upsilon} \\ \sqrt{\bar{\gamma}_1} F_\varepsilon C_1 \\ \sqrt{(1-\bar{\gamma}_1)\bar{\gamma}_2} F_\varepsilon C_2 \\ \vdots \\ \sqrt{\prod_{j=1}^{N-1}(1-\bar{\gamma}_j)\bar{\gamma}_N F_\varepsilon C_N} \end{bmatrix}^{\mathrm{T}}, \quad \hat{\Xi} \triangleq \begin{bmatrix} \varsigma_1 (U^{\mathrm{T}} + \bar{\varepsilon} T^{\mathrm{T}}) \\ \varsigma_2 F_\varepsilon \bar{\Upsilon} \\ \sqrt{\bar{\gamma}_1} F_\varepsilon C_1 \\ \sqrt{(1-\bar{\gamma}_1)\bar{\gamma}_2} F_\varepsilon C_2 \\ \vdots \\ \sqrt{\prod_{j=1}^{N-1}(1-\bar{\gamma}_j)\bar{\gamma}_N F_\varepsilon C_N} \end{bmatrix}^{\mathrm{T}},$$

$$\tilde{\Lambda} \triangleq \operatorname{diag}\{P_0, B_2^{\mathrm{T}} P_{22} B_2, B_2^{\mathrm{T}} P_{22} B_2, \underbrace{B_2^{\mathrm{T}} P_{22} B_2, \cdots, B_2^{\mathrm{T}} P_{22} B_2}_{N-1}\},$$

$$\hat{\Lambda} \triangleq \operatorname{diag}\{P_{\bar{\varepsilon}}, H, H, \underbrace{H, \cdots, H}_{N-1}\}, \check{\Lambda} \triangleq \operatorname{diag}\{P_{\bar{\varepsilon}}, \check{H}, \check{H}, \underbrace{\check{H}, \cdots, \check{H}}_{N-1}\},$$

$$H \triangleq B_2^{\mathrm{T}} P_{22} B_2 + \bar{\varepsilon}(B_2^{\mathrm{T}} P_{12}^{\mathrm{T}} B_1 + B_1^{\mathrm{T}} P_{12} B_2 + B_2^{\mathrm{T}} Q B_2),$$

$$\check{H} \triangleq B_2^{\mathrm{T}} P_{22} B_2 + \bar{\varepsilon}(B_2^{\mathrm{T}} P_{12}^{\mathrm{T}} B_1 + B_1^{\mathrm{T}} P_{12} B_2 + B_2^{\mathrm{T}} Q B_2) + \bar{\varepsilon}^2 B_1^{\mathrm{T}} P_{11} B_1,$$

then, the closed-loop system (7.13) is MSEUB for any $\varepsilon \in (0, \bar{\varepsilon}]$.

Proof *We select the Lyapunov function candidate as* $V(k) \triangleq x^{\mathrm{T}}(k) P_\varepsilon x(k)$. *Along the closed-loop system (7.13), the difference of* $V(k)$ *is obtained by using Lemma 7.2 as follows:*

$$\mathbb{E}\{\Delta V(k) \mid x(k)\}$$
$$= \mathbb{E}\{x^{\mathrm{T}}(k+1) P_\varepsilon x(k+1) \mid x(k)\} - x^{\mathrm{T}}(k) P_\varepsilon x(k)$$
$$= x^{\mathrm{T}}(k) \left\{ \left[A_\varepsilon - B_\varepsilon \left(G_\varepsilon \bar{\Upsilon} B_\varepsilon\right)^{-1} F_\varepsilon \bar{\Upsilon} \right]^{\mathrm{T}} P_\varepsilon \left[A_\varepsilon - B_\varepsilon \left(G_\varepsilon \bar{\Upsilon} B_\varepsilon\right)^{-1} F_\varepsilon \bar{\Upsilon} \right] \right.$$
$$- \bar{\Upsilon}^{\mathrm{T}} F_\varepsilon^{\mathrm{T}} (B_\varepsilon^{\mathrm{T}} P_\varepsilon B_\varepsilon)^{-1} F_\varepsilon \bar{\Upsilon} + \bar{\gamma}_1 C_1^{\mathrm{T}} F_\varepsilon^{\mathrm{T}} (B_\varepsilon^{\mathrm{T}} P_\varepsilon B_\varepsilon)^{-1} F_\varepsilon C_1$$
$$+ \sum_{i=2}^{N} \left\{ \prod_{j=1}^{i-1} (1-\bar{\gamma}_j)\bar{\gamma}_i C_i^{\mathrm{T}} F_\varepsilon^{\mathrm{T}} (B_\varepsilon^{\mathrm{T}} P_\varepsilon B_\varepsilon)^{-1} F_\varepsilon C_i \right\} - P_\varepsilon \left. \right\} x(k)$$
$$+ 2x^{\mathrm{T}}(k) \left[A_\varepsilon - B_\varepsilon \left(G_\varepsilon \bar{\Upsilon} B_\varepsilon\right)^{-1} F_\varepsilon \bar{\Upsilon} \right]^{\mathrm{T}} P_\varepsilon B_\varepsilon \left(G_\varepsilon \bar{\Upsilon} B_\varepsilon\right)^{-1} [D_\varepsilon(k) - D_o$$
$$- D_s \operatorname{sgn}(s(k))] + [D_\varepsilon(k) - D_o - D_s \operatorname{sgn}(s(k))]^{\mathrm{T}} \left[B_\varepsilon \left(G_\varepsilon \bar{\Upsilon} B_\varepsilon\right)^{-1} \right]^{\mathrm{T}} P_\varepsilon$$
$$\times B_\varepsilon \left(G_\varepsilon \bar{\Upsilon} B_\varepsilon\right)^{-1} [D_\varepsilon(k) - D_o - D_s \operatorname{sgn}(s(k))]. \tag{7.23}$$

For two adjustable scalars $\mu > 0$ *and* $\delta > 0$, *in light of the equality condition (7.7), one has*

$$2x^{\mathrm{T}}(k) \left[A_\varepsilon - B_\varepsilon \left(G_\varepsilon \bar{\Upsilon} B_\varepsilon\right)^{-1} F_\varepsilon \bar{\Upsilon} \right]^{\mathrm{T}} P_\varepsilon B_\varepsilon \left(G_\varepsilon \bar{\Upsilon} B_\varepsilon\right)^{-1} [D_\varepsilon(k) - D_o - D_s \operatorname{sgn}(s(k))]$$

$$\leq \delta x^{\mathrm{T}}(k) \left[A_\varepsilon - B_\varepsilon \left(G_\varepsilon \bar{\Upsilon} B_\varepsilon\right)^{-1} F_\varepsilon \bar{\Upsilon} \right]^{\mathrm{T}} P_\varepsilon \left[A_\varepsilon - B_\varepsilon \left(G_\varepsilon \bar{\Upsilon} B_\varepsilon\right)^{-1} F_\varepsilon \bar{\Upsilon} \right] x(k)$$

Main Results

$$+ \delta^{-1} \left[D_\varepsilon(k) - D_o - D_s \mathrm{sgn}(s(k)) \right]^{\mathrm{T}} \left(B_\varepsilon P_\varepsilon B_\varepsilon \right)^{-1} \left[D_\varepsilon(k) - D_o - D_s \mathrm{sgn}(s(k)) \right],$$
$$(7.24)$$

$$\left[A_\varepsilon - B_\varepsilon \left(G_\varepsilon \bar{\Upsilon} B_\varepsilon \right)^{-1} F_\varepsilon \bar{\Upsilon} \right]^{\mathrm{T}} P_\varepsilon \left[A_\varepsilon - B_\varepsilon \left(G_\varepsilon \bar{\Upsilon} B_\varepsilon \right)^{-1} F_\varepsilon \bar{\Upsilon} \right]$$
$$\leq (1+\mu) A_\varepsilon^{\mathrm{T}} P_\varepsilon A_\varepsilon + (1+\mu^{-1}) \bar{\Upsilon}^{\mathrm{T}} F_\varepsilon^{\mathrm{T}} \left(B_\varepsilon^{\mathrm{T}} P_\varepsilon B_\varepsilon \right)^{-1} F_\varepsilon \bar{\Upsilon}. \tag{7.25}$$

By combining (7.23)–(7.25), it follows from (7.16) that for any scalar $\xi > 0$:

$$\mathbb{E} \left\{ \Delta V(k) \mid x(k) \right\}$$
$$\leq \mathbb{E} \left\{ x^{\mathrm{T}}(k+1) P_\varepsilon x(k+1) \mid x(k) \right\} - x^{\mathrm{T}}(k) P_\varepsilon x(k) + \xi [4\|D_s\|^2 - \|D_\varepsilon(k) - D_o$$
$$- D_s \mathrm{sgn}(s(k))\|^2]$$
$$\leq x^{\mathrm{T}}(k) \Sigma_1(\varepsilon) x(k) + [D_\varepsilon(k) - D_o - D_s \mathrm{sgn}(s(k))]^{\mathrm{T}} \Sigma_2(\varepsilon) [D_\varepsilon(k) - D_o$$
$$\times - D_s \mathrm{sgn}(s(k))] + 4\xi \|D_s\|^2, \tag{7.26}$$

where $\Sigma_1(\varepsilon) \triangleq (1+\delta)(1+\mu) A_\varepsilon^{\mathrm{T}} P_\varepsilon A_\varepsilon + (1+\delta)(1+\mu^{-1}) \bar{\Upsilon}^{\mathrm{T}} F_\varepsilon^{\mathrm{T}} \left(B_\varepsilon^{\mathrm{T}} P_\varepsilon B_\varepsilon \right)^{-1} F_\varepsilon \bar{\Upsilon} - \bar{\Upsilon}^{\mathrm{T}} F_\varepsilon^{\mathrm{T}} \left(B_\varepsilon^{\mathrm{T}} P_\varepsilon B_\varepsilon \right)^{-1} F_\varepsilon \bar{\Upsilon} + \bar{\gamma}_1 C_1^{\mathrm{T}} F_\varepsilon^{\mathrm{T}} (B_\varepsilon^{\mathrm{T}} P_\varepsilon B_\varepsilon)^{-1} F_\varepsilon C_1 + \sum_{i=2}^{N} \left\{ \prod_{j=1}^{i-1} (1 - \bar{\gamma}_j) \bar{\gamma}_i C_i^{\mathrm{T}} F_\varepsilon^{\mathrm{T}} \right.$
$\times (B_\varepsilon^{\mathrm{T}} P_\varepsilon B_\varepsilon)^{-1} F_\varepsilon C_i \left. \right\} - P_\varepsilon$ *and* $\Sigma_2(\varepsilon) \triangleq -\xi I + (1 + \delta^{-1}) \left(B_\varepsilon^{\mathrm{T}} P_\varepsilon B_\varepsilon \right)^{-1}.$

As per the Schur complement, it is known that $\Sigma_1(\varepsilon) < 0$ is equivalent to

$$\tilde{\Sigma}_1(\varepsilon) \triangleq \begin{bmatrix} -P_\varepsilon & \Xi(\varepsilon) \\ \star & -\Lambda(\varepsilon) \end{bmatrix} < 0, \tag{7.27}$$

where

$$\Xi(\varepsilon) \triangleq \begin{bmatrix} \varsigma_1 A_\varepsilon^{\mathrm{T}} P_\varepsilon & \varsigma_2 \bar{\Upsilon}^{\mathrm{T}} F_\varepsilon^{\mathrm{T}} & \sqrt{\bar{\gamma}_1} C_1^{\mathrm{T}} F_\varepsilon^{\mathrm{T}} \end{bmatrix}$$
$$\sqrt{(1 - \bar{\gamma}_1)\bar{\gamma}_2} C_2^{\mathrm{T}} F_\varepsilon^{\mathrm{T}} \quad \cdots \quad \sqrt{\prod_{j=1}^{N-1}(1 - \bar{\gamma}_j)\bar{\gamma}_N} C_N^{\mathrm{T}} F_\varepsilon^{\mathrm{T}} \Big],$$
$$\Lambda(\varepsilon) \triangleq \mathrm{diag}\{ P_\varepsilon, B_\varepsilon^{\mathrm{T}} P_\varepsilon B_\varepsilon, B_\varepsilon^{\mathrm{T}} P_\varepsilon B_\varepsilon, \underbrace{B_\varepsilon^{\mathrm{T}} P_\varepsilon B_\varepsilon, \cdots, B_\varepsilon^{\mathrm{T}} P_\varepsilon B_\varepsilon}_{N-1} \},$$

and $\Sigma_2(\varepsilon) < 0$ is equivalent to

$$\tilde{\Sigma}_2(\varepsilon) \triangleq \begin{bmatrix} -\xi I & \varsigma_3 I \\ \star & -B_\varepsilon^{\mathrm{T}} P_\varepsilon B_\varepsilon \end{bmatrix} < 0. \tag{7.28}$$

Let $P_\varepsilon \triangleq \begin{bmatrix} P_{11} & P_{12} \\ P_{12}^{\mathrm{T}} & P_{22} + \varepsilon Q \end{bmatrix}$. *It is noted that* $B_\varepsilon^{\mathrm{T}} P_\varepsilon B_\varepsilon = B_2^{\mathrm{T}} P_{22} B_2 + \varepsilon (B_2^{\mathrm{T}} P_{12}^{\mathrm{T}} B_1 + B_1^{\mathrm{T}} P_{12} B_2 + B_2^{\mathrm{T}} Q B_2) + \varepsilon^2 B_1^{\mathrm{T}} P_{11} B_1$ *and* $A_\varepsilon^{\mathrm{T}} P_\varepsilon = U + \varepsilon T$. *With the aid of the Lemma 7.1, given any $\varepsilon \in (0, \bar{\varepsilon}]$, the conditions (7.17)–(7.19) ensure $\tilde{\Sigma}_1(\varepsilon) < 0$ and the conditions (7.20)–(7.22) guarantee $\tilde{\Sigma}_2(\varepsilon) < 0$, which* *imply that*

$$\mathbb{E}\left\{x^{\mathrm{T}}(k+1)P_\varepsilon x(k+1) \mid x(k)\right\} - x^{\mathrm{T}}(k)P_\varepsilon x(k)$$
$$\leq x^{\mathrm{T}}(k)\Sigma_1(\varepsilon)x(k) + 4\xi\|D_s\|^2 \leq -\lambda_{\min}(-\Sigma_1(\varepsilon))\|x(k)\|^2 + 4\xi\|D_s\|^2$$
$$\leq -\psi x^{\mathrm{T}}(k)P_\varepsilon x(k) + 4\xi\|D_s\|^2, \tag{7.29}$$

where $\psi \triangleq \frac{\lambda_{\min}(-\Sigma_1(\varepsilon))}{\lambda_{\max}(P_\varepsilon)} < 1$.
Then, it is obtained that

$$\mathbb{E}\left\{x^{\mathrm{T}}(k)P_\varepsilon x(k) \mid x(k-1)\right\}$$
$$\leq (1-\psi)x^{\mathrm{T}}(k-1)P_\varepsilon x(k-1) + 4\xi\|D_s\|^2$$
$$\leq (1-\psi)^k x^{\mathrm{T}}(0)P_\varepsilon x(0) + 4\xi\|D_s\|^2\frac{1-(1-\psi)^k}{\psi},$$

and by taking mathematical expectation again, it further leads to

$$\mathbb{E}\left\{\|x(k)\|^2\right\} \leq \frac{\lambda_{\max}(P_\varepsilon)}{\lambda_{\min}(P_\varepsilon)}(1-\psi)^k\|x(0)\|^2 + 4\xi\|D_s\|^2\frac{1-(1-\psi)^k}{\psi\lambda_{\min}(P_\varepsilon)}. \tag{7.30}$$

This means that for any singular perturbed parameter $\varepsilon \in (0, \bar\varepsilon]$, *the conditions* *(7.17)–(7.22) ensure that the closed-loop system (7.13) is MSEUB with the ultimate bound:*

$$\bar\chi = \lim_{k\to\infty}\left\{4\xi\|D_s\|^2\frac{1-(1-\psi)^k}{\psi\lambda_{\min}(P_\varepsilon)}\right\} = \frac{4\xi\|D_s\|^2}{\psi\lambda_{\min}(P_\varepsilon)}.$$

This completes the proof.

7.2.3 The reachability of sliding surface

This subsection carries out the reachability analysis for the specified sliding surface $s(k) = 0$ under the output-feedback SMC law (7.12). From the sliding function (7.6) and the closed-loop system (7.13), we have

$$s(k+1) = \left[B_\varepsilon^{\mathrm{T}}P_\varepsilon A_\varepsilon - F_\varepsilon\Upsilon(k)\right]x(k) + \left[D_\varepsilon(k) - D_o - D_s\mathrm{sgn}(s(k))\right]. \tag{7.31}$$

The following theorem establishes a sufficient condition to the reachability of the specified sliding surface (7.6).

Theorem 7.2 *Consider the fast sampling SPS (7.1) and the output-feedback SMC law (7.12) with the RCTP (7.2)–(7.3). For prescribed scalars* $\delta > 0$, $\mu > 0$ *and* $\bar\varepsilon > 0$, *if there exist symmetric matrices* $P_{11} > 0$, $P_{22} > 0$, $Q > 0$, $W > 0$, *matrices* P_{12} *and* $F_\varepsilon \in \mathbb{R}^{m\times p}$, *and a scalar* $\xi > 0$ *satisfying the*

Main Results 127

following LMIs:

$$\begin{bmatrix} \bar{\Theta} & \bar{\Psi} \\ \star & \bar{\Gamma} \end{bmatrix} < 0, \tag{7.32}$$

$$\begin{bmatrix} \tilde{\Theta} & \tilde{\Psi} \\ \star & \tilde{\Gamma} \end{bmatrix} < 0, \tag{7.33}$$

$$\begin{bmatrix} \tilde{\Theta} & \check{\Psi} \\ \star & \check{\Gamma} \end{bmatrix} < 0, \tag{7.34}$$

where

$$L \triangleq \begin{bmatrix} B_2^\mathrm{T} P_{12}^\mathrm{T} + B_2^\mathrm{T} P_{22} A_{21} & B_2^\mathrm{T} P_{22} A_{22} \end{bmatrix},$$

$$R \triangleq \begin{bmatrix} B_1^\mathrm{T} P_{11} + B_2^\mathrm{T} P_{12}^\mathrm{T} A_{11} + B_1^\mathrm{T} P_{12} + B_2^\mathrm{T} Q A_{21} \end{bmatrix}$$
$$\qquad B_2^\mathrm{T} P_{12}^\mathrm{T} A_{12} + B_1^\mathrm{T} P_{12} A_{22} + B_2^\mathrm{T} Q A_{22} \end{bmatrix},$$

$$J \triangleq \begin{bmatrix} B_1^\mathrm{T} P_{11} A_{11} & B_1^\mathrm{T} P_{11} A_{12} \end{bmatrix}, \bar{\Theta} \triangleq -\mathrm{diag}\{P_0, \xi I\},$$

$$\bar{\Gamma} \triangleq -\mathrm{diag}\{W, W, \underbrace{W, \ldots, W}_{N-1}, P_0, B_2^\mathrm{T} P_{22} B_2, B_2^\mathrm{T} P_{22} B_2,$$
$$\qquad \underbrace{B_2^\mathrm{T} P_{22} B_2, \ldots, B_2^\mathrm{T} P_{22} B_2}_{N-1}, B_2^\mathrm{T} P_{22} B_2\},$$

$$\bar{\Psi} \triangleq \begin{bmatrix} [L - F_\varepsilon \bar{\Upsilon}]^\mathrm{T} & \sqrt{\bar{\gamma}_1} C_1^\mathrm{T} F_\varepsilon^\mathrm{T} & \sqrt{(1-\bar{\gamma}_1)\bar{\gamma}_2} C_2^\mathrm{T} F_\varepsilon^\mathrm{T} & \cdots & \sqrt{\prod_{j=1}^{N-1}(1-\bar{\gamma}_j)\bar{\gamma}_N} C_N^\mathrm{T} F_\varepsilon^\mathrm{T} \\[2ex] I & 0 & 0 & \cdots & 0 \\[2ex] \varsigma_1 U & \varsigma_2 \bar{\Upsilon}^\mathrm{T} F_\varepsilon^\mathrm{T} & \sqrt{\bar{\gamma}_1} C_1^\mathrm{T} F_\varepsilon^\mathrm{T} & \sqrt{(1-\bar{\gamma}_1)\bar{\gamma}_2} C_2^\mathrm{T} F_\varepsilon^\mathrm{T} & \cdots & \sqrt{\prod_{j=1}^{N-1}(1-\bar{\gamma}_j)\bar{\gamma}_N} C_N^\mathrm{T} F_\varepsilon^\mathrm{T} \quad 0 \\[2ex] 0 & 0 & 0 & 0 & \cdots & 0 & \varsigma_3 I \end{bmatrix},$$

$$\tilde{\Theta} \triangleq -\mathrm{diag}\{P_{\bar{\varepsilon}}, \xi I\}, \tilde{\Gamma} \triangleq -\mathrm{diag}\{W, W, \underbrace{W, \ldots, W}_{N-1}, P_{\bar{\varepsilon}}, H, H, \underbrace{H, \ldots, H}_{N-1}, H\},$$

$$\tilde{\Psi} \triangleq \begin{bmatrix} [L + \varepsilon R - F_\varepsilon \bar{\Upsilon}]^\mathrm{T} & \sqrt{\bar{\gamma}_1} C_1^\mathrm{T} F_\varepsilon^\mathrm{T} & \sqrt{(1-\bar{\gamma}_1)\bar{\gamma}_2} C_2^\mathrm{T} F_\varepsilon^\mathrm{T} & \cdots & \sqrt{\prod_{j=1}^{N-1}(1-\bar{\gamma}_j)\bar{\gamma}_N} C_N^\mathrm{T} F_\varepsilon^\mathrm{T} \\[2ex] I & 0 & 0 & \cdots & 0 \\[2ex] \varsigma_1(U + \bar{\varepsilon} T) & \varsigma_2 \bar{\Upsilon}^\mathrm{T} F_\varepsilon^\mathrm{T} & \sqrt{\bar{\gamma}_1} C_1^\mathrm{T} F_\varepsilon^\mathrm{T} & \sqrt{(1-\bar{\gamma}_1)\bar{\gamma}_2} C_2^\mathrm{T} F_\varepsilon^\mathrm{T} & \cdots & \sqrt{\prod_{j=1}^{N-1}(1-\bar{\gamma}_j)\bar{\gamma}_N} C_N^\mathrm{T} F_\varepsilon^\mathrm{T} \quad 0 \\[2ex] 0 & 0 & 0 & 0 & \cdots & 0 & \varsigma_3 I \end{bmatrix},$$

$$\check{\Gamma} \triangleq -\mathrm{diag}\{W, W, \underbrace{W, \ldots, W}_{N-1}, P_{\bar{\varepsilon}}, \check{H}, \check{H}, \underbrace{\check{H}, \ldots, \check{H}}_{N-1}, \check{H}\},$$

$$\check{\Psi} \triangleq \begin{bmatrix} [L + \varepsilon R + \bar{\varepsilon}^2 J - F_\varepsilon \bar{\Upsilon}]^\mathrm{T} & \sqrt{\bar{\gamma}_1} C_1^\mathrm{T} F_\varepsilon^\mathrm{T} & \sqrt{(1-\bar{\gamma}_1)\bar{\gamma}_2} C_2^\mathrm{T} F_\varepsilon^\mathrm{T} & \cdots & \sqrt{\prod_{j=1}^{N-1}(1-\bar{\gamma}_j)\bar{\gamma}_N} C_N^\mathrm{T} F_\varepsilon^\mathrm{T} \\[2ex] I & 0 & 0 & \cdots & 0 \end{bmatrix}$$

$$\left. \begin{matrix} \varsigma_1(U+\bar{\varepsilon}T) & \varsigma_2\bar{\Upsilon}^{\mathrm{T}}F_\varepsilon^{\mathrm{T}} & \sqrt{\bar{\gamma}_1}C_1^{\mathrm{T}}F_\varepsilon^{\mathrm{T}} & \sqrt{(1-\bar{\gamma}_1)\bar{\gamma}_2}C_2^{\mathrm{T}}F_\varepsilon^{\mathrm{T}} & \cdots & \sqrt{\prod_{j=1}^{N-1}(1-\bar{\gamma}_j)\bar{\gamma}_N}C_N^{\mathrm{T}}F_\varepsilon^{\mathrm{T}} & 0 \\ 0 & 0 & 0 & 0 & \cdots & 0 & \varsigma_3 I \end{matrix} \right],$$

and the other matrices are defined in Theorem 7.1, then for the singularly perturbed parameter $\varepsilon \in (0, \bar{\varepsilon}]$, the sliding variable $s(k)$ will be driven into the following sliding region $\mathbf{O}$ in mean-square sense by the output-feedback SMC law (7.12):

$$\mathbf{O} \triangleq \left\{ s(k) \mid \|s(k)\| \le \bar{\xi} \right\}, \tag{7.35}$$

where $\bar{\xi} \triangleq \sqrt{\frac{4\xi\|D_s\|^2}{\lambda_{\min}(W^{-1})}}$ is the bound of the sliding region.

Proof *Consider the following Lyapunov function candidate:*

$$\tilde{V}(k) \triangleq x^{\mathrm{T}}(k)P_\varepsilon x(k) + s^{\mathrm{T}}(k)W^{-1}s(k).$$

By utilizing the relationship (7.26) and the solution of (7.31), we evaluate the difference of $\tilde{V}(k)$ as follows:

$$\mathbb{E}\left\{ \Delta\tilde{V}(k) \mid x(k) \right\}$$

$$=\mathbb{E}\left\{ x^{\mathrm{T}}(k+1)P_\varepsilon x(k+1) \mid x(k) \right\} - x^{\mathrm{T}}(k)P_\varepsilon x(k) + s^{\mathrm{T}}(k+1)W^{-1}s(k+1)$$
$$- s^{\mathrm{T}}(k)W^{-1}s(k)$$

$$\le x^{\mathrm{T}}(k)\Sigma_1(\varepsilon)x(k) + [D_\varepsilon(k) - D_o - D_s\mathrm{sgn}(s(k))]^{\mathrm{T}} \Sigma_2(\varepsilon) [D_\varepsilon(k) - D_o$$

$$- D_s\mathrm{sgn}(s(k))] + 4\xi\|D_s\|^2 + x^{\mathrm{T}}(k)\left\{ \left[B_\varepsilon^{\mathrm{T}}P_\varepsilon A_\varepsilon - F_\varepsilon\bar{\Upsilon} \right]^{\mathrm{T}} W^{-1} \right.$$

$$\times \left[B_\varepsilon^{\mathrm{T}}P_\varepsilon A_\varepsilon - F_\varepsilon\bar{\Upsilon} \right] - \bar{\Upsilon}^{\mathrm{T}}F_\varepsilon^{\mathrm{T}}W^{-1}F_\varepsilon\bar{\Upsilon} + \bar{\gamma}_1 C_1^{\mathrm{T}}F_\varepsilon^{\mathrm{T}}W^{-1}F_\varepsilon C_1$$

$$+ \sum_{i=2}^{N}\left\{ \prod_{j=1}^{i-1}(1-\bar{\gamma}_j)\bar{\gamma}_i C_i^{\mathrm{T}}F_\varepsilon^{\mathrm{T}}W^{-1}F_\varepsilon C_i \right\}\left. \right\}x(k)$$

$$+ 2x^{\mathrm{T}}(k) \left[B_\varepsilon^{\mathrm{T}}P_\varepsilon A_\varepsilon - F_\varepsilon\bar{\Upsilon} \right]^{\mathrm{T}} W^{-1} [D_\varepsilon(k) - D_o - D_s\mathrm{sgn}(s(k))]$$

$$+ [D_\varepsilon(k) - D_o - D_s\mathrm{sgn}(s(k))]^{\mathrm{T}} W^{-1} [D_\varepsilon(k) - D_o - D_s\mathrm{sgn}(s(k))]$$

$$- s^{\mathrm{T}}(k)W^{-1}s(k)$$

$$\le \eta^{\mathrm{T}}(k)\acute{\Sigma}(\varepsilon)\eta(k) - \left[\lambda_{\min}(W^{-1})\|s(k)\|^2 - 4\xi\|D_s\|^2 \right], \tag{7.36}$$

Main Results 129

where $\eta(k) \triangleq \begin{bmatrix} x^{\mathrm{T}}(k) & [D_\varepsilon(k) - D_o - D_s\mathrm{sgn}(s(k))]^{\mathrm{T}} \end{bmatrix}^{\mathrm{T}}$, and

$$\acute{\Sigma}(\varepsilon) \triangleq \begin{bmatrix} \acute{\Sigma}_1(\varepsilon) & [B_\varepsilon^{\mathrm{T}} P_\varepsilon A_\varepsilon - F_\varepsilon \bar{\Upsilon}]^{\mathrm{T}} W^{-1} \\ \star & \Sigma_2(\varepsilon) + W^{-1} \end{bmatrix},$$

$$\acute{\Sigma}_1(\varepsilon) \triangleq \Sigma_1(\varepsilon) + [B_\varepsilon^{\mathrm{T}} P_\varepsilon A_\varepsilon - F_\varepsilon \bar{\Upsilon}]^{\mathrm{T}} W^{-1} [B_\varepsilon^{\mathrm{T}} P_\varepsilon A_\varepsilon - F_\varepsilon \bar{\Upsilon}]$$
$$+ \bar{\gamma}_1 C_1^{\mathrm{T}} F_\varepsilon^{\mathrm{T}} W^{-1} F_\varepsilon C_1 + \sum_{i=2}^{N} \left\{ \prod_{j=1}^{i-1} (1 - \bar{\gamma}_j) \bar{\gamma}_i C_i^{\mathrm{T}} F_\varepsilon^{\mathrm{T}} W^{-1} F_\varepsilon C_i \right\}.$$

By resorting to the Schur complement, it is shown that $\acute{\Sigma}(\varepsilon) < 0$ is equivalent to the following inequality hold:

$$\dot{\Sigma}(\varepsilon) \triangleq \begin{bmatrix} \Theta(\varepsilon) & \Psi(\varepsilon) \\ \star & \Gamma(\varepsilon) \end{bmatrix} < 0, \tag{7.37}$$

where

$$\Theta(\varepsilon) \triangleq -\mathrm{diag}\left\{ P_\varepsilon, \xi I \right\},$$
$$\Gamma(\varepsilon) \triangleq -\mathrm{diag}\{ W, W, \underbrace{W, \dots, W}_{N-1}, P_\varepsilon, B_\varepsilon^{\mathrm{T}} P_\varepsilon B_\varepsilon, B_\varepsilon^{\mathrm{T}} P_\varepsilon B_\varepsilon,$$
$$\underbrace{B_\varepsilon^{\mathrm{T}} P_\varepsilon B_\varepsilon, \dots, B_\varepsilon^{\mathrm{T}} P_\varepsilon B_\varepsilon}_{N-1}, B_\varepsilon^{\mathrm{T}} P_\varepsilon B_\varepsilon \},$$

$$\Psi(\varepsilon) \triangleq \begin{bmatrix} [B_\varepsilon^{\mathrm{T}} P_\varepsilon A_\varepsilon - F_\varepsilon \bar{\Upsilon}]^{\mathrm{T}} & \sqrt{\bar{\gamma}_1} C_1^{\mathrm{T}} F_\varepsilon^{\mathrm{T}} & \sqrt{(1-\bar{\gamma}_1)\bar{\gamma}_2} C_2^{\mathrm{T}} F_\varepsilon^{\mathrm{T}} & \cdots & \sqrt{\prod_{j=1}^{N-1}(1-\bar{\gamma}_j)\bar{\gamma}_N} C_N^{\mathrm{T}} F_\varepsilon^{\mathrm{T}} \\ I & 0 & 0 & \cdots & 0 \\ \varsigma_1 A_\varepsilon^{\mathrm{T}} P_\varepsilon & \varsigma_2 \bar{\Upsilon}^{\mathrm{T}} F_\varepsilon^{\mathrm{T}} & \sqrt{\bar{\gamma}_1} C_1^{\mathrm{T}} F_\varepsilon^{\mathrm{T}} & \sqrt{(1-\bar{\gamma}_1)\bar{\gamma}_2} C_2^{\mathrm{T}} F_\varepsilon^{\mathrm{T}} & \cdots & \sqrt{\prod_{j=1}^{N-1}(1-\bar{\gamma}_j)\bar{\gamma}_N} C_N^{\mathrm{T}} F_\varepsilon^{\mathrm{T}} & 0 \\ 0 & 0 & 0 & 0 & \cdots & 0 & \varsigma_3 I \end{bmatrix}.$$

Let $P_\varepsilon \triangleq \begin{bmatrix} P_{11} & P_{12} \\ P_{12}^{\mathrm{T}} & P_{22} + \varepsilon Q \end{bmatrix}$. *We have the following relationships:*

$$B_\varepsilon^{\mathrm{T}} P_\varepsilon B_\varepsilon = B_2^{\mathrm{T}} P_{22} B_2 + \varepsilon(B_2^{\mathrm{T}} P_{12}^{\mathrm{T}} B_1 + B_1^{\mathrm{T}} P_{12} B_2 + B_2^{\mathrm{T}} Q B_2) + \varepsilon^2 B_1^{\mathrm{T}} P_{11} B_1,$$
$$A_\varepsilon^{\mathrm{T}} P_\varepsilon = U + \varepsilon T, B_\varepsilon^{\mathrm{T}} P_\varepsilon A_\varepsilon = L + \varepsilon R + \varepsilon^2 J.$$

Thus, according to Lemma 7.1, it is obtained that the conditions (7.32)–(7.34) guarantee $\dot{\Sigma}(\varepsilon) < 0$ for any $\varepsilon \in (0, \bar{\varepsilon}]$, which implies that when the state trajectories escape from the region $\mathbf{O}$ around the specified sliding surface (7.6) (i.e., $\|s(k)\| \geq \sqrt{\frac{4\xi\|D_s\|^2}{\lambda_{\min}(W^{-1})}}$), it renders $\mathbb{E}\left\{ \Delta\tilde{V}(k) \right\} < 0$ in (7.36). Therefore, the sliding variable $s(k)$ is strictly deceasing outside the region $\mathbf{O}$, and the reachability of the sliding region $\mathbf{O}$ is ensured in mean-square sense.

130   *Reliable SMC under RCTP*

It is seen from Theorems 7.1 and 7.2 that under the output-feedback SMC law (7.12) with the RCTP (7.2)–(7.3), the state trajectories of the closed-loop system (7.13) will not only be driven into the sliding region $\mathbf{O}$ around the sliding surface (7.6) in a finite time but also be guaranteed as MSEUB over the sliding region $\mathbf{O}$.

7.2.4   Solving algorithm

In practical applications, it is very meaningful to find the upper-bound $\bar{\varepsilon}$ for the singular perturbation parameter, which reflects the conservativeness of the proposed SMC strategy in stabilizing fast-sampling SPSs (7.1). Based on Theorems 7.1 and 7.2, we develop an algorithm to design the sliding surface (7.6) and the SMC law (7.12) with estimating the ε-bound as follows.

- **Step 1.** Choose a sufficiently small scalar $\Delta\varepsilon$, and set the scalars $\bar{\varepsilon} = \Delta\varepsilon$ and $t = 1$.

- **Step 2.** Let $\delta = \frac{\zeta}{1-\zeta}$ and $\mu = \frac{\iota}{1-\iota}$ with $\zeta \in (0,1)$ and $\iota \in (0,1)$. Check the feasibility of the LMIs in Theorems 7.1 and 7.2 simultaneously for prescribed adjustable scalars $\zeta = i\Delta\nu_1$ and $\iota = j\Delta\nu_2$, where $i = j = 1$, $\Delta\nu_1 > 0$ and $\Delta\nu_2 > 0$ are preassigned small fixed step sizes.

- **Step 3.** If the conditions have feasible solutions for $t \leq L$ (L denotes the given maximum iterative steps), set $\bar{\varepsilon} = \bar{\varepsilon} + \Delta\varepsilon$ and $t = t+1$, go to Step 2; otherwise, modify the scalars δ and μ by adjusting i and j incrementally over $i \in \left[1, \lceil \frac{1}{\Delta\nu_1} \rceil - 1\right]$ and $j \in \left[1, \lceil \frac{1}{\Delta\nu_2} \rceil - 1\right]$ until no feasible solutions can be found, go to Step 4.

- **Step 4.** Output the maximum feasible bound $\bar{\varepsilon}$, and get the matrices F_ε, P_{11}, P_{12}, P_{22} and Q. For a known singular perturbation parameter $\varepsilon \in (0, \bar{\varepsilon}]$, solve the LMI (7.8) with a sufficiently small scalar $\kappa > 0$ and the matrix $P_\varepsilon = \begin{bmatrix} P_{11} & P_{12} \\ P_{12}^{\mathrm{T}} & P_{22} + \varepsilon Q \end{bmatrix}$, and thus obtain the gain matrix G_ε in sliding function (7.6).

- **Step 5.** Produce the output-feedback SMC law (7.12) with matrices F_ε, G_ε, D_o and D_s, and apply it to the fast-sampling SPSs (7.1) with the singularly perturbed parameter $\varepsilon \in (0, \bar{\varepsilon}]$ and the RCTP (7.2)–(7.3).

Remark 7.4 *For the first attempt, this chapter investigates the SMC problem for the fast-sampling SPS (7.1) under the RCTP (7.2)–(7.3). The proposed output-feedback SMC scheme exhibits the following distinct features: 1) In order to ensure the engineering implementation of the SMC, the sliding function (7.6) is designed by considering the structural characteristics of the measurement output model (7.3) under the RCTP. 2) The established sufficient conditions for guaranteeing the MSEUB and the reachability of the SMC system are dependent on the packet loss probabilities in the RCTP (7.2)–(7.3) and*

Example

the available upper bound of the singularly perturbed parameter ε. If the upper bound ϖ of the external disturbance $w(k)$ is unknown, then the stabilization problem for the fast-sampling SPS (7.1) under the RCTP (7.2)-(7.3) still can be solved readily by the proposed SMC scheme with combining the adaptive technique as in [21].

7.3 Example

As shown in Fig. 7.2, we conduct a simulation for remote SMC of an operational amplifier (OPA) circuit under the RCTP. Owing to the virtual short circuit between OPA input terminals, the voltage at the inverting terminal will be equal to v_2. Thus, the current through R_2 will be $\frac{u-v_2}{R_2}$. Meanwhile, owing to the infinite input impedance of the OPA, the current through R_3 will be also $\frac{u-v_2}{R_2}$. Furthermore, it is concluded that the current through R_4 will be $\frac{(v_2-u)R_3}{R_2 R_4}$. Now, by applying the Kirchoff current law to v_1 and v_2, we obtain the following state equations:

$$\dot{v}_1 = -\frac{1}{R_1 \mathbf{C}_1} v_1 + \frac{1}{R_1 \mathbf{C}_1} v_2, \tag{7.38}$$

$$\mathbf{C}_2 \dot{v}_2 = \frac{1}{R_1} v_1 + \left(\frac{R_3}{R_2 R_4} - \frac{1}{R_1} \right) v_2 - \frac{R_3}{R_2 R_4} u. \tag{7.39}$$

Let $x_1(t) = v_1$ and $x_2(t) = v_2$. Similar to [1], it is supposed that $\mathbf{C}_2$ is a

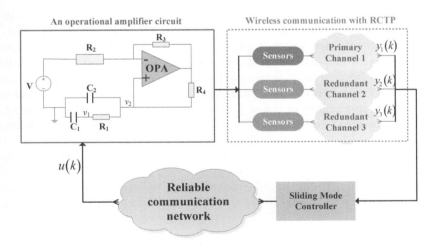

FIGURE 7.2: Remote SMC of an operational amplifier circuit under the RCTP.

TABLE 7.1: RCTP ($N = 3$) in the example.

Protocol Settings	Measurement matrix C_i	Probability $\bar{\gamma}_i$
Primary channel ($i = 1$)	$C_1 = \begin{bmatrix} 1 & 0 \\ 0 & 1 \end{bmatrix}$	$\bar{\gamma}_1 = 0.7$
Redundant channel(a) ($i = 2$)	$C_2 = \begin{bmatrix} 0.8 & 0 \\ 0 & 0.8 \end{bmatrix}$	$\bar{\gamma}_2 = 0.6$
Redundant channel(b) ($i = 3$)	$C_3 = \begin{bmatrix} 0.8 & 0.1 \\ 0 & 0.7 \end{bmatrix}$	$\bar{\gamma}_3 = 0.5$

small "parasitic" capacitor, which can be regarded as a singularly perturbed parameter, i.e., $\mathbf{C}_2 = \varepsilon$. Other parameters are given as $R_1 = R_4 = 2\ \Omega$, $R_2 = 3\ \Omega$, $R_3 = 1\ \Omega$, and $\mathbf{C}_1 = 0.3$ F. By using the discretizing approach in [70], we obtain the following discrete-time fast-sampling SPS:

$$\begin{bmatrix} x_1(k+1) \\ x_2(k+1) \end{bmatrix} = \begin{bmatrix} 1-1.2927\varepsilon & 0.9921\varepsilon \\ 0.4252 & 0.7165 \end{bmatrix} \begin{bmatrix} x_1(k) \\ x_2(k) \end{bmatrix} + \begin{bmatrix} -0.1247\varepsilon \\ -0.1417 \end{bmatrix} (u(k)+w(k)), \tag{7.40}$$

where the matched disturbance is $w(k) = \varpi \cos(20k)$ with $\varpi = 0.1$.

The simulation experiment is conducted by MATLAB (R2014a) for the SMC of the fast-sampling SPS under the RCTR (7.2)–(7.3) with three channels (i.e., one primary channel and two redundant channels) as shown in Fig. 7.2, where the implementation of the OPA circuit is based on the state equation (7.40) by setting the fast sampling interval $T_f = \varepsilon$ as per [70] and the communication between sensors and controller is executed via IEC 62439-3-based industrial WiFi network [22]. The number of redundant channels is $N = 3$ with the measurement matrices and the probabilities of successfully transmitted packets given as in Table 7.1.

Now, we further explore the SMC performance under different RCTP cases. By applying the algorithm in Section 7.2.4, four cases are considered as shown in Table 7.2, from which it is observed that for Case 4, i.e. the primary channel (PC) and the two redundant channels (RC(a) and RC(b)) are exploited simultaneously, the largest upper bound of the singular perturbation parameter ε can be obtained, while for Case 1, i.e., only PC is utilized, the smallest $\bar{\varepsilon}$ just be ensured.

In the simulation, we choose the singular perturbation parameter $\varepsilon = 0.3$ and the bounds in (7.11) are $\bar{d}_i = -\underline{d}_i = \varpi \|G_\varepsilon \tilde{\Upsilon} B_\varepsilon\|$, that is, $D_o = 0$ and $D_s = \varpi \|G_\varepsilon \tilde{\Upsilon} B_\varepsilon\|$. Then, the following reliable SMC laws are designed for the above mentioned four cases:

- Case 1 (PC):

$$\begin{aligned} u(k) &= \begin{bmatrix} -1.6194 & 2.5658 \end{bmatrix} y(k) - 0.1\mathrm{sgn}(s(k)), \\ y(k) &= \gamma_1(k)C_1 x(k), \\ s(k) &= \bar{\gamma}_1 G_\varepsilon y_1(k), \end{aligned}$$

Example   133

TABLE 7.2: Four different RCTP cases.

Cases and Solutions	$\bar{\varepsilon}$	Feasible solutions
Case 1: PC	0.3401	$F_\varepsilon = \begin{bmatrix} 0.3304 & -0.5235 \end{bmatrix},$ $P_\varepsilon = \begin{bmatrix} 11.3105 & -14.4738 \\ -14.4738 & 21.1932 + 17.8084\varepsilon \end{bmatrix}$ $W = 11.1571,\ \xi = 2.9725$
Case 2: PC+RC(a)	0.8194	$F_\varepsilon = \begin{bmatrix} 0.1017 & -0.1578 \end{bmatrix},$ $P_\varepsilon = \begin{bmatrix} 5.8083 & -4.7922 \\ -4.7922 & 9.8067 + 2.9061\varepsilon \end{bmatrix}$ $W = 11.2992,\ \xi = 2.9048$
Case 3: PC+RC(b)	0.5436	$F_\varepsilon = \begin{bmatrix} 0.2677 & -0.4166 \end{bmatrix},$ $P_\varepsilon = \begin{bmatrix} 11.1213 & -13.1923 \\ -13.1923 & 20.6578 + 5.7722\varepsilon \end{bmatrix},$ $W = 11.2425,\ \xi = 2.8263$
Case 4: PC+RC(a)+RC(b)	0.9079	$F_\varepsilon = \begin{bmatrix} 0.0757 & -0.1223 \end{bmatrix},$ $P_\varepsilon = \begin{bmatrix} 5.0070 & -3.5575 \\ -3.5575 & 8.3842 + 2.8645\varepsilon \end{bmatrix},$ $W = 11.1636,\ \xi = 2.9724$

- Case 2 (PC+RC(a)):

$$u(k) = \begin{bmatrix} -0.6931 & 1.0754 \end{bmatrix} y(k) - 0.1\mathrm{sgn}(s(k)),$$
$$y(k) = \gamma_1(k)C_1 x(k) + (1 - \gamma_1(k))\gamma_2(k)C_2 x(k),$$
$$s(k) = G_\varepsilon \left[\bar{\gamma}_1 y_1(k) + (1 - \bar{\gamma}_1)\bar{\gamma}_2 y_2(k) \right],$$

- Case 3 (PC+RC(b)):

$$u(k) = \begin{bmatrix} -1.4574 & 2.2680 \end{bmatrix} y(k) - 0.1\mathrm{sgn}(s(k)),$$
$$y(k) = \gamma_1(k)C_1 x(k) + (1 - \gamma_1(k))\gamma_3(k)C_3 x(k),$$
$$s(k) = G_\varepsilon \left[\bar{\gamma}_1 y_1(k) + (1 - \bar{\gamma}_1)\bar{\gamma}_3 y_3(k) \right],$$

- Case 4 (PC+RC(a)+RC(b)):

$$u(k) = \begin{bmatrix} -0.9563 & 1.5214 \end{bmatrix} y(k) - 0.1\mathrm{sgn}(s(k)),$$
$$y(k) = \gamma_1(k)C_1 x(k) + (1 - \gamma_1(k))\gamma_2(k)C_2 x(k)$$
$$+ (1 - \gamma_1(k))(1 - \gamma_2(k))\gamma_3(k)C_3 x(k),$$
$$s(k) = G_\varepsilon \left[\bar{\gamma}_1 y_1(k) + (1 - \bar{\gamma}_1)\bar{\gamma}_2 y_2(k) + (1 - \bar{\gamma}_1)(1 - \bar{\gamma}_2)\bar{\gamma}_3 y_3(k) \right].$$

Under the initial condition $x(0) = \begin{bmatrix} -3 & 5 \end{bmatrix}^{\mathrm{T}}$, the simulation results shown in Figs. 7.3–7.8 are obtained from MATLAB (R2014a) by simulating

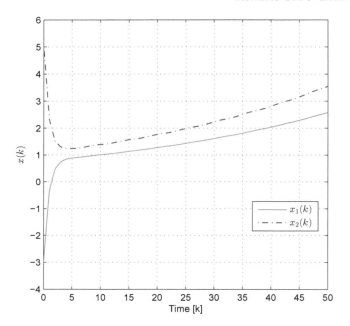

FIGURE 7.3: State trajectories $x(k)$ in open-loop case.

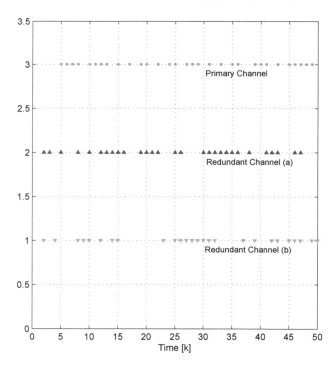

FIGURE 7.4: Random packet dropouts in three channels.

Example

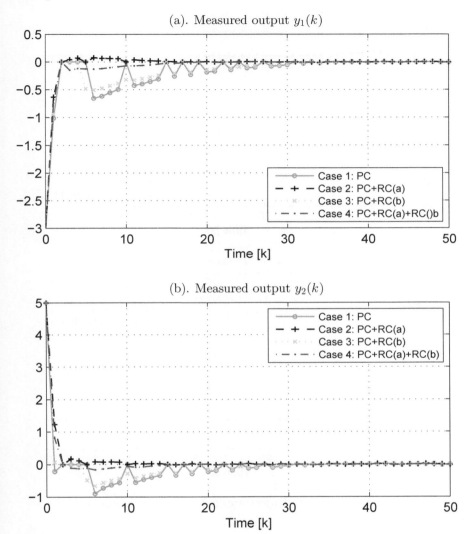

FIGURE 7.5: Measured outputs $y(k)$ in four cases.

OPA circuit based on the state equation (7.40) with the fast sampling interval $T_f = \varepsilon$, where the voltages v_1 and v_2 are captured from the simulation. In practical application, the voltages v_1 and v_2 can be measured directly via voltmeters or voltage sensors. Fig. 7.3 shows the fast-sampling SPS (7.40) with the singular perturbation parameter $\varepsilon = 0.3$ is unstable. Fig. 7.4 depicts the random packet dropouts in three channels. The measured outputs in four cases are plotted in Fig. 7.5. By using the above measurement outputs, the state

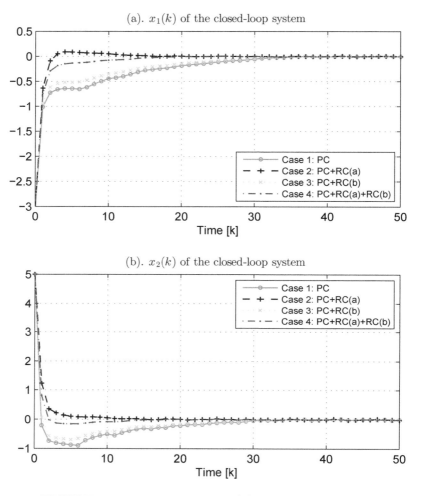

FIGURE 7.6: State trajectories $x(k)$ in closed-loop cases.

trajectories $x(k)$ and the sliding variable $s(k)$ of the closed-loop system are shown in Figs. 7.6–7.7, where one can observe that in Case 4, the convergence speed of $x(k)$ is fastest and the bound $\bar{\xi}$ of the sliding region is smallest. The SMC input $u(k)$ in four cases are given in Fig. 7.8. All simulation results illustrate that the utilization of the redundant channels can substantially improve the output-feedback SMC performance for the fast-sampling SPSs (7.40).

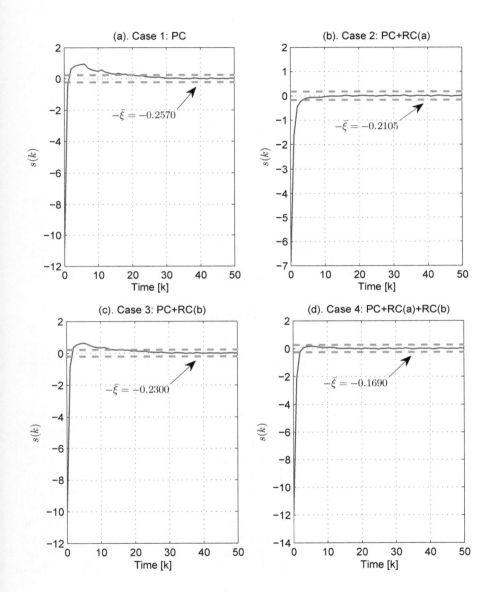

FIGURE 7.7: Sliding variables $s(k)$ in four cases.

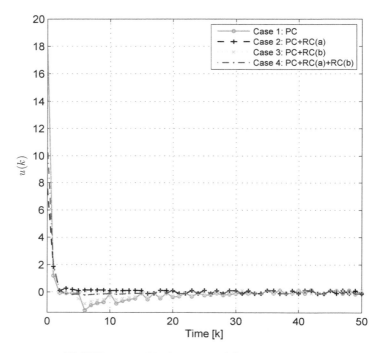

FIGURE 7.8: SMC input $u(k)$ in four cases.

7.4 Conclusion

This chapter has addressed the output-feedback SMC problem of the fast-sampling singularly perturbed systems. The redundant channels transmission protocol has been employed to improve the reliability of the network transmission between the sensors and the controller. An output-feedback sliding function has been constructed based on the structure characteristic of the redundant channels transmission protocol. By resorting to some appropriate Lyapunov functions, the MSEUB of the closed-loop system and the reachability of the specified sliding surface have been analyzed with estimating the available upper bound for the singularly perturbed parameter. It has been shown from a numerical example that the more redundant channels are applied, the better output-feedback SMC performance are achieved for the fast-sampling SPSs.

8

State-Saturated Sliding Mode Control Under Multiple-Packet Transmission Protocol

As a typical kind of nonlinear systems, state-saturated systems possess the distinctive characteristics that the states are constrained into a prescribed bounded area quantified by *saturation level*. The state saturation phenomenon occurs frequently in engineering practice owing to physical limitations of the devices and/or protection equipment's. By introducing a polytopic representation of saturation nonlinearities, some representative results have been acquired on various dynamics analysis issues (e.g. stability, robustness, filtering, and control) for state-saturated systems. Nonetheless, the sliding mode control (SMC) problem for state-saturated systems remains an open yet interesting research problems.

In the real world, the channel fading phenomenon occurs typically in a *time-varying* fashion. In other words, the amplitude and/or phase of the transmitted signal may experience fadings caused by the changing communication environment. It has been shown that the wireless flat-fading channels can be well described by the model of *finite-state* Markov fading channels (FSMFCs), where the stochastic time-varying feature is captured by a finite-state Markov process. From the viewpoint of practical applications, a key problem in employing the FSMFCs model is how to estimate the actual network mode based on the measured network information. A seemingly interesting question is whether it is possible to develop a more general yet simple estimation model to not only cover the above three special cases but also observe the *multi*-step delayed network mode, and one of the motivations of this chapter is therefore to answer such a question.

According to the above discussion, it makes both theoretical and practical sense to study the SMC design for networked state-saturated systems under MPTP with FSMFCs. Very recently, the mode detector approach has been put forward to describe the asynchronous phenomena between the Markovian jump system modes and the controller/filtering modes. Inspired by such an approach, in this chapter, we focus our attention on the design task of the SMC scheme for state-saturated systems over *hidden Markov fading channels* (HMFCs), in which the jumping network mode of the FSMFCs is estimated/observed by a mode detector via a hidden Markov model (HMM). *The novelties of this chapter lie in the following four aspects: 1) Based on*

DOI: 10.1201/9781003309499-8

139

140 State-Saturated SMC under MPTP

a novel fading channels model of HMFCs, the SMC problem is, for the first time, investigated for the networked state-saturated systems over the fading channels; 2) Based on a linear sliding surface, a switching SMC strategy is proposed by using the estimated network mode only; 3) By exploiting the polytopic representation of the state saturation and the HMM approach, both the mean-square stability and the reachability are studied for the SMC system; and 4) With the aid of the Hadamard product, a genetic algorithm (GA) is formulated to solve the proposed SMC design problem subject to some nonconvex constraint conditions resulting from the state saturation and the fading channels.

8.1 Problem Formulation

This chapter is concerned with the problem of sliding mode stabilization over HMFCs for the following state-saturated systems:

$$x(k+1) = \sigma\Big((A + \Delta A(k))\,x(k) + B\,(u(k) + \phi(x(k), k))\Big), \qquad (8.1)$$

where $x(k) \in \mathbb{R}^n$ is the state vector, $u(k) \in \mathbb{R}^m$ is the actuator input vector, and $\phi(x(k), k) \in \mathbb{R}^m$ is the matched input disturbance satisfying $\|\phi(x(k), k)\| \le \varphi\|x(k)\|$ with a known scalar $\varphi \ge 0$. The norm-bounded parameter uncertainty $\Delta A(k)$ satisfies $\Delta A(k) = MF(k)H$ with $F^{\mathrm{T}}(k)F(k) \le I$. Here, A, B, M, and H are known constant matrices, and the input matrix B satisfies $\mathrm{rank}(B) = m$.

The state-saturated function $\sigma(\cdot) : \mathbb{R}^n \to \mathbb{R}^n$ in (8.1) is defined as follows:

$$\sigma(v) \triangleq \big[\ \sigma_1(v_1) \quad \sigma_2(v_2) \quad \cdots \quad \sigma_n(v_n)\ \big]^{\mathrm{T}}, \ \forall v \in \mathbb{R}^n, \qquad (8.2)$$

where the scalar-valued saturation function $\sigma_i(v_i)$ (shown in Fig. 8.1) is defined by

$$\sigma_i(v_i) \triangleq \mathrm{sgn}(v_i) \min\{\bar{\varpi}_i, |v_i|\}, \qquad (8.3)$$

with v_i representing the ith element of the vector v and $\bar{\varpi}_i$ standing for the saturation level [127]. To help with the analysis later, we denote

$$\Lambda \triangleq \mathrm{diag}\{\bar{\varpi}_1, \bar{\varpi}_2, \ldots, \bar{\varpi}_n\}. \qquad (8.4)$$

In this chapter, it is supposed that control signal $u(k)$ is transmitted to the actuators via a Multiple-Packet Transmission Protocol (MPTP) due to the packet size constraint. Specifically, MPTP is executed over a set of FSMFCs. Let the network mode $\varsigma(k) \in \Theta \triangleq \{1, 2, \ldots, N\}$ obey a discrete-time Markov process with the following transition probabilities:

$$\pi_{ij} \triangleq \mathrm{Pr}\{\varsigma(k+1) = j \mid \varsigma(k) = i\}, \ k \in \mathbb{N}_0, \ \forall i, j \in \Theta \qquad (8.5)$$

Problem Formulation

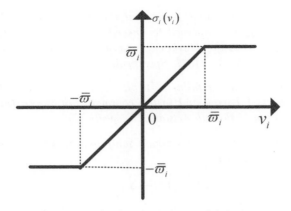

FIGURE 8.1: The saturation function $\sigma_i(v_i)$.

where $\pi_{ij} \in [0,1]$ and $\sum_{j=1}^{N} \pi_{ij} = 1$. The transition probability matrix is defined as $\Pi \triangleq [\pi_{ij}]_{i,j \in \Theta}$. In practical application, the network modes $\varsigma(k)$ correspond to different channel fading amplitudes or different configurations of the overall physical environment (e.g., positions of mobile agents) [118, 122, 176].

Now, at the network mode $\varsigma(k)$, the network input-output behaviour can be characterized by

$$u(k) = \vartheta_{\varsigma(k)}(k) v(k), \tag{8.6}$$

where the channel fading phenomenon is reflected in $\vartheta_{\varsigma(k)}(k) \in \mathbb{R}^{m \times m}$ that has the diagonal structure

$$\vartheta_{\varsigma(k)}(k) = \text{diag}\left\{\vartheta_{1,\varsigma(k)}(k), \vartheta_{2,\varsigma(k)}(k), \ldots, \vartheta_{m,\varsigma(k)}(k)\right\} \tag{8.7}$$

and, for any $b = 1, 2, \ldots, m$, $\vartheta_{b,\varsigma(k)}(k)$ is scalar-valued random process with

$$\begin{aligned} \mu_{b,\varsigma(k)} &\triangleq \mathbb{E}\left\{\vartheta_{b,\varsigma(k)}(k)\right\}, \\ \xi_{bl,\varsigma(k)} &\triangleq \mathbb{E}\left\{\left(\vartheta_{b,\varsigma(k)} - \mu_{b,\varsigma(k)}\right)\left(\vartheta_{l,\varsigma(k)} - \mu_{l,\varsigma(k)}\right)\right\} \end{aligned} \tag{8.8}$$

which satisfy $\mu_{b,\varsigma(k)} > 0$, $\xi_{bb,\varsigma(k)} > 0$, and $\xi_{bl,\varsigma(k)} = \xi_{lb,\varsigma(k)}$ for all $\varsigma(k) \in \Theta$ and $b, l = 1, 2, \ldots, m$.

We further write

$$\begin{aligned} \Gamma_{\varsigma(k)} &\triangleq \text{diag}\left\{\mu_{1,\varsigma(k)}, \mu_{2,\varsigma(k)}, \ldots, \mu_{m,\varsigma(k)}\right\}, \\ \Phi_{\varsigma(k)} &\triangleq \left[\xi_{bl,\varsigma(k)}\right]_{b,l=1,2,\ldots,m}, \\ \Psi_{\varsigma(k)} &\triangleq \text{diag}\left\{\xi_{11,\varsigma(k)}, \xi_{22,\varsigma(k)}, \ldots, \xi_{mm,\varsigma(k)}\right\}. \end{aligned}$$

It is easily seen that $\Phi_{\varsigma(k)}$ is positive semidefinite.

142   *State-Saturated SMC under MPTP*

Remark 8.1 *It is worth mentioning that the model (8.6) with the Markov process (8.5) can also be employed to describe the jumping actuator failures [115], the quantizer with the Markovian jump quantization density [108] and the Markovian packet losses [131]. Specifically,*

- *for the jumping actuator failures [115], $\vartheta_{b,\varsigma(k)}(k) = 0$, $0 < \vartheta_{b,\varsigma(k)}(k) < 1$, and $\vartheta_{b,\varsigma(k)}(k) = 1$ correspond to the complete failure, partial failure, and failure-free cases of the rth actuator, respectively;*

- *for the quantizer with the Markovian jump quantization density [108], the model (8.6) is specified as*

$$u(k) = \left(I_m + \bar{\Delta}_{\varsigma(k)}(k)\right) v(k),$$

where $\bar{\Delta}_{\varsigma(k)}(k) \triangleq \mathrm{diag}\left\{\Delta_{1,\varsigma(k)}(k), \ldots, \Delta_{m,\varsigma(k)}(k)\right\}$, $\Delta_{l,\varsigma(k)}(k) \in \left[-\zeta_{l,\varsigma(k)}, \zeta_{l,\varsigma(k)}\right]$, $\zeta_{l,\varsigma(k)} \triangleq \frac{1-\rho_{l,\varsigma(k)}}{1+\rho_{l,\varsigma(k)}}$, and $\rho_{l,\varsigma(k)} \triangleq e^{-\frac{2}{\varrho_{l,\varsigma(k)}}}$ $(l = 1, 2, \ldots, m)$ with the Markovian jump quantization density $\varrho_{l,\varsigma(k)}$;

- *for the Markovian packet losses [131], the model (8.6) is set as $\varsigma(k) \in \{1, 2\}$, where the mode $\varsigma(k) = 1$ means that the packet is received, i.e., $\Gamma_1 = I_m$, $\Phi_1 = 0_m$, and the mode $\varsigma(k) = 2$ implies that the packet is lost, i.e., $\Gamma_2 = 0_m$, $\Phi_2 = 0_m$.*

Remark 8.2 *Over the past decade, the SMC problem subject to actuator degradation has been addressed well as in [207]. Unfortunately, in order to ensure the full column rank condition of control gain matrix under the actuator degradation, the actuator outage case is always excluded in the existing literature [207]. In this chapter, the fading model (8.6) covers the actuator outage as a special case in a stochastic setting, that is, the stochastic variable $\vartheta_{\varsigma(k)}(k)$ may be equal to zero at some sampling instants as shown in the simulation example later. This fact just shows a key contribution of this chapter.*

As depicted in Fig. 8.2, the mode detector is utilized to estimate the actual network mode $\varsigma(k)$ and also emit an estimated mode signal $\theta(k) \in \Xi \triangleq \{1, 2, \ldots, L\}$ to the controller with the mode detection probability δ_{iq} given by

$$\delta_{iq} \triangleq \Pr\left\{\theta(k) = q \mid \varsigma(k) = i\right\}, \tag{8.9}$$

where $\delta_{iq} \in [0, 1]$ and $\sum_{q=1}^{L} \delta_{iq} = 1$. We denote the mode detection probability matrix as $\Omega \triangleq [\delta_{iq}]_{i\in\Theta, q\in\Xi}$. It is easy to see that the detector (8.9) covers the following three special cases:

- Synchronous observation of network mode: $\Theta = \Xi$, and $\delta_{ii} = 1$ ($\theta(k) = \varsigma(k)$ in probability 1);

- Asynchronous observation of network mode: $\Xi \neq I$ ($\theta(k) = \varsigma(k)$ with some probabilities $0 < \delta_{ii} < 1$);

Main Results 143

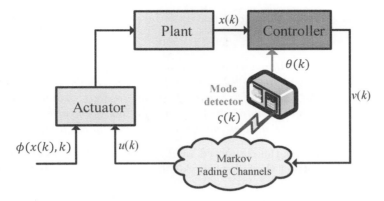

FIGURE 8.2: Reliable SMC over hidden Markov fading channels.

- No observation/estimation of network mode: $L = 1$ ($\theta(k) = 1$ in probability 1).

Remark 8.3 *In fact, the asynchronous network mode observation can be interpreted as the* multi-*step delayed network mode observation in the sense of probability, where the mode detection probabilities* $\{\delta_{iq}\}$ *are obtained through the following statistical method via the channel field measurements [122]:*

$$\delta_{iq} = \lim_{n_i \to \infty} \frac{l_{iq}}{n_i}, \; i \in \Theta, \; q \in \Xi,$$

where $n_i \in \mathbb{N}$ *is the times of network mode* $\varsigma(k) = i$, *and* $l_{iq} \in \mathbb{N}$ *is the times of detector mode* $\theta(k) = q$ *under the case of system mode* $\varsigma(k) = i$. *The utilization of the hidden Markov mode detector (8.9) to the FSMFCs is motivated by [27, 112, 137]. For presentation convenience, the novel FSMFCs (8.5)–(8.9) is named as the HMFCs hereafter.*

In this chapter, we are interested in developing a SMC input $v(k)$ by just using the estimated network mode signal $\theta(k)$ such that the resultant closed-loop system is mean-square stable subject to the HMFCs (8.5)–(8.9).

8.2 Main Results

8.2.1 Sliding function and sliding mode controller

In this chapter, the following linear sliding function is utilized:

$$s(k) = Tx(k) \tag{8.10}$$

144 *State-Saturated SMC under MPTP*

where $T \in \mathbb{R}^{m \times n}$ is a sliding gain matrix to be determined later such that

$$\det(TB) \neq 0. \tag{8.11}$$

Remark 8.4 *It is worth mentioning that the mode-dependent sliding surfaces as in [97] may provide more freedom to design the mode-dependent controller. However, due to the switching frequently from one mode to another, the reachability of mode-dependent sliding surfaces may not always be attained actually. To this end, this chapter utilizes the common sliding function (8.10).*

By only employing the estimated mode signal $\theta(k) \in \Xi$ to the network mode $\varsigma(k)$, we construct the following SMC law for the state-saturated system (8.1):

$$v(k) = K_{\theta(k)}x(k) - \varphi\|x(k)\| \cdot \text{sgn}(s(k)) \tag{8.12}$$

where the matrices $K_{\theta(k)} \in \mathbb{R}^{m \times n}$ will be designed later.

Remark 8.5 *It is well recognized that the chattering is the main drawback of the SMC approach. Notice that the matched external disturbance $\phi(x(k), k)$ in the system (8.1) is supposed to satisfy Lipchitz condition so that the discontinuous term can have an adaptive gain. This feature is actually benefit for chattering alleviation.*

Combining (8.1), (8.6) and (8.12), we have the following closed-loop system:

$$x(k+1) = \sigma\Big((A + \Delta A(k) + B\vartheta_i(k)K_q)\, x(k) + B\varepsilon_i(k)\Big), \tag{8.13}$$

where $i \in \Theta$, $q \in \Xi$ and

$$\varepsilon_i(k) \triangleq \phi(x(k), k) - \varphi\|x(k)\| \cdot \vartheta_i(k)\text{sgn}(s(k)). \tag{8.14}$$

Definition 8.1 *[27] The closed-loop system (8.13) is said to be mean-square stable if, for any initial condition $x(0) \neq 0$, the condition $\lim_{k \to \infty} \mathbb{E}\left\{\|x(k)\|^2\right\}|_{x(0)} = 0$ holds.*

The following three technical lemmas will be useful in the subsequent investigation on the mean-square stability of the closed-loop system (8.13) and the reachability for the specified sliding surface (8.10).

Lemma 8.1 *[31, 68] Let $\mathbb{Z}_n$ be the set of $n \times n$ diagonal matrices whose diagonal elements are either 1 or 0. i) There are 2^n elements in $\mathbb{Z}_n$ where its rth element is denoted as Z_r, $r \in \mathbb{M}_n \triangleq \{1, 2 \ldots, 2^n\}$. ii) By defining $Z_r^- \triangleq I - Z_r$ and letting $G \in \mathbb{R}^{n \times n}$ satisfy $\|G\|_\infty \leq 1$, for any vector $w \in \mathbb{R}^n$, we have*

$$\Lambda^{-1}\sigma\Big(Ax(k) + w\Big) \in \text{co}\Big\{Z_r\Lambda^{-1}\big(Ax(k) + w\big) + Z_r^- G\Lambda^{-1}x(k)\Big\}, \quad r \in \mathbb{M}_n$$

where the matrix Λ is defined in (8.4), and $\text{co}\{\cdot\}$ denotes the convex hull of a set.

Main Results 145

Lemma 8.2 *For any* $\varsigma(k) = i \in \Theta$*, the stochastic variable* $\varepsilon_i(k)$ *in (8.14)*
satisfies

$$\mathbb{E}\left\{\varepsilon_i^{\mathrm{T}}(k)\varepsilon_i(k) \mid x(k)\right\} \leq \left(2 + 2\|\Gamma_i\|^2 + \mathrm{tr}(\Psi_i)\right)\varphi^2 x^{\mathrm{T}}(k)x(k).$$

Proof *For any* $i \in \Theta$*, it yields from (8.14) that*

$$\begin{aligned}
&\mathbb{E}\left\{\varepsilon_i^{\mathrm{T}}(k)\varepsilon_i(k) \mid x(k)\right\}\\
=&\|\phi(x(k),k) - \varphi\|x(k)\| \cdot \Gamma_i\mathrm{sgn}(s(k))\|^2 + \varphi^2\|x(k)\|^2\mathrm{sgn}^{\mathrm{T}}(s(k))\Psi_i\mathrm{sgn}(s(k))\\
\leq&\left(2 + 2\|\Gamma_i\|^2 + \mathrm{tr}(\Psi_i)\right)\varphi^2 x^{\mathrm{T}}(k)x(k),
\end{aligned}$$

which completes the proof.

Lemma 8.3 *Let* X*,* Y *and* Z *be positive semidefinite matrices. If* $X \leq Y$*,*
then the inequality $Z \circ X \leq Z \circ Y$ *holds.*

Proof *Clearly,* $Y - X \geq 0$*. Then, according to Theorem 7.5.3 in [60], we*
have $Z \circ (Y - X) \geq 0$*, which is equivalent to* $Z \circ X \leq Z \circ Y$*. The proof is*
complete.

8.2.2 Analysis of sliding mode dynamics

In the following theorem, we analyze the mean-square stability of the closed-
loop system (8.13) by utilizing a fading-related state-dependent Lyapunov
function.

Theorem 8.1 *Consider the networked state-saturated system (8.1) and the*
SMC law (8.12). The closed-loop system (8.13) is mean-square stable if, for
any $i \in \Theta$*,* $q \in \Xi$ *and* $r \in \mathrm{M}_n$*, there exist matrices* $P_i > 0$*,* $K_q \in \mathbb{R}^{m \times n}$*,*
$G \in \mathbb{R}^{n \times n}$*, and scalars* $\gamma_{ir} > 0$ *such that the following matrix inequalities*
hold:

$$\|G\|_\infty \leq 1, \tag{8.15}$$

$$B^{\mathrm{T}}\Lambda^{-1}Z_r\Lambda\mathbb{P}_i\Lambda Z_r\Lambda^{-1}B \leq \gamma_{ir}I, \tag{8.16}$$

$$\begin{aligned}
&- P_i + 2\gamma_{ir}\left(2 + 2\|\Gamma_i\|^2 + \mathrm{tr}(\Psi_i)\right)\varphi^2 I + 2\sum_{q=1}^{L}\delta_{iq}\Big\{\mathbf{Z}_{irq}^{\mathrm{T}}\Lambda\mathbb{P}_i\Lambda\mathbf{Z}_{irq}\\
&+ K_q^{\mathrm{T}}\left(\Phi_i \circ B^{\mathrm{T}}\Lambda^{-1}Z_r\Lambda\mathbb{P}_i\Lambda Z_r\Lambda^{-1}B\right)K_q\Big\} < 0, \tag{8.17}
\end{aligned}$$

where $\mathbb{P}_i \triangleq \sum_{j=1}^{N}\pi_{ij}P_j$*, and* $\mathbf{Z}_{irq} \triangleq Z_r\Lambda^{-1}\left(A + \Delta A(k)\right) + Z_r^-G\Lambda^{-1} +$
$Z_r\Lambda^{-1}B\Gamma_iK_q$*.*

Proof *For the network mode* $\varsigma(k) = i \in \Theta$*, we consider the following*
Lyapunov function:

$$V(x(k),\varsigma(k)) \triangleq x^{\mathrm{T}}(k)P_{\varsigma(k)}x(k). \tag{8.18}$$

146   *State-Saturated SMC under MPTP*

Exploiting the conditional expectation and Lemma 8.1, we have the following relationships for $\varsigma(k) = i$:

$$\mathbb{E}\{\Delta V(x(k), \varsigma(k)) \mid x(k)\}$$
$$\triangleq \mathbb{E}\{x^{\mathrm{T}}(k+1)P_{\varsigma(k+1)}x(k+1) \mid x(k)\} - x^{\mathrm{T}}(k)P_i x(k)$$
$$= \mathbb{E}\{x^{\mathrm{T}}(k+1)\mathbb{P}_i x(k+1) \mid x(k)\} - x^{\mathrm{T}}(k)P_i x(k)$$
$$= \sum_{q=1}^{L} \delta_{iq} \mathbb{E}\Big\{\sigma^{\mathrm{T}}\Big(\big(A + \Delta A(k) + B\vartheta_i(k)K_q\big) x(k) + B\varepsilon_i(k)\Big)$$
$$\times \mathbb{P}_i \sigma\Big(\big(A + \Delta A(k) + B\vartheta_i(k)K_q\big) x(k) + B\varepsilon_i(k)\Big) \mid x(k)\Big\} - x^{\mathrm{T}}(k)P_i x(k)$$
$$= \sum_{q=1}^{L} \delta_{iq} \mathbb{E}\Big\{\Big\{ \sum_{r=1}^{2^n} \alpha_r \Big[Z_r \Lambda^{-1}\Big(\big(A + \Delta A(k) + B\vartheta_i(k)K_q\big) x(k) + B\varepsilon_i(k)\Big)$$
$$+ Z_r^- G\Lambda^{-1} x(k)\Big]^{\mathrm{T}}\Big\}\Lambda\mathbb{P}_i\Lambda\Big\{ \sum_{r=1}^{2^n} \alpha_r \Big[Z_r \Lambda^{-1}\Big(\big(A + \Delta A(k)$$
$$+ B\vartheta_i(k)K_q\big) x(k) + B\varepsilon_i(k)\Big) + Z_r^- G\Lambda^{-1} x(k)\Big]\Big\} \mid x(k)\Big\} - x^{\mathrm{T}}(k)P_i x(k)$$
$$\leq \max_{r \in \mathbb{M}_n} \Big\{ \sum_{q=1}^{L} \delta_{iq} \mathbb{E}\Big\{ \Big[Z_r \Lambda^{-1}\Big(\big(A + \Delta A(k) + B\vartheta_i(k)K_q\big) x(k) + B\varepsilon_i(k)\Big)$$
$$+ Z_r^- G\Lambda^{-1} x(k)\Big]^{\mathrm{T}} \Lambda\mathbb{P}_i\Lambda\Big[Z_r\Lambda^{-1}\Big(\big(A + \Delta A(k) + B\vartheta_i(k)K_q\big) x(k) + B\varepsilon_i(k)\Big)$$
$$+ Z_r^- G\Lambda^{-1} x(k)\Big] \mid x(k)\Big\}\Big\} - x^{\mathrm{T}}(k)P_i x(k)$$
$$\leq \max_{r \in \mathbb{M}_n} \Big\{ \sum_{q=1}^{L} \delta_{iq} \Big\{ 2\mathbb{E}\Big\{ x^{\mathrm{T}}(k)\Big[Z_r\Lambda^{-1}\big(A + \Delta A(k)\big) + Z_r^- G\Lambda^{-1}$$
$$+ Z_r\Lambda^{-1}B\vartheta_i(k)K_q\Big]^{\mathrm{T}} \Lambda\mathbb{P}_i\Lambda\Big[Z_r\Lambda^{-1}\big(A + \Delta A(k)\big) + Z_r^- G\Lambda^{-1}$$
$$+ Z_r\Lambda^{-1}B\vartheta_i(k)K_q\Big] x(k) \mid x(k)\Big\}$$
$$+ 2\mathbb{E}\Big\{ \varepsilon_i^{\mathrm{T}}(k)B^{\mathrm{T}}\Lambda^{-1}Z_r\Lambda\mathbb{P}_i\Lambda Z_r\Lambda^{-1}B\varepsilon_i(k) \mid x(k)\Big\}\Big\} - x^{\mathrm{T}}(k)P_i x(k)\Big\},$$

$$(8.19)$$

where $\alpha_r \geq 0$ and $\sum_{r=1}^{2^n} \alpha_r = 1$.

Main Results 147

It follows from (8.16) and Lemma 8.2 that

$$\mathbb{E}\left\{\varepsilon_i^{\mathrm{T}}(k)B^{\mathrm{T}}\Lambda^{-1}Z_r\Lambda\mathbb{P}_i\Lambda Z_r\Lambda^{-1}B\varepsilon_i(k)\mid x(k)\right\}$$
$$\leq\gamma_{ir}\left(2+2\|\Gamma_i\|^2+\mathrm{tr}(\Psi_i)\right)\varphi^2 x^{\mathrm{T}}(k)x(k). \tag{8.20}$$

Then, for $\varsigma(k)=i$, substituting (8.20) into (8.19) gives

$$\mathbb{E}\left\{\Delta V(x(k),\varsigma(k))\mid x(k)\right\}$$

$$\leq \max_{r\in\mathbb{M}_n}\left\{\sum_{q=1}^{L}\delta_{iq}\left\{2\left\{x^{\mathrm{T}}(k)\left[Z_r\Lambda^{-1}\left(A+\Delta A(k)\right)+Z_r^- G\Lambda^{-1}+Z_r\Lambda^{-1}B\Gamma_i K_q\right]^{\mathrm{T}}\right.\right.$$

$$\times \Lambda\mathbb{P}_i\Lambda\left[Z_r\Lambda^{-1}\left(A+\Delta A(k)\right)+Z_r^- G\Lambda^{-1}+Z_r\Lambda^{-1}B\Gamma_i K_q\right]x(k)$$

$$\left.+x^{\mathrm{T}}(k)K_q^{\mathrm{T}}\left(\Phi_i\circ B^{\mathrm{T}}\Lambda^{-1}Z_r\Lambda\mathbb{P}_i\Lambda Z_r\Lambda^{-1}B\right)K_q x(k)\right\}$$

$$\left.\left.+2\gamma_{ir}\left(2+2\|\Gamma_i\|^2+\mathrm{tr}(\Psi_i)\right)\varphi^2 x^{\mathrm{T}}(k)x(k)\right\}-x^{\mathrm{T}}(k)P_i x(k)\right\}. \tag{8.21}$$

It is easily shown that, for any $\varsigma(k)=i\in\Theta$, $\mathbf{E}\{\Delta V(x(k),i)\mid x(k)\}<0$ if the condition (8.17) holds, and this completes the proof.

Remark 8.6 *In the discrete-time setting of the SMC theory, the equivalent control law has been widely employed to analyze the stability of the sliding mode dynamics, see [111, 133] for example. Unfortunately, it follows from (8.1), (8.6), and (8.10) that*

$$s(k+1)=T\sigma\left(\left(A+\Delta A(k)\right)x(k)+B\left(\vartheta_{\varsigma(k)}(k)v(k)+\phi(x(k),k)\right)\right),$$

from which one can see that the equivalent control law approach cannot be employed in this chapter due to the co-existence of the state-saturation function $\sigma(\cdot)$ and the channel fading variable $\vartheta_{\varsigma(k)}(k)$, and this gives rise to substantial challenges regarding the stability analysis of the SMC systems subject to state saturation and HMFCs. With the aid of Lemmas 8.1 and 8.2 as well as the HMM approach, sufficient conditions are established such that the closed-loop system (8.13) is mean-square stable under the SMC law (8.12).

8.2.3 Analysis of reachability

This subsection is devoted to the analysis of the reachability for the specified sliding surface (8.10). First, it follows from (8.10) and (8.13) that

$$s(k+1)=T\sigma\left(\left(A+\Delta A(k)+B\vartheta_i(k)K_q\right)x(k)+B\varepsilon_i(k)\right). \tag{8.22}$$

Next, define the following *time-varying* sliding domain:

$$\mathcal{O}\triangleq\left\{s(k)\mid \|s(k)\|\leq\tilde{\epsilon}(k)\right\} \tag{8.23}$$

where

$$\tilde{\epsilon}(k) \triangleq \max_{i\in\Theta, r\in\mathbb{M}_n} \left\{ \sqrt{\frac{\epsilon_{ir}(k)}{\lambda_{\min}(W_i)}} \right\},$$

$$\epsilon_{ir}(k) \triangleq 2\Big[\lambda_{\max}\left(B^{\mathrm{T}}\Lambda^{-1}Z_r\Lambda\mathbb{P}_i\Lambda Z_r\Lambda^{-1}B\right)$$
$$+ \lambda_{\max}\left(B^{\mathrm{T}}\Lambda^{-1}Z_r\Lambda T^{\mathrm{T}}\mathbb{W}_iT\Lambda Z_r\Lambda^{-1}B\right)\Big]$$
$$\times \left(2 + 2\|\Gamma_i\|^2 + \mathrm{tr}(\Psi_i)\right)\varphi^2\|x(k)\|^2$$

with the matrix $\mathbb{P}_i$ being defined in Theorem 8.1, and the matrices W_i and $\mathbb{W}_i$ being defined in Theorem 8.2.

Clearly, $\mathcal{O}$ is just a *vicinity* of the sliding surface (8.10), i.e., the quasi-sliding mode (QSM) around the sliding surface $s(k) = 0$. In the following theorem, we will propose a sufficient condition to ensure the above QSM motion for the state-saturated system (8.1) subject to the HMFCs (8.5)–(8.9).

Theorem 8.2 *Consider the networked state-saturated system (8.1) and the SMC law (8.12). For any $i \in \Theta$, $q \in \Xi$ and $r \in \mathbb{M}_n$, assume that there exist matrices $P_i > 0$, $W_i > 0$, $K_q \in \mathbb{R}^{m\times n}$, $T \in \mathbb{R}^{m\times n}$, and $G \in \mathbb{R}^{n\times n}$ satisfying the condition (8.15) and the following matrix inequalities:*

$$- P_i + 2\sum_{q=1}^{L} \delta_{iq}\Big\{\mathbf{Z}_{irq}^{\mathrm{T}}\Lambda\mathbb{P}_i\Lambda\mathbf{Z}_{irq} + \mathbf{Z}_{irq}^{\mathrm{T}}\Lambda T^{\mathrm{T}}\mathbb{W}_iT\Lambda\mathbf{Z}_{irq}$$
$$+ K_q^{\mathrm{T}}\left(\Phi_i \circ B^{\mathrm{T}}\Lambda^{-1}Z_r\Lambda\mathbb{P}_i\Lambda Z_r\Lambda^{-1}B\right)K_q$$
$$+ K_q^{\mathrm{T}}\left(\Phi_i \circ B^{\mathrm{T}}\Lambda^{-1}Z_r\Lambda T^{\mathrm{T}}\mathbb{W}_iT\Lambda Z_r\Lambda^{-1}B\right)K_q\Big\} < 0 \qquad (8.24)$$

where $\mathbb{W}_i \triangleq \sum_{j=1}^{N} \pi_{ij}W_j$, and $\mathbb{P}_i$, $\mathbf{Z}_{irq}$ are defined in Theorem 8.1. Then, the SMC law (8.12) can force the state trajectories of the closed-loop system (8.13) into the sliding region $\mathcal{O}$ in mean-square sense.

Proof *Consider the following extended Lyapunov functional:*

$$\tilde{V}(x(k), \varsigma(k)) \triangleq x^{\mathrm{T}}(k)P_{\varsigma(k)}x(k) + s^{\mathrm{T}}(k)W_{\varsigma(k)}s(k). \qquad (8.25)$$

Along with the solution of the sliding function (8.22), for any $\varsigma(k) = i \in \Theta$,

Main Results 149

one has from Lemma 8.1 that

$$\mathbb{E}\left\{s^{\mathrm{T}}(k+1)W(\varsigma(k+1))s(k+1) \mid x(k)\right\}$$

$$=\sum_{q=1}^{L}\delta_{iq}\mathbb{E}\left\{\left\{\sum_{r=1}^{2^n}\alpha_r\left[Z_r\Lambda^{-1}\left((A+\Delta A(k)+B\vartheta_i(k)K_q)\,x(k)+B\varepsilon_i(k)\right)\right.\right.\right.$$

$$\left.\left.+Z_r^-G\Lambda^{-1}x(k)\right]^{\mathrm{T}}\right\}\Lambda T^{\mathrm{T}}\mathbb{W}_iT\Lambda\left\{\sum_{r=1}^{2^n}\alpha_r\left[Z_r\Lambda^{-1}\left((A+\Delta A(k)\right.\right.\right.$$

$$\left.\left.\left.+B\vartheta_i(k)K_q)\,x(k)+B\varepsilon_i(k)\right)+Z_r^-G\Lambda^{-1}x(k)\right]\right\}\mid x(k)\right\}$$

$$\leq \max_{r\in\mathbb{M}_n}\left\{\sum_{q=1}^{L}\delta_{iq}\left\{2\mathbb{E}\left\{x^{\mathrm{T}}(k)\left[Z_r\Lambda^{-1}(A+\Delta A(k))+Z_r^-G\Lambda^{-1}\right.\right.\right.\right.$$

$$\left.+Z_r\Lambda^{-1}B\vartheta_i(k)K_q\right]^{\mathrm{T}}\Lambda T^{\mathrm{T}}\mathbb{W}_iT\Lambda\left[Z_r\Lambda^{-1}(A+\Delta A(k))+Z_r^-G\Lambda^{-1}\right.$$

$$\left.\left.+Z_r\Lambda^{-1}B\vartheta_i(k)K_q\right]x(k)\mid x(k)\right\}$$

$$\left.\left.+2\mathbb{E}\left\{\varepsilon_i^{\mathrm{T}}(k)B^{\mathrm{T}}\Lambda^{-1}Z_r\Lambda T^{\mathrm{T}}\mathbb{W}_iT\Lambda Z_r\Lambda^{-1}B\varepsilon_i(k)\mid x(k)\right\}\right\}\right\}$$

$$\leq \max_{r\in\mathbb{M}_n}\left\{\sum_{q=1}^{L}\delta_{iq}\left\{2\left\{x^{\mathrm{T}}(k)\left[Z_r\Lambda^{-1}(A+\Delta A(k))+Z_r^-G\Lambda^{-1}\right.\right.\right.\right.$$

$$\left.+Z_r\Lambda^{-1}B\Gamma_iK_q\right]^{\mathrm{T}}\Lambda T^{\mathrm{T}}\mathbb{W}_iT\Lambda\left[Z_r\Lambda^{-1}(A+\Delta A(k))+Z_r^-G\Lambda^{-1}\right.$$

$$\left.+Z_r\Lambda^{-1}B\Gamma_iK_q\right]x(k)+x^{\mathrm{T}}(k)K_q^{\mathrm{T}}\left(\Phi_i\circ B^{\mathrm{T}}\Lambda^{-1}Z_r\Lambda T^{\mathrm{T}}\mathbb{W}_iT\Lambda Z_r\Lambda^{-1}B\right)$$

$$\left.\times K_qx(k)\right\}+2\lambda_{\max}\left(B^{\mathrm{T}}\Lambda^{-1}Z_r\Lambda T^{\mathrm{T}}\mathbb{W}_iT\Lambda Z_r\Lambda^{-1}B\right)$$

$$\times \left(2+2\|\Gamma_i\|^2+\mathrm{tr}(\Psi_i)\right)\varphi^2\|x(k)\|^2\right\}\right\}, \tag{8.26}$$

where $\alpha_r\geq 0$ and $\sum_{r=1}^{2^n}\alpha_r=1$.

Bearing (8.19) and (8.26) in mind, we obtain the following relationships for any $\varsigma(k)=i\in\Theta$:

$$\mathbb{E}\left\{\Delta\tilde{V}(x(k),\varsigma(k)) \mid x(k)\right\}$$

$$\triangleq\mathbb{E}\left\{\Delta V(x(k),\varsigma(k)) \mid x(k)\right\}+\mathbb{E}\left\{s^{\mathrm{T}}(k+1)W(\varsigma(k+1))s(k+1) \mid x(k)\right\}$$

$$-s^{\mathrm{T}}(k)W_{\varsigma(k)}s(k)$$

$$\leq \max_{r\in\mathbb{M}_n}\left\{\sum_{q=1}^{L}\delta_{iq}\left\{2\left\{x^{\mathrm{T}}(k)\left[Z_r\Lambda^{-1}(A+\Delta A(k))+Z_r^-G\Lambda^{-1}\right.\right.\right.\right.$$

$$
+ Z_r\Lambda^{-1}B\Gamma_iK_q\Big]^{\mathrm{T}}\Lambda\mathbb{P}_i\Lambda\Big[Z_r\Lambda^{-1}\big(A+\Delta A(k)\big)+Z_r^-G\Lambda^{-1}
$$

$$
+ Z_r\Lambda^{-1}B\Gamma_iK_q\Big]x(k)+x^{\mathrm{T}}(k)K_q^{\mathrm{T}}\big(\Phi_i\circ B^{\mathrm{T}}\Lambda^{-1}Z_r\Lambda\mathbb{P}_i\Lambda Z_r\Lambda^{-1}B\big)K_qx(k)\bigg\}
$$

$$
+ 2\bigg\{x^{\mathrm{T}}(k)\Big[Z_r\Lambda^{-1}\big(A+\Delta A(k)\big)+Z_r^-G\Lambda^{-1}+Z_r\Lambda^{-1}B\Gamma_iK_q\Big]^{\mathrm{T}}
$$

$$
\times \Lambda T^{\mathrm{T}}\mathbb{W}_iT\Lambda\Big[Z_r\Lambda^{-1}\big(A+\Delta A(k)\big)+Z_r^-G\Lambda^{-1}+Z_r\Lambda^{-1}B\Gamma_iK_q\Big]x(k)
$$

$$
+ x^{\mathrm{T}}(k)K_q^{\mathrm{T}}\big(\Phi_i\circ B^{\mathrm{T}}\Lambda^{-1}Z_r\Lambda T^{\mathrm{T}}\mathbb{W}_iT\Lambda Z_r\Lambda^{-1}B\big)K_qx(k)\bigg\}\bigg\}
$$

$$
- x^{\mathrm{T}}(k)P_ix(k)-\lambda_{\min}\big(W_i\big)\|s(k)\|^2+\epsilon_{ir}(k)\bigg\}. \tag{8.27}
$$

If the following is true:

$$
\|s(k)\|>\tilde{\epsilon}(k)\geq\sqrt{\frac{\epsilon_{ir}(k)}{\lambda_{\min}(W_i)}},\ \forall i\in\Theta,\ r\in\mathbb{M}_n,
$$

that is, the state trajectories escape the region $\mathcal{O}$, then we can have from (8.24) and (8.27) that

$$
\mathbb{E}\Big\{\Delta\tilde{V}(x(k),\varsigma(k))\mid x(k)\Big\}
$$

$$
\leq \max_{r\in\mathbb{M}_n}\bigg\{\sum_{q=1}^{L}\delta_{iq}\bigg\{2\Big\{x^{\mathrm{T}}(k)\Big[Z_r\Lambda^{-1}\big(A+\Delta A(k)\big)+Z_r^-G\Lambda^{-1}
$$

$$
+ Z_r\Lambda^{-1}B\Gamma_iK_q\Big]^{\mathrm{T}}\Lambda\mathbb{P}_i\Lambda\Big[Z_r\Lambda^{-1}\big(A+\Delta A(k)\big)+Z_r^-G\Lambda^{-1}
$$

$$
+ Z_r\Lambda^{-1}B\Gamma_iK_q\Big]x(k)+x^{\mathrm{T}}(k)K_q^{\mathrm{T}}\big(\Phi_i\circ B^{\mathrm{T}}\Lambda^{-1}Z_r\Lambda\mathbb{P}_i\Lambda Z_r\Lambda^{-1}B\big)K_qx(k)\Big\}
$$

$$
+ 2\Big\{x^{\mathrm{T}}(k)\Big[Z_r\Lambda^{-1}\big(A+\Delta A(k)\big)+Z_r^-G\Lambda^{-1}+Z_r\Lambda^{-1}B\Gamma_iK_q\Big]^{\mathrm{T}}
$$

$$
\times \Lambda T^{\mathrm{T}}\mathbb{W}_iT\Lambda\Big[Z_r\Lambda^{-1}\big(A+\Delta A(k)\big)+Z_r^-G\Lambda^{-1}+Z_r\Lambda^{-1}B\Gamma_iK_q\Big]x(k)
$$

$$
+ x^{\mathrm{T}}(k)K_q^{\mathrm{T}}\big(\Phi_i\circ B^{\mathrm{T}}\Lambda^{-1}Z_r\Lambda T^{\mathrm{T}}\mathbb{W}_iT\Lambda Z_r\Lambda^{-1}B\big)K_qx(k)\Big\}\bigg\}-x^{\mathrm{T}}(k)P_ix(k)\bigg\}
$$

$$
<0,\ \forall i\in\Theta, \tag{8.28}
$$

which means that, outside the region $\mathcal{O}$ defined in (8.23), the state trajectories of the closed-loop system (8.13) are strictly deceasing in mean-square sense. *The proof is now complete.*

8.2.4 Synthesis of SMC law

Actually, it is quite difficult to directly solve the inequalities in Theorems 8.1 and 8.2 due to the coupling terms and the Hadamard product terms in (8.17) and (8.24). In the following theorem, by means of the property of the Hadamard product in Lemma 8.3, we establish a sufficient condition to guarantee the mean-square stability of the closed-loop system (8.13) and the reachability of the specified sliding surface (8.10) simultaneously.

Theorem 8.3 *For any $i \in \Theta$, $q \in \Xi$ and $r \in \mathbb{M}_n$, assume that there exist $\bar{P}_i > 0$, $\bar{W}_i > 0$, $\bar{Q}_{irq} > 0$, $\mathcal{K}_q \in \mathbb{R}^{m \times n}$, $J_q \in \mathbb{R}^{n \times n}$, $T \in \mathbb{R}^{m \times n}$, $G \in \mathbb{R}^{n \times n}$, and scalars $\bar{\gamma}_{ir} > 0$, $\bar{\zeta}_{ir} > 0$, $\chi > 0$ such that the following coupled matrix inequalities hold:*

$$\|G\|_\infty \leq 1, \tag{8.29}$$

$$\begin{bmatrix} -\bar{\gamma}_{ir}I & \mathcal{B}_{ir} \\ \mathcal{B}_{ir}^{\mathrm{T}} & -\mathcal{P} \end{bmatrix} \leq 0, \tag{8.30}$$

$$\begin{bmatrix} -\bar{\zeta}_{ir}I & \mathcal{T}_{ir} \\ \mathcal{T}_{ir}^{\mathrm{T}} & -\mathcal{W} \end{bmatrix} \leq 0, \tag{8.31}$$

$$\begin{bmatrix} \tilde{\mathcal{Q}}_{irq} & \Sigma_{irq} \\ \Sigma_{irq}^{\mathrm{T}} & -\mathcal{I}_{ir} \end{bmatrix} \leq 0, \tag{8.32}$$

$$\begin{bmatrix} -\bar{P}_i & \tilde{\varphi}_i \bar{P}_i & \tilde{\mathcal{P}}_i \\ \tilde{\varphi}_i \bar{P}_i & -\bar{\gamma}_{ir}I & 0 \\ \tilde{\mathcal{P}}_i & 0 & -\mathcal{Q}_{ir} \end{bmatrix} < 0, \tag{8.33}$$

where

$$\mathcal{P} \triangleq \mathrm{diag}\left\{\bar{P}_1, \bar{P}_2, \ldots, \bar{P}_N\right\}, \mathcal{W} \triangleq \mathrm{diag}\left\{\bar{W}_1, \bar{W}_2, \ldots, \bar{W}_N\right\},$$

$$\mathcal{B}_{ir} \triangleq \begin{bmatrix} \bar{\gamma}_{ir}\sqrt{\pi_{i1}}B^{\mathrm{T}}\Lambda^{-1}Z_r\Lambda & \cdots & \bar{\gamma}_{ir}\sqrt{\pi_{iN}}B^{\mathrm{T}}\Lambda^{-1}Z_r\Lambda \end{bmatrix},$$

$$\mathcal{T}_{ir} \triangleq \begin{bmatrix} \bar{\zeta}_{ir}\sqrt{\pi_{i1}}B^{\mathrm{T}}\Lambda^{-1}Z_r\Lambda T^{\mathrm{T}} & \cdots \bar{\zeta}_{ir}\sqrt{\pi_{iN}}B^{\mathrm{T}}\Lambda^{-1}Z_r\Lambda T^{\mathrm{T}} \end{bmatrix},$$

$$\tilde{\varphi}_i \triangleq \varphi\sqrt{2\left(2 + 2\|\Gamma_i\|^2 + \mathrm{tr}(\Psi_i)\right)},$$

$$\tilde{\mathcal{P}}_i \triangleq \begin{bmatrix} \sqrt{2\delta_{i1}}\bar{P}_i & \sqrt{2\delta_{i2}}\bar{P}_i & \cdots & \sqrt{2\delta_{iL}}\bar{P}_i \end{bmatrix},$$

$$\mathcal{Q}_{ir} \triangleq \mathrm{diag}\left\{\bar{Q}_{ir1}, \bar{Q}_{ir2}, \ldots, \bar{Q}_{irL}\right\},$$

$$\tilde{\mathcal{Q}}_{irq} \triangleq \begin{bmatrix} \bar{Q}_{irq} - J_q^{\mathrm{T}} - J_q & \mathcal{Z}_{irq} & \tilde{\mathcal{Z}}_{irq} \\ \mathcal{Z}_{irq}^{\mathrm{T}} & -\mathcal{P} & 0 \\ \tilde{\mathcal{Z}}_{irq}^{\mathrm{T}} & 0 & -\mathcal{W} \end{bmatrix},$$

$$\Sigma_{irq} \triangleq \begin{bmatrix} \mathcal{K}_q^{\mathrm{T}}\Psi_i & \mathcal{K}_q^{\mathrm{T}}\Psi_i & J_q^{\mathrm{T}}N^{\mathrm{T}} & 0 \\ 0 & 0 & 0 & \chi\mathcal{M}_{ir} \\ 0 & 0 & 0 & \chi\tilde{\mathcal{M}}_{ir} \end{bmatrix},$$

$$\mathcal{I}_{ir} \triangleq \mathrm{diag}\left\{\bar{\gamma}_{ir}\Psi_i, \bar{\zeta}_{ir}\Psi_i, \chi I, \chi I\right\},$$

$$\mathcal{Z}_{irq} \triangleq \begin{bmatrix} \sqrt{\pi_{i1}}\tilde{\mathbf{Z}}_{irq} & \sqrt{\pi_{i2}}\tilde{\mathbf{Z}}_{irq} & \cdots & \sqrt{\pi_{iN}}\tilde{\mathbf{Z}}_{irq} \end{bmatrix},$$

$$\tilde{\mathbf{Z}}_{irq} \triangleq \left(Z_r\Lambda^{-1}AJ_q + Z_r^- G\Lambda^{-1}J_q + Z_r\Lambda^{-1}B\Gamma_i\mathcal{K}_q \right)^{\mathrm{T}} \Lambda,$$

$$\vec{\tilde{\mathcal{Z}}}_{irq} \triangleq \left[\ \sqrt{\pi_{i1}}\tilde{\mathbf{Z}}_{irq} \quad \sqrt{\pi_{i2}}\tilde{\mathbf{Z}}_{irq} \quad \cdots \quad \sqrt{\pi_{iN}}\tilde{\mathbf{Z}}_{irq} \ \right],$$

$$\mathbf{Z}_{irq} \triangleq \left(Z_r\Lambda^{-1}AJ_q + Z_r^- G\Lambda^{-1}J_q + Z_r\Lambda^{-1}B\Gamma_i\mathcal{K}_q \right)^{\mathrm{T}} \Lambda T^{\mathrm{T}},$$

$$\mathcal{M}_{ir} \triangleq \left[\ \sqrt{\pi_{i1}}M^{\mathrm{T}}\Lambda^{-1}Z_r\Lambda \quad \cdots \quad \sqrt{\pi_{iN}}M^{\mathrm{T}}\Lambda^{-1}Z_r\Lambda \ \right]^{\mathrm{T}},$$

$$\tilde{\mathcal{M}}_{ir} \triangleq \left[\ \sqrt{\pi_{i1}}M^{\mathrm{T}}\Lambda^{-1}Z_r\Lambda T^{\mathrm{T}} \quad \cdots \sqrt{\pi_{iN}}M^{\mathrm{T}}\Lambda^{-1}Z_r\Lambda T^{\mathrm{T}} \ \right]^{\mathrm{T}}.$$

Then, the SMC law (8.12) with $K_q = \mathcal{K}_q J_q^{-1}$ can guarantee both the mean-square stability of the closed-loop system (8.13) and the QSM with the sliding region $\mathcal{O}$ define in (8.23).

Proof *For any $i \in \Theta$ and $r \in \mathbb{M}_n$, we consider the following inequalities:*

$$\|G\|_\infty \leq 1, \tag{8.34}$$

$$B^{\mathrm{T}}\Lambda^{-1}Z_r\Lambda\mathbb{P}_i\Lambda Z_r\Lambda^{-1}B \leq \gamma_{ir}I, \tag{8.35}$$

$$B^{\mathrm{T}}\Lambda^{-1}Z_r\Lambda T^{\mathrm{T}}\mathbb{W}_iT\Lambda Z_r\Lambda^{-1}B \leq \zeta_{ir}I, \tag{8.36}$$

$$- P_i + 2\gamma_{ir}\left(2 + 2\|\Gamma_i\|^2 + \mathrm{tr}(\Psi_i)\right)\varphi^2 I + 2\sum_{q=1}^{L}\delta_{iq}\Big\{\mathbf{Z}_{irq}^{\mathrm{T}}\Lambda\mathbb{P}_i\Lambda\mathbf{Z}_{irq}$$
$$+ \mathbf{Z}_{irq}^{\mathrm{T}}\Lambda T^{\mathrm{T}}\mathbb{W}_iT\Lambda\mathbf{Z}_{irq} + K_q^{\mathrm{T}}\left(\Phi_i \circ B^{\mathrm{T}}\Lambda^{-1}Z_r\Lambda\mathbb{P}_i\Lambda Z_r\Lambda^{-1}B\right)K_q$$
$$+ K_q^{\mathrm{T}}\left(\Phi_i \circ B^{\mathrm{T}}\Lambda^{-1}Z_r\Lambda T^{\mathrm{T}}\mathbb{W}_iT\Lambda Z_r\Lambda^{-1}B\right)K_q\Big\} < 0. \tag{8.37}$$

It is straightforward to find that the conditions (8.17) and (8.24) are guaranteed by the inequality (8.37), which implies that Theorems 8.1 and 8.2 are ensured by the inequalities (8.34), (8.35), and (8.37).

By using Lemma 8.3 and the definition of the Hadamard product, it follows from the inequalities (8.35) and (8.36) that

$$\Phi_i \circ B^{\mathrm{T}}\Lambda^{-1}Z_r\Lambda\mathbb{P}_i\Lambda Z_r\Lambda^{-1}B \leq \Phi_i \circ \gamma_{ir}I = \gamma_{ir}\Psi_i,$$
$$\Phi_i \circ B^{\mathrm{T}}\Lambda^{-1}Z_r\Lambda T^{\mathrm{T}}\mathbb{W}_iT\Lambda Z_r\Lambda^{-1}B \leq \Phi_i \circ \zeta_{ir}I = \zeta_{ir}\Psi_i.$$

Thus, the inequality (8.37) is ensured by

$$- P_i + 2\gamma_{ir}\left(2 + 2\|\Gamma_i\|^2 + \mathrm{tr}(\Psi_i)\right)\varphi^2 I + 2\sum_{q=1}^{L}\delta_{iq}\Big\{\mathbf{Z}_{irq}^{\mathrm{T}}\Lambda\mathbb{P}_i\Lambda\mathbf{Z}_{irq}$$
$$+ \mathbf{Z}_{irq}^{\mathrm{T}}\Lambda T^{\mathrm{T}}\mathbb{W}_iT\Lambda\mathbf{Z}_{irq} + \gamma_{ir}K_q^{\mathrm{T}}\Psi_iK_q + \zeta_{ir}K_q^{\mathrm{T}}\Psi_iK_q\Big\} < 0. \tag{8.38}$$

Clearly, the following inequalities are sufficient conditions for the inequality

Main Results 153

(8.38) to hold:

$$\mathbf{Z}_{irq}^{\mathrm{T}}\Lambda\mathbb{P}_i\Lambda\mathbf{Z}_{irq} + \mathbf{Z}_{irq}^{\mathrm{T}}\Lambda T^{\mathrm{T}}\mathbb{W}_i T\Lambda\mathbf{Z}_{irq} + \gamma_{ir}K_q^{\mathrm{T}}\Psi_i K_q + \zeta_{ir}K_q^{\mathrm{T}}\Psi_i K_q \leq Q_{irq},$$

$$(8.39)$$

$$- P_i + 2\gamma_{ir}\left(2 + 2\|\Gamma_i\|^2 + \mathrm{tr}(\Psi_i)\right)\varphi^2 I + 2\sum_{q=1}^{L}\delta_{iq}Q_{irq} < 0. \qquad (8.40)$$

Next, we denote $\bar{P}_i \triangleq P_i^{-1}$, $\bar{W}_i \triangleq W_i^{-1}$, $\bar{Q}_{irq} \triangleq Q_{irq}^{-1}$, $\bar{\gamma}_{ir} \triangleq \gamma_{ir}^{-1}$, $\bar{\zeta}_{ir} \triangleq \zeta_{ir}^{-1}$, *and* $\mathcal{K}_q \triangleq K_q J_q$. *By resorting to the Schur complement and the following matrix inequality:*

$$\left(J_q^{\mathrm{T}} - Q_{irq}^{-1}\right)Q_{irq}\left(J_q - Q_{irq}^{-1}\right) \geq 0, i \in \Theta, r \in \mathbb{M}_n, q \in \Xi,$$

it is concluded that the inequalities (8.34)–(8.36) hold because of the conditions (8.29)–(8.31), and the inequalities (8.39)–(8.40) are guaranteed by the conditions (8.32)–(8.33), respectively. The proof is now complete.

8.2.5 Solving algorithm

Notice that it is difficult to solve the proposed SMC design problem by directly exploiting Theorem 8.3 due to the nonlinear constraint conditions (8.11) and (8.29) as well as the nonconvex coupling terms in the condition (8.32), e.g., $J_q^{\mathrm{T}}\Lambda^{-1}G^{\mathrm{T}}Z_r^- \Lambda$, $J_q^{\mathrm{T}}\Lambda^{-1}G^{\mathrm{T}}Z_r^- \Lambda T^{\mathrm{T}}$.

Fortunately, the coupled matrix inequalities (8.30)–(8.33) reduce to be coupled LMIs if the sliding mode matrix $T \in \mathbb{R}^{m \times n}$ and the matrix $G \in \mathbb{R}^{n \times n}$ are specified *a priori*, and the corresponding numerical difficulty scales down to the feasible solution to certain coupled LMIs. On the other hand, it is widely known that many existing evolutionary algorithms are able to effectively deal with various optimization problems subject to *nonlinear* constraints. Some popular evolutionary algorithms include genetic algorithms (GAs) [37, 134] and particle swarm optimization (PSO) algorithms [87]. As such, it is quite natural to combine GAs and LMIs to solve the proposed SMC design problem under the nonlinear constraint conditions (8.11) and (8.29).

In order to take advantage of the GAs, an adequate objective function should be given for a certain formulated optimization problem. Notice that the reachability of the sliding region $\mathcal{O}$ in (8.23) just reflects the SMC performance. Therefore, we consider the following minimization problem for the proposed SMC design:

$$\min_{T,G} \check{\epsilon}$$

subject to: LMIs (8.30)–(8.33),

and constraints (8.11), (8.29), $\qquad$ (8.41)

where

$$\check{\epsilon} \triangleq \max_{i\in\Theta, r\in\mathbb{M}_n} \left\{ \sqrt{\frac{\bar{\epsilon}_{ir}}{\lambda_{\min}(W_i)}} \right\},$$

$$\bar{\epsilon}_{ir} \triangleq 2\Big[\lambda_{\max}\left(B^{\mathrm{T}}\Lambda^{-1}Z_r\Lambda\mathbb{P}_i\Lambda Z_r\Lambda^{-1}B\right) + \lambda_{\max}\left(B^{\mathrm{T}}\Lambda^{-1}Z_r\Lambda T^{\mathrm{T}}\mathbb{W}_i T\Lambda Z_r\Lambda^{-1}B\right) \Big]$$
$$\times \left(2 + 2\|\Gamma_i\|^2 + \mathrm{tr}(\Psi_i)\right)\varphi^2.$$

Remark 8.7 *It is worth mentioning that the nonconvex conditions (8.11) and (8.29) are actually difficulty to be solved directly. In the existing literature, the nonconvex condition (8.11) is always addressed via trial-and-error way or some specific structures, e.g. $L = B^{\mathrm{T}}P$ with Lyapunov matrix P to be determined [111] or $L = B^{\mathrm{T}}X$ with a given matrix $X > 0$ [133, 135]. Besides, the nonconvex condition (8.29) is overcome in the existing literature by trial-and-error way or iterative LMI approach [31, 68]. However, these existing methods may be conservative to the controller design. Compared with the aforementioned results, the present method of introducing GA to the design in this chapter can not only reduce the conservative but also produce an optimized sliding mode controller in the sense of the optimization problem (8.41). This fact actually shows that the proposed GA-assisted SMC design approach is more "smarter" than the ones in the existing literature [111, 133–135].*

Based upon the above objective function, the binary-based GA is formulated in Algorithm 2 to solve the proposed SMC design problem with coupled LMIs.

Remark 8.8 *In this chapter, the SMC problem is investigated for state-saturated systems over the HMFCs. Actually, there are two technical obstacles to solve the specified SMC problem, that is, the state saturation reflecting in the non-convex condition (8.29) and the HMFCs leading to the asynchronous SMC law (8.12). The main results proposed in Theorems 8.1–8.3 exhibit the following distinctive merits: 1) the designed SMC law is dependent on the estimated network mode, which is a kind of non-synchronization with the actual network mode via a HMM; 2) the reachability of a sliding domain around the specified sliding surface is analyzed by a stochastic Lyapunov function; 3) the HMFCs and the state saturation nonlinearities are tackled by using the HMM approach and some properties of the Hadamard product; and 4) the proposed SMC problem subject to nonconvex constraint conditions can be solved effectively by the developed GA in combination with the coupled LMIs.*

Main Results

Algorithm 2 GA-Assisted State-Saturated SMC Design

▶ **Step 1: Parameter encoding.** Denote the sliding gain matrix $T = [t_{ij}]_{m \times n}$ and the matrix $G = [g_{ij}]_{n \times n}$, so there are $mn + n^2$ independent variables. To this end, the phenotype in the search space is expressed as a row vector $w \in \mathbb{R}^{1 \times (mn+n^2)}$:

$$[T, G] \to w \triangleq [t_{11} \ \dots \ t_{1n} \ t_{21} \ \dots \ t_{mn}$$
$$g_{11} \ \cdots \ g_{1n} \ g_{21} \ \cdots \ g_{nn}].$$

In w, each element t_{ij} is coded as a binary string with the length $\ell_{t_{ij}}$ over the range of $t_{ij} \in [\underline{t}_{ij}, \overline{t}_{ij}]$. The precision $q_{t_{ij}}$ under the linearly-mapped coding can be obtained by $q_{t_{ij}} = \frac{\overline{t}_{ij} - \underline{t}_{ij}}{2^{\ell_{t_{ij}}} - 1}$. Furthermore, it is noted that the nonlinear constraint condition (8.29) implies that $g_{ij} \in [-1, 1]$. Hence, for each element g_{ij} in w is coded as a binary string of length $\ell_{g_{ij}}$, the precision $q_{g_{ij}}$ can be computed as $q_{g_{ij}} = \frac{2}{2^{\ell_{g_{ij}}} - 1}$.

▶ **Step 2: Population initialization.** Initial population of Nc chromosomes w_l, $l = 1, 2, \dots, Nc$, is generated randomly.

▶ **Step 3: Fitness function and assignment.** Decode the initial population produced in Step 2 into a real values for every phenotype, and then the fitness function **Fitness**$(T_l, G_l) \triangleq \frac{1}{\tilde{\epsilon}}$ is computed for every T_l and G_l via solving coupled LMIs (8.30)–(8.33). If either the coupled LMIs (8.30)–(8.33) are infeasible or the constraint conditions (8.11), (8.29) are not hold, then the fitness function **Fitness**(T_l, G_l) will be artificially assigned a sufficiently small value (10^{-6} in this chapter) for reducing its opportunity to survive in the next generation.

▶ **Step 4: Performing genetic operations.** According to the assigned fitness in Step 3, we obtain the next population by executing the sequence of genetic operations *Selection*, *Crossover*, and *Mutation*, respectively (more details can be found in [134,143]). Here, we denote the single-point crossover probability as p_c, and a single bit mutation probability as p_m.

▶ **Step 5: Stop criterion.** The evolution process will be repeated from Step 3 to 4 in each generation until the maximum generations $N_{\max}$ is reached. And then, decode the *best* chromosome w_l into real values with producing the sliding gain matrix T and the matrix G.

▶ **Step 6: Design of SMC laws.** Produce the SMC law (8.12) by using the sliding gain matrix T and the gain matrices $K_q = \mathcal{K}_q J_q^{-1}$ obtained from Step 5, and then apply it to stabilize the state-saturated system (8.1) subject to the HMFCs (8.5)–(8.9).

8.3 Example

Consider a state-saturated system in form of (8.1) with

$$A = \begin{bmatrix} -1.5 & 1 \\ -0.3 & 1.3 \end{bmatrix}, \ B = \begin{bmatrix} 1 \\ 1 \end{bmatrix},$$

$$M = \begin{bmatrix} 0.2 & -0.3 \end{bmatrix}^{\mathrm{T}}, \ H = \begin{bmatrix} -0.1 & 0.09 \end{bmatrix},$$

$$F(k) = \cos(k), \ \phi(x(k), k) = 0.2\sqrt{x_1^2 + x_2^2}.$$

The saturation levels are taken as $\bar{\varpi}_1 = \bar{\varpi}_2 = 1$ and the bound coefficient of nonlinear function $\phi(x(k), k)$ is set to be $\varphi = 0.2$. In this example, we utilize a two-mode ("good" and "bad") Markov fading channel with the transition probability matrix:

$$\Pi = \begin{bmatrix} 0.6 & 0.4 \\ 0.4 & 0.6 \end{bmatrix}.$$

When the network channel is in the "good" mode ($\varsigma(k) = 1$), the channel fading variable $\vartheta_1(k)$ has the mathematical expectation $\Gamma_1 = \mu_{1,1} = 0.8$ and variance $\Phi_1 = \Psi_1 = \xi_{11,1} = 0.09$ while, if the network channel is in the "bad" mode ($\varsigma(k) = 2$), the channel fading variable $\vartheta_2(k)$ has the mathematical expectation $\Gamma_2 = \mu_{1,2} = 0.2$ and variance $\Phi_2 = \Psi_2 = \xi_{11,2} = 0.1$.

Now, we utilize a mode detector to estimate the states of the Markov fading channel, where the mode detection probability matrix is given as follows:

$$\Omega = \begin{bmatrix} 0.5 & 0.5 \\ 0.3 & 0.7 \end{bmatrix}.$$

Specifically, when the network channel is in the "good" mode ($\varsigma(k) = 1$), the mode detector has 0.5 probability to emit the "good" mode ($\theta(k) = 1$) or the "bad" mode ($\theta(k) = 2$) to the controller. This is similar to the case that the network channel is in the "bad" mode ($\varsigma(k) = 2$).

The objective of this example is to synthesize an SMC law (8.12) to stabilize the networked state-saturated system subject to the above HMFC. GA is recalled to solve the proposed SMC design problem. In GA, the parameters are set as follows: population size $Nc = 50$; maximum of generations $T_{\max} = 100$; crossover probability $p_c = 0.8$; mutation probability $p_m = 0.1$; bounds of the elements $\bar{t}_{11} = \bar{t}_{12} = 5$ and $\underline{t}_{11} = \underline{t}_{12} = -5$; and lengths of binary strings $\ell_{t_{11}} = \ell_{t_{12}} = 12$ and $\ell_{g_{11}} = \ell_{g_{12}} = \ell_{g_{21}} = \ell_{g_{22}} = 10$. By executing Algorithm

Example

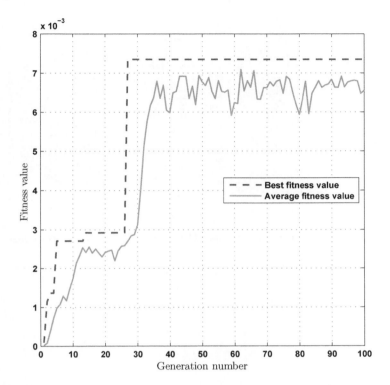

FIGURE 8.3: Fitness value of each generation in solving Algorithm 2.

2, we obtain the following optimized solutions:

$$T = \begin{bmatrix} 3.8938 & 4.5092 \end{bmatrix},$$
$$G = \begin{bmatrix} -0.3568 & 0.2356 \\ 0.6872 & 0.1711 \end{bmatrix},$$
$$v(k) = \begin{cases} \begin{bmatrix} 1.4847 & -1.5954 \end{bmatrix} x(k) - 0.2\|x(k)\| \cdot \mathrm{sgn}(s(k)), & \theta(k) = 1; \\ \begin{bmatrix} 1.6680 & -1.8379 \end{bmatrix} x(k) - 0.2\|x(k)\| \cdot \mathrm{sgn}(s(k)), & \theta(k) = 2. \end{cases}$$
(8.42)

Fig. 8.3 depicts the best and average fitness values of each generation in the computation process of GA.

The simulation results for the initial condition $x(0) = \begin{bmatrix} 1 & -1 \end{bmatrix}^\mathrm{T}$ are given in Figs. 8.4–8.8. It is clear from the Fig. 8.4(a) that the state-saturated system in open-loop case is unstable. As expected, as shown in Fig. 8.4(b), the designed SMC law (8.42) stabilizes the state-saturated system under a possible fading state and detection mode sequence in Fig. 8.5 and the channel fading variable $\vartheta_{\varsigma(k)}(k)$ in Fig. 8.6. The reachability of the sliding region $\mathcal{O}$ in (8.23) is clearly illustrated in Fig. 8.7. Fig. 8.8 shows the SMC signal $v(k)$ and

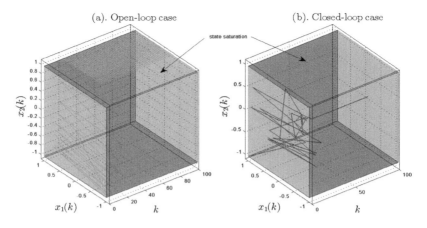

FIGURE 8.4: State trajectories $x(k)$ in open- and closed-loop cases.

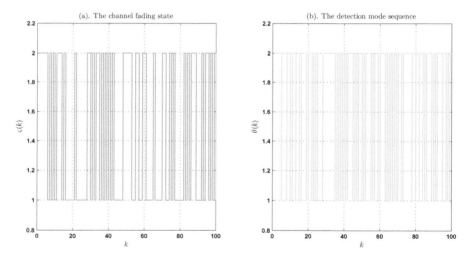

FIGURE 8.5: A possible fading state and detection mode sequence.

the actual control signal $u(k)$ simultaneously. To this end, the effectiveness of the proposed GA-based state-saturated SMC scheme subject to the HMFCs is validated from all obtained simulation results.

Example

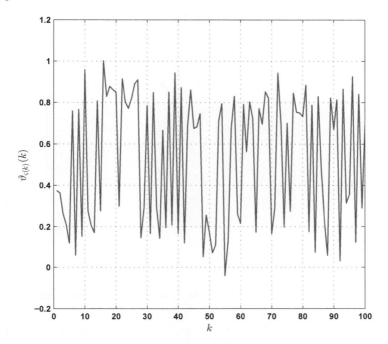

FIGURE 8.6: The channel fading variable $\vartheta_{\varsigma(k)}(k)$.

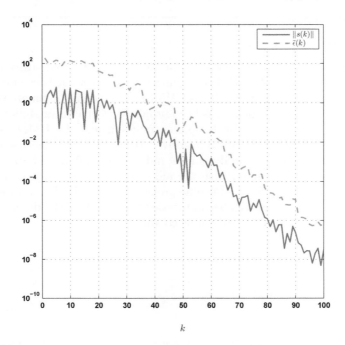

FIGURE 8.7: Illustration of the reachability to the sliding region $\mathcal{O}$ in (8.23).

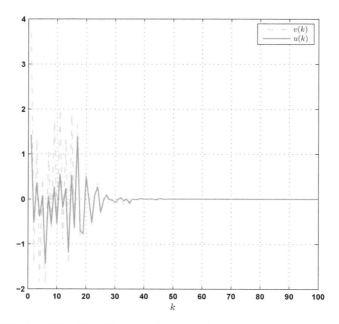

FIGURE 8.8: The SMC input $v(k)$ and the actual control input $u(k)$.

8.4 Conclusion

This chapter has dealt with the SMC problem for networked state-saturated systems over the finite-state HMFCs. The mode detector has been employed to estimate the actual network mode via a HMM. Based on the estimated network modes, a suitable SMC scheme has been proposed with a linear sliding surface. By resorting to the stochastic Lyapunov function and the HMM approach, the sufficient conditions have been derived to ensure both the mean-square stability of the sliding mode dynamics and the reachability of the specified sliding surface. Furthermore, with the aid of the polytopic representation of the saturation nonlinearities and some properties of the Hadamard product, a GA combining with LMIs has been proposed to solve the SMC design problem subject to some nonconvex constraint conditions.

9

ESO-Based Terminal Sliding Mode Control Under Periodic Event-Triggered Protocol

Permanent magnet synchronous motor (PMSM) has been widely used in various fields due to its excellent performance such as high power density, low inertia, and high efficiency. As a conventional linear control scheme, proportional integral differential (PID) control is widely used in PMSM system due to its simple implementation. However, in practical application, the PMSM system is a highly coupled and nonlinear system with unmeasured disturbances. It means that the PID control scheme can hardly achieve the desired high performance control. Therefore, various nonlinear control theories have been adopted to improve the control performances of PMSM system.

Inspired by the aforementioned discussions, in this chapter, we endeavour to proposed a novel *periodic event-triggered terminal sliding mode control (TSMC)* approach and implement it to a speed regulation problem in PMSM. In periodic event-triggered method, the continuous state measurement is no longer required due to the periodic evaluation of the triggering rule. This just means that the triggering time is always an integral multiple of the sampling period which avoids the Zeno phenomenon. Besides, an extended-state-observer (ESO) is introduced to estimate the unknown perturbations in PMSM mainly arising from the external disturbance and the parameter uncertainties. *The main contributions of this chapter can be concluded as follows: 1) Based on a nonsingular terminal sliding function, a novel ESO-based periodic event-triggered TSMC is designed for the networked PMSM speed regulation system, in which the explicit upper-bound of the ESO estimation error is given; 2) The actual bound of the triggering error is analyzed with proposing a proper selection criterion of the periodic sampling period; 3) A design condition for the controller gain is established for ensuring the reachability to a sliding domain and the ultimate boundedness of the closed-loop PMSM speed regulation system; 4) In order to reduce the chattering and the communication burden simultaneously, a binary-based genetic algorithm (GA) is formulated to get the optimized ESO with the ideal estimation error; and 5) The effectiveness of the proposed novel periodic event-triggered TSMC scheme with GA-optimized ESO is demonstrated in both of the simulation and experiment results.*

DOI: 10.1201/9781003309499-9

9.1 Problem Formulation

In this chapter, we consider a surface mounted PMSM. Assume that the rotor of the motor has no damping winding and the permanent magnet has no damping effect, magnetic circuit saturation and the effect of eddy current and hysteresis are neglected, and the distribution of permanent magnetic field in air gap space is sinusoidal. We can obtain the model of PMSM on d-q coordinates as follows [84]:

$$\begin{aligned}
\frac{\mathrm{d}i_d(t)}{\mathrm{d}t} &= \frac{1}{L_d}\left(u_d(t) - R_s i_d(t) + n_p \omega(t) L_q i_q(t)\right) \\
\frac{\mathrm{d}i_q(t)}{\mathrm{d}t} &= \frac{1}{L_q}\left(u_q(t) - R_s i_q(t) - n_p \omega(t) L_d i_d(t) - n_p \omega(t) \psi_f\right) \\
\frac{\mathrm{d}\omega(t)}{\mathrm{d}t} &= \frac{1}{J}\left(1.5 n_p \psi_f i_q(t) - T_L(t) - B_v \omega(t)\right),
\end{aligned} \qquad (9.1)$$

where L_d and L_q with $L_d = L_q$ are the stator inductances of d and q axes, respectively; R_s is the stator resistance; $u_d(t)$, $u_q(t)$ and $i_d(t)$, $i_q(t)$ are the stator voltages and currents of d, q axes, respectively; $\omega(t)$, n_p, and ψ_f are the rotor angular velocity, number of pole pairs, and permanent magnet flux linkage, respectively; and J, $T_L(t)$, and B_v denote the moment of inertia, load torque, and viscous friction coefficient, respectively.

The vector control, which includes a speed loop and two current loops, is applied to the PMSM, and PI controllers are adopted to eliminate the tracking error of the two current loops. In vector control scheme, the output of the speed loop is used as the reference current $i_q^*(t)$ of the current loop $i_q(t)$. The principle of vector control of PMSM is shown in Fig. 9.1.

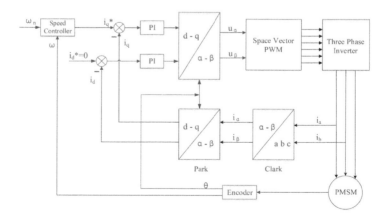

FIGURE 9.1: Schematic diagram of PMSM speed regulation system based on vector control.

Main Results 163

Here, we assume that the dynamic response speed of the current loop is faster than the speed loop. Now, we can use $i_q(t)$ to replace reference current $i_q^*(t)$ approximately. By using the vector control strategy of $i_d^*(t) \equiv 0$, the model can be simplified as follows:

$$
\begin{cases}
\dfrac{\mathrm{d}i_q(t)}{\mathrm{d}t} = -\dfrac{R_s}{L_q}i_q(t) - \dfrac{n_p\psi_f}{L_q}\omega(t) + \dfrac{u_q(t)}{L_q}, \\
\dfrac{\mathrm{d}\omega(t)}{\mathrm{d}t} = ai_q(t) - b\omega(t) - \dfrac{T_L(t)}{J},
\end{cases}
\tag{9.2}
$$

where $a \triangleq \frac{1.5n_p\psi_f}{J}$, $b \triangleq \frac{B_v}{J}$.

Define the following state variables:

$$
x_1(t) = \omega_n(t) - \omega(t), \quad x_2(t) = \dot{x}_1(t) = \dot{\omega}_n(t) - \dot{\omega}(t),
\tag{9.3}
$$

where $\omega_n(t)$ is the given reference speed. Then, we can rewrite the PMSM speed regulation system (9.2) as follows:

$$
\begin{cases}
\dot{x}_1(t) = x_2(t), \\
\dot{x}_2(t) = -au(t) + d(t),
\end{cases}
\tag{9.4}
$$

where $u(t) \triangleq i_q(t)$ is the control input and $d(t) \triangleq \ddot{\omega}_n(t) + b\dot{\omega}(t) + \frac{\dot{T}_L(t)}{J}$, which is handled as the matched disturbance.

In this chapter, our aim is to design a remote speed control scheme such that the rotor angular velocity $\omega(t)$ tracks the given reference speed $\omega_n(t)$ within finite time under the limited communication resource. To achieve the above objective, we develop a novel ESO-based periodic event-triggered TSMC approach, in which the ESO is employed to estimate the disturbance $d(t)$ for reducing the chattering in the designed TSMC and the periodic event-triggered mechanism is utilized to reduce the communication burden between the speed sensor and the remote controller.

9.2 Main Results

9.2.1 The design of ESO

It is worth mentioning that the design of TSMC scheme requires the upper-bound $d_{\max}$ of the disturbance $d(t)$, which is difficulty obtained readily in the practical applications. To this end, we introduce the ESO to estimate the upper-bound $d_{\max}$, where the disturbance $d(t)$ is regarded as an extended state. Then, we rearrange the PMSM speed regulation system (9.2) as follows:

$$
\begin{cases}
\dot{x}_1(t) = x_2(t), \\
\dot{x}_2(t) = -au(t) + x_3(t), \\
\dot{x}_3(t) = D(t),
\end{cases}
\tag{9.5}
$$

where $x_3(t) = d(t)$, $D(t) = \dot{d}(t)$. It is assumed that the $D(t)$ is bounded by $|D(t)| \leq L_0$, where $L_0 \geq 0$ is a known constant.

For the system (9.5), an ESO is designed as follows:

$$
\begin{cases}
\dot{z}_1(t) = z_2(t) + \beta_1(x_1(t) - z_1(t)), \\
\dot{z}_2(t) = z_3(t) + \beta_2(x_1(t) - z_1(t)) - au(t), \\
\dot{z}_3(t) = \beta_3(x_1(t) - z_1(t)),
\end{cases} \tag{9.6}
$$

where β_1, β_2, β_3 are the observer parameters to be determined later. $z_1(t)$, $z_2(t)$, and $z_3(t)$ are the estimates of $x_1(t)$, $x_2(t)$, and $x_3(t)$, respectively. It is noticed that $z_3(t)$ is an estimate of the disturbance $d(t)$. Thus, one can utilize $z_3(t)$ in the designed TSMC for compensating the disturbance $d(t)$.

By combining (9.5) and (9.6), the estimation error dynamics can be obtained as follows:

$$
\dot{\tilde{x}}(t) = A\tilde{x}(t) + BD(t), \tag{9.7}
$$

where $\tilde{x}(t) \triangleq \begin{bmatrix} x_1(t) - z_1(t) & x_2(t) - z_2(t) & x_3(t) - z_3(t) \end{bmatrix}^{\mathrm{T}}$, and

$$
A \triangleq \begin{bmatrix} -\beta_1 & 1 & 0 \\ -\beta_2 & 0 & 1 \\ -\beta_3 & 0 & 0 \end{bmatrix}, \quad B \triangleq \begin{bmatrix} 0 \\ 0 \\ 1 \end{bmatrix}.
$$

In what follows, we will analyze the exponentially ultimate boundedness of the estimation error dynamics (9.7). Firstly, the following technique lemma is introduced.

Lemma 9.1 *[200] For any positive definite matrix Q and matrices E and F, there exists*

$$
E^{\mathrm{T}}QF + F^{\mathrm{T}}QE \leq E^{\mathrm{T}}QE + F^{\mathrm{T}}QF.
$$

Theorem 9.1 *If there exist a matrix $P > 0$ and a decay rate $\gamma > 0$ satisfies*

$$
PA + A^{\mathrm{T}}P + (1 + \gamma)P < 0, \tag{9.8}
$$

then the estimation error dynamics (9.7) is exponentially ultimately bounded with satisfying

$$
\|\tilde{x}(t)\| \leq \sqrt{\frac{V(0)e^{-\gamma t}}{\lambda_{\min}\{P\}}} + \sqrt{\frac{\lambda_{\max}\{P\}L_0^2}{\gamma \lambda_{\min}\{P\}}}, \tag{9.9}
$$

where $V(0) \triangleq \tilde{x}^T(0)P\tilde{x}(0)$, and $\lambda_{\min}\{P\}$ans $\lambda_{\max}\{P\}$ represent the smallest and largest eigenvalues of the matrix P, respectively.

Proof *Choose the Lyapunov function as $V(t) = \tilde{x}^T(t)P\tilde{x}(t)$. Then, it gets*

$$
\dot{V}(t) = \tilde{x}^{\mathrm{T}}(t)\left(PA + A^{\mathrm{T}}P\right)\tilde{x}(t) + \tilde{x}^{\mathrm{T}}(t)PBD(t) + (BD(t))^{\mathrm{T}}P\tilde{x}(t). \tag{9.10}
$$

Main Results 165

By resorting to Lemma 9.1, we further have

$$\dot{V}(t) \le \tilde{x}^{\mathrm{T}}(t)\left(PA + A^{\mathrm{T}}P + P\right)\tilde{x}(t) + (BD(t))^{\mathrm{T}}PBD(t)$$
$$\le \tilde{x}^{\mathrm{T}}(t)\left(PA + A^{\mathrm{T}}P + P\right)\tilde{x}(t) + \lambda_{\max}\{P\}L_0^2.$$

Furthermore, the condition (9.8) gives

$$\dot{V}(t) + \gamma V(t)$$
$$\le \tilde{x}^{\mathrm{T}}(t)\left[PA + A^{\mathrm{T}}P + (1+\gamma)P\right]\tilde{x}(t) + \lambda_{\max}\{P\}L_0^2$$
$$\le \lambda_{\max}\{P\}L_0^2. \tag{9.11}$$

Then, one can obtain the following solution of (9.11):

$$V(t) \le V(0)e^{-\gamma t} + \frac{\lambda_{\max}\{P\}L_0^2}{\gamma}\left(1 - e^{-\gamma t}\right)$$
$$\le V(0)e^{-\gamma t} + \frac{\lambda_{\max}\{P\}L_0^2}{\gamma}. \tag{9.12}$$

Clearly, the ultimate bound (9.9) can be obtained directly by using the fact that $V(t) \ge \lambda_{\min}\{P\}\|\tilde{x}(t)\|^2$. This completes the proof.

Notice that the function $e^{-\gamma t}$ decreases monotonically. Thus, we can get the upper bound of the estimation error vector $\tilde{x}(t)$, i.e.

$$\|\tilde{x}(t)\| \le \sqrt{\frac{V(0)}{\lambda_{\min}\{P\}}} + \sqrt{\frac{\lambda_{\max}\{P\}L_0^2}{\gamma\lambda_{\min}\{P\}}} \triangleq \bar{\gamma}, \ \forall t \ge 0. \tag{9.13}$$

The above upper bound $\bar{\gamma}$ will be utilized in the design of periodic event-triggered TSMC later.

9.2.2 Design of periodic event-triggered TSMC based on ESO

The nonsingular terminal sliding function is chosen as follows:

$$s(t) = x_1(t) + \frac{1}{\beta}x_2^{\frac{p}{q}}(t), \tag{9.14}$$

where $\beta > 0$ is a given scalar, and p, q are two assigned odd integers satisfying $1 < \frac{p}{q} < 2$.

The TSMC scheme is composed of equivalent control law $u_{eq}(t)$ and discontinuous control law $u_{sw}(t)$, in which the equivalent control law $u_{eq}(t)$ is used to drive the sliding mode dynamics converging to the equilibrium point in finite time and the discontinuous control law $u_{sw}(t)$ is selected to force the system states converging to the sliding surface in finite time. First, setting $\dot{s}(t) = 0$, it follows from (9.4) that

$$\dot{s}(t) = x_2(t) + \frac{p}{\beta q}x_2^{\frac{p}{q}-1}(t) \cdot (-au_{eq}(t) + d(t)) = 0. \tag{9.15}$$

166   *ESO-Based TSMC under PETP*

Furthermore, by applying the ESO estimation $z_3(t)$ of the unknown disturbance $d(t)$, the following equivalent control law $u_{eq}(t)$ is derived from (9.15):

$$u_{eq}(t) = a^{-1}\left[f(x_2(t)) + z_3(t)\right], \tag{9.16}$$

with

$$f(x_2(t)) \triangleq \frac{q\beta}{p} x_2^{2-\frac{p}{q}}(t). \tag{9.17}$$

On the other hand, the discontinuous control law $u_{sw}(t)$ is generally chosen in the following form:

$$u_{sw}(t) = a^{-1}[k \cdot \operatorname{sgn}(s(t))]. \tag{9.18}$$

Now, combining (9.16) and (9.18) gives the following TSMC scheme:

$$u(t) = u_{eq}(t) + u_{sw}(t) = a^{-1}[f(x_2(t)) + z_3(t) + k \cdot \operatorname{sgn}(s(t))]. \tag{9.19}$$

Remark 9.1 *It is worth pointing out that the choice of different parameters β, p, and q will offer different control performance. In the case that β is prespecified, $\frac{p}{q}$ is chosen more closer to 2, then a faster convergence of sliding function $s(t)$ will be achieved. Conversely, if $\frac{p}{q}$ is preassigned, the smaller β is selected, then the slower convergence of sliding mode dynamics will be attained.*

In this chapter, the implementation of controller is based on the periodic event-triggered mechanism for the purpose of reducing the communication burden between the PMSM and the remote controller. Fig. 9.2 depicts the framework of the proposed ESO-based periodic event-triggered TSMC, where $z_3(t)$, $x_1(t)$, $x_2(t)$ will be sampled periodically with a proper periodic time λ and then a preassigned event generator will determine whether or not the transmission happens in this sampling instant. Now, during a consecutive triggering interval $[t_i, t_{i+1})$, the TSMC law (9.19) actually becomes the following periodic event-triggered TSMC scheme:

$$u(t_i) = a^{-1}\left[f(x_2(t_i)) + z_3(t_i) + k \cdot \operatorname{sgn}(s(t_i))\right]. \tag{9.20}$$

Now, the PMSM speed regulation system (9.4) under the designed periodic event-triggered TSMC law (9.20) becomes:

$$\begin{cases} \dot{x}_1(t) = x_2(t), \\ \dot{x}_2(t) = -f(x_2(t_i)) - z_3(t_i) - k \cdot \operatorname{sgn}(s(t_i)) + d(t). \end{cases} \tag{9.21}$$

Define the following triggering error vector $e(t)$:

$$e(t) = \begin{bmatrix} x_1(t) - x_1(t_i) \\ x_2(t) - x_2(t_i) \\ z_3(t) - z_3(t_i) \end{bmatrix}.$$

Main Results

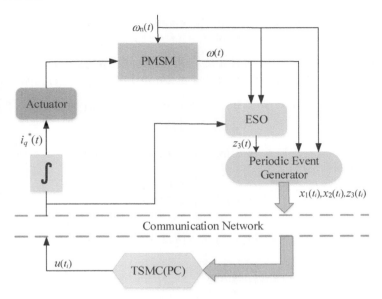

FIGURE 9.2: The framework of the ESO-based periodic event-triggered TSMC.

In (9.20), the triggering instant t_i is determined by the following periodic event-triggered mechanism:

$$t_{i+1} = \inf_{j \in \mathbb{Z}} \{t = t_i + j\lambda \mid \|e(t)\| \geq \alpha\}, \qquad (9.22)$$

where $\alpha > 0$ is an assigned event-triggered parameter, λ is a sampling period to be determined properly (see Eq. (9.32) later). Clearly, the Zeno phenomenon is avoided since the inter-event time $T_i = t_{i+1} - t_i$ is always integer multiple of λ [13].

9.2.3 Estimation of actual bound for $\|e(t)\|$

It is noticed that the periodic event-triggered condition (9.22) can only evaluate the triggering condition periodically with a given sampling period λ. This implies that the actual bound of triggering error $\|e(t)\|$ may exceed α. In this section, we give an estimation for the actual bound of $\|e(t)\|$ by considering the following worst case:

- Suppose that the $(i+1)$th triggering time $t_{i+1} = t_i + j^*\lambda$, and the event-triggering condition $\|e(t)\| \geq \alpha$ is violated immediately after the instant $t_i + (j^* - 1)\lambda$. Under the periodic assessment of the triggering rule, the above violation will only be detected at time $t_i + j^*\lambda$.

168 *ESO-Based TSMC under PETP*

Clearly, the worst case renders the maximum estimation of actual bound for $\|e(t)\|$ during the time interval $[t_i + (j^* - 1)\lambda, t_i + j^*\lambda)$.

Lemma 9.2 *Consider the closed-loop PMSM speed regulation system (9.21) and periodic event-triggered triggering mechanism (9.22), $\|e(t)\|$ satisfies*

$$\|e(t)\| < (h(x_2(t_i)) + \varrho\bar{\gamma} + k)\left(e^\lambda - 1\right) + \alpha \cdot e^\lambda,$$
$$\forall t \in [t_i + (j^* - 1)\lambda, t_i + j^*\lambda), \tag{9.23}$$

where $\varrho \triangleq \max\{1, \beta_3\}$, $\bar{\gamma}$ is defined in (9.13) and $h(x_2(t_i))$ is a real-valued function defined as

$$h(x_2(t_i)) = \left\|\begin{array}{c} x_2(t_i) \\ -f(x_2(t_i)) \\ 0 \end{array}\right\|. \tag{9.24}$$

Consider the time derivative of $\|e(t)\|$ on time interval $[t_i + (j^* - 1)\lambda, t_i + j^*\lambda)$, we obtain

$$\frac{\mathrm{d}}{\mathrm{d}t}\|e(t)\| \leq \|\dot{e}(t)\| = \left\|\begin{array}{c} \dot{x}_1(t) \\ \dot{x}_2(t) \\ \dot{z}_3(t) \end{array}\right\|$$

$$\leq \left\|\begin{array}{c} x_2(t) - x_2(t_i) \\ z_3(t) - z_3(t_i) \\ 0 \end{array}\right\| + \left\|\begin{array}{c} x_2(t_i) \\ -f(x_2(t_i)) \\ 0 \end{array}\right\| + \left\|\begin{array}{c} 0 \\ x_3(t) - z_3(t) \\ \beta_3(x_1(t) - z_1(t)) \end{array}\right\| + k$$

$$\leq \|e(t)\| + h(x_2(t_i)) + \varrho\bar{\gamma} + k. \tag{9.25}$$

Notice that the triggering condition is violated immediately after the instant $t_i + (j^* - 1)\lambda$, that is, it has

$$e(t_i + (j^* - 1)\lambda) = \alpha. \tag{9.26}$$

Under the initial condition (9.26), the following solution of (9.25) can be obtained by resorting to Comparison Lemma:

$$\|e(t)\| \leq (h(x_2(t_i)) + \varrho\bar{\gamma} + k)\left(e^{[t - t_i - (j^* - 1)\lambda]} - 1\right) + \alpha \cdot e^{[t - t_i - (j^* - 1)\lambda]}. \tag{9.27}$$

Notice that the function $e^{[t - t_i - (j^* - 1)\lambda]}$ increases monotonously. One can conclude that the upper bound of the triggering error term $\|e(t)\|$ reaches the maximum value when $t = t_i + j^*\lambda$. Consequently, the relationship (9.23) is attained, which completes the proof.

Remark 9.2 *In the case that the periodic sampling period $\lambda = 0$, the periodic event-triggered mechanism (9.22) will be reduced to the static event-triggered protocol as in [45, 184, 214]. One can also find the feature in (9.23), where the actual upper bound of $\|e(t)\|$ will be reduced to α while $\lambda = 0$. This shows that the periodic event-triggered mechanism is more general than the static one.*

Main Results 169

9.2.4 Selection criterion for periodic sampling period λ

It is noticed that the actual upper bound (9.23) of $\|e(t)\|$ depends on the controller gain k and the periodic sampling period λ. In order to obtain an explicit design condition for the controller gain k (see Theorem 11.2 later), it is required to decouple parameters k and λ in (9.23) by choosing a proper λ.

First, we analyze the upper bound of the function $h(x_2(t_i))$ in (9.24). According to (9.17) and (9.24), one has

$$h(x_2(t_i)) = \sqrt{|x_2(t_i)|^2 + \frac{q^2\beta^2}{p^2}|x_2(t_i)|^{4-\frac{2p}{q}}}. \tag{9.28}$$

Furthermore, it is found from (9.2) that

$$|x_2(t)| \leq |\dot{\omega}_n(t)| + |\dot{\omega}(t)| \leq |\dot{\omega}_n(t)| + a|i_q(t)| + b|\omega(t)| + \frac{1}{J}|T_L(t)|. \tag{9.29}$$

In the practical application, to guarantee the safe operation of the motor, there always exist the known rated current $\bar{i}_q$ (i.e., $|i_q(t)| \leq \bar{i}_q$) and the known rated speed $\bar{\omega}$ (i.e., $|\omega(t)| \leq \bar{\omega}$) of the PMSM. With the meanings of these known information, we further have from (9.29) that

$$|x_2(t)| \leq |\dot{\omega}_n(t)| + a\bar{i}_q + b\bar{\omega} + \frac{1}{J}|T_L(t)| \leq \mu, \tag{9.30}$$

where $\mu \triangleq \bar{\omega}_n + a\bar{i}_q + b\bar{\omega} + \frac{\bar{T}_L}{J}$, $\bar{\omega}_n \triangleq \max\{|\dot{\omega}_n(t)|\}$ and $\bar{T}_L \triangleq \max\{|T_L(t)|\}$.

Since the reference speed signal $\omega_n(t)$ and the load torque $T_L(t)$ are given by the designer, thus the parameters $\bar{\omega}_n$ and $\bar{T}_L$ are known. Substituting (9.30) into (9.28), one can has

$$h(x_2(t_i)) \leq h_0 \triangleq \sqrt{\mu^2 + \frac{q^2\beta^2}{p^2}\mu^{4-\frac{2p}{q}}}. \tag{9.31}$$

Now, we propose the following selection criterion for the periodic sampling period λ:

$$0 < \lambda \leq \lambda^*, \tag{9.32}$$

where

$$\lambda^* \triangleq \ln\left(\frac{k+\rho}{h_0 + k + \varrho\bar{\gamma} + \alpha}\right) = \ln\left(1 + \frac{\sigma}{h_0 + k + \varrho\bar{\gamma} + \alpha}\right) \tag{9.33}$$

with a specified parameter $\sigma > 0$ and the scalar ρ satisfying

$$\rho = h_0 + \varrho\bar{\gamma} + \alpha + \sigma. \tag{9.34}$$

Now, by resorting to the selection criterion (9.32), one can further obtain from (9.23) that

$$\begin{aligned}
\|e(t)\| &< (h(x_2(t_i)) + k + \varrho\bar{\gamma})\left(e^\lambda - 1\right) + \alpha \cdot e^\lambda \\
&\leq (h_0 + k + \varrho\bar{\gamma} + \alpha)e^\lambda - h_0 - \varrho\bar{\gamma} - k \\
&\leq (h_0 + k + \varrho\bar{\gamma} + \alpha)e^{\lambda^*} - h_0 - \varrho\bar{\gamma} - k \\
&\leq \alpha + \sigma.
\end{aligned} \tag{9.35}$$

Clearly, with the help of the selection criterion (9.32), the upper bound of $\|e(t)\|$ under the periodic event-triggered protocol (9.22) is reduced to be an constant $\alpha + \sigma$, which is independent of k and λ. This feature is beneficial to derive an explicit design condition for the controller gain k in Theorem 11.2 later.

9.2.5 Reachability and stability

This subsection analyzes the reachability of the nonsingular terminal sliding surface (9.14) and the stability of the closed-loop system (9.21).

For the specified nonsingular terminal sliding function (9.14), the following establishes a sufficient condition for the existence of the controller gain k, which ensures the practical sliding mode under the TSMC law (9.20) with the ESO (9.6) and the periodic event-triggered mechanism (9.22) with the selection criterion (9.32).

Theorem 9.2 *Consider the closed-loop PMSM speed regulation system (9.21) and periodic event-triggered triggering mechanism (9.22), if the periodic sampling period is selected according to (9.32) and the controller gain k is chosen to satisfy*

$$k = \eta + \alpha + \sigma + c(\alpha + \sigma)^r + \bar{\gamma}, \tag{9.36}$$

where $\eta > 0$ is an assigned small parameter, then the practical sliding mode of nonsingular terminal sliding function (9.14) occurs in the following region:

$$\Omega \triangleq \left\{ s(t) \in \mathbb{R} \mid |s(t)| < \left(1 + \frac{c_2}{\beta}\right)(\alpha + \sigma) \right\}, \tag{9.37}$$

where $f(x_2(t))$ satisfies the Hölder continuity with the constant c ($c > 0$) and the order $r \in (0,1)$; $c_2 > 0$ is the Lipschitz constant of $x_2^{\frac{p}{q}}(t)$ on compact set; h_0 and ρ are defined in (9.31) and (9.34), respectively; and $\bar{\gamma}$ is the upper bound of the ESO estimation error defined in (9.13).

Proof *Consider the Lyapunov function candidate $V_1(t) = \frac{1}{2}s(t)^2$. The time derivative of $V_1(t)$ is expressed as follows:*

$$\begin{aligned}
\dot{V}_1(t) &= s(t)\dot{s}(t) \\
&= s(t)\left[x_2(t) + g(x_2(t))\left(-au(t_i) + d(t)\right)\right] \\
&= s(t)g(x_2(t))\left[f(x_2(t)) - f(x_2(t_i)) + d(t)\right. \\
&\quad \left. - z_3(t) + z_3(t) - z_3(t_i) - k \cdot \mathrm{sgn}(s(t_i))\right],
\end{aligned} \tag{9.38}$$

Main Results 171

where $f(x_2(t))$ is defined in (9.17) and

$$g(x_2(t)) \triangleq \frac{p}{q\beta} x_2^{\frac{p}{q}-1}(t) \geq 0.$$

Considering $f(x_2(t))$ is Hölder continuous, we obtain the following relationship:

$$|f(x_2(t)) - f(x_2(t_i))| \leq c|x_2(t) - x_2(t_i)|^r \leq c\|e(t)\|^r. \tag{9.39}$$

- **Case 1**: *When $\text{sgn}(s(t)) = \text{sgn}(s(t_i))$, it has $\text{sgn}(s(t_i)) \cdot s(t) = |s(t)|$. Now, the time derivative of $V_1(t)$ is further derived as follows:*

$$\dot{V}_1(t) \leq |s(t)|g(x_2(t))\left[c\|e(t)\|^r + \|\tilde{x}(t)\| + \|e(t)\| - k\right]. \tag{9.40}$$

In what follows, two subcases are analyzed, respectively:

1) *$x_2(t) \neq 0$. In this case, by resorting to (9.36), it follows from (9.35) and (9.40) that*

$$\dot{V}_1(t) \leq -\eta g(x_2(t))|s(t)|, \tag{9.41}$$

which just implies that the sliding variable $s(t)$ is attracted to the sliding surface $s(t) = 0$ within finite time.

2) *$x_2(t) = 0$. Actually, it is trivial to consider the case that $s(t) = 0$ since it just means the sliding variable $s(t)$ has achieved to the sliding surface. Now, we consider $s(t) \neq 0$, which gives from (9.14) that $x_1(t) \neq 0$. Notice that the point $(x_1(t) \neq 0, x_2(t) = 0)$ is not the equilibrium point of the system. This means that the state trajectories will continue to cross the axis $x_2(t) = 0$ in the phase plane of $(x_1(t), x_2(t))$, that is, $\dot{V}_1(t) = 0$ is not always hold for $s(t) \neq 0$.*

While $x_2(t) = 0$ changes to be $x_2(t) \neq 0$, one can conclude from the above first case that the sliding variable $s(t)$ will be forced to the sliding surface $s(t) = 0$ by the periodic event-triggered TSMC law (9.20).

Based upon the above discussions, we obtain that the sliding variable $s(t)$ is forced to the sliding surface in this case under the periodic event-triggered TSMC law (9.20) with the condition (9.36).

- **Case 2**: *When $\text{sgn}(s(t)) \neq \text{sgn}(s(t_i))$, we obtain from (9.14) that*

$$|s(t) - s(t_i)| = \left|x_1(t) - x_1(t_i) + \frac{1}{\beta}\left(x_2^{\frac{p}{q}}(t) - x_2^{\frac{p}{q}}(t_i)\right)\right|. \tag{9.42}$$

Notice that $x_2^{\frac{p}{q}}(t)$ is Lipschitz continuous on compact set with constant c_2. Thus, it has

$$\left|x_2^{\frac{p}{q}}(t) - x_2^{\frac{p}{q}}(t_i)\right| \leq c_2|x_2(t) - x_2(t_i)|. \tag{9.43}$$

172   *ESO-Based TSMC under PETP*

By means of (9.43), one can further gets from (9.42) that

$$|s(t) - s(t_i)| \leq \|e(t)\| \left(1 + \frac{c_2}{\beta}\right) \leq \left(1 + \frac{c_2}{\beta}\right)(\alpha + \sigma). \qquad (9.44)$$

Since the signs of $s(t)$ and $s(t_i)$ are different in this case, one can easily find in (9.44) that the bound of $s(t)$ reaches the maximum value when $s(t_i) = 0$. This completes the proof.

In the sequel, the stability of the closed-loop PMSM speed regulation system (9.21) is analyzed. Firstly, from (9.4) and (9.14), one has

$$\dot{x}_1(t) = x_2(t) = [\beta(s(t) - x_1(t))]^{\frac{q}{p}}. \qquad (9.45)$$

By using the practical sliding mode region (9.37), we analyze the ultimately boundedness of the closed-loop PMSM speed regulation system (9.21) in the following theorem.

Theorem 9.3 *Under the designed TSMC law (9.20) with the periodic event-triggered mechanism (9.22), the state trajectories of the closed-loop PMSM speed regulation system (9.21) are ultimately bounded by the following domains:*

$$\Theta_1 \triangleq \left\{x_1(t) \mid |x_1(t)| < (\alpha + \sigma)\left(1 + \frac{c_2}{\beta}\right)\right\}, \qquad (9.46)$$

$$\Theta_2 \triangleq \left\{x_2(t) \mid |x_2(t)| < \left[2\beta(\alpha + \sigma)\left(1 + \frac{c_2}{\beta}\right)\right]^{\frac{q}{p}}\right\}. \qquad (9.47)$$

Proof *Choose the Lyapunov function candidate as $V_2(t) = \frac{1}{2}x_1^2(t)$. Then, the time derivative of $V_2(t)$ is obtained by*

$$\dot{V}_2(t) = \dot{x}_1(t)x_1(t) = x_1(t)[\beta(s(t) - x_1(t))]^{\frac{q}{p}}. \qquad (9.48)$$

Notice that $\dot{V}_2(t) < 0$ can be attained if and only if the state $x_1(t)$ stays in the two domains: $\Lambda_1 \triangleq \{x_1(t) > 0 \mid x_1(t) > |s(t)|\}$ and $\Lambda_2 \triangleq \{x_1(t) < 0 \mid |s(t)| < -x_1(t)\}$. That is to say, we have $\dot{V}_2(t) < 0$ for any $x_1(t) \in \Lambda_3$, where $\Lambda_3 \triangleq \Lambda_1 \cup \Lambda_2 = \{x_1(t) \mid |x_1(t)| > |s(t)|\}$.

Clearly, the system trajectory $x_1(t)$ is strictly decreasing in the domain Λ_3. This just means that $x_1(t)$ converges to Θ_1 defined in (9.46).

Furthermore, by resorting to (9.37) and (9.46), one has

$$|s(t) - x_1(t)| \leq |s(t)| + |x_1(t)| < 2(\alpha + \sigma)\left(1 + \frac{c_2}{\beta}\right). \qquad (9.49)$$

As per (9.45), one can yield readily that the state trajectory $x_2(t)$ is ultimately bounded by the domain Θ_2 defined in (9.47). The proof is completed.

Main Results 173

Remark 9.3 *It is seen from (9.33) that the upper bound of sampling period* λ *increases as* ρ *increases. Meanwhile, as* ρ *increases, the regulation error* $|x_1(t)|$ *increases as shown in (9.46). The above facts mean that the larger* ρ *renders the less communication times as well as the worse speed regulation performance. In the practical applications, one may choose a proper scalar* ρ *by considering the tradeoff between the regulation performance and the communication burden.*

9.2.6 Solving algorithm

In this subsection, a practical implementation algorithm of the periodic event-triggered TSMC law (9.20) is proposed. Firstly, a GA-optimized ESO is developed to reduce the chattering in the proposed periodic event-triggered TSMC law (9.20). Secondly, the implementation steps of the proposed control strategy are formulated.

Assume that the disturbance $d(t)$ satisfies $|d(t)| \leq d_{\max}$, where $d_{\max} \geq 0$ is the actual bound but *unknown* to the designer. For the case that without ESO, with the similar lines to Theorem 11.2, the controller gain k^* will be chosen as

$$k^* = \eta + c(\alpha + \sigma)^r + d_{\max}, \tag{9.50}$$

where $\eta > 0$ is an assigned small parameter.

By choosing appropriate parameters ρ, γ, and A in Theorem 11.1, it is possible to get $\bar{\gamma} \leq d_{\max} - (\alpha + \sigma)$. This means that the introduction of ESO with some appropriate parameters may render a smaller controller gain $k \leq k^*$, which implies the chattering is reduced by means of ESO.

For given parameters α, η, σ, as shown in (9.36), the controller gain k now only depends on the upper bound $\bar{\gamma}$ of the ESO estimation error. The smaller $\bar{\gamma}$ renders the less chattering. Besides, as shown in (9.33), the smaller $(1+\varrho)\bar{\gamma}$ renders the larger upper bound λ^*, which may imply the less communication transmissions. To this end, we introduce the following minimization objective problem

$$\min_{\gamma,\beta_1,\beta_2,\beta_3} \zeta(\theta_1,\theta_2)$$
$$\text{subject to: Eq. (9.8),} \tag{9.51}$$

where

$$\zeta(\theta_1,\theta_2) \triangleq \theta_1\bar{\gamma} + \theta_2(1+\varrho)\bar{\gamma}, \tag{9.52}$$

with two assigned weighting parameters $\theta_1 \geq 0$, $\theta_2 \geq 0$, and $\theta_1 + \theta_2 = 1$. Here, θ_1 and θ_2 reflect the expectation for reducing the chattering and the communication burden, respectively.

It is worth pointing out that the matrix inequality (9.8) is *nonlinear* for the decision variable $\gamma > 0$ and decision matrices $P > 0$ and A. Clearly, it is difficult for us to employ the existing convex algorithm to solve the optimization problem (9.51). Fortunately, the matrix inequality (9.8) can be reduced to be an LMI for the decision matrix P if the parameters $\beta_1, \beta_2, \beta_3$, and γ are preassigned. This fact motivates us to combine GA with LMI technique to overcome the above obstacle [71, 134, 143].

In what follows, we develop a binary-based GA to solve the optimization problem (9.51).

- (Parameter Encoding): The phenotype in the search space is expressed as the following row vector v:

$$v = [v_1, v_2, v_3, v_4],$$

where v_1, v_2, v_3, and v_4 represent β_1, β_2, β_3, and γ, respectively. The element v_i ($i \in \{1, 2, 3, 4\}$) is coded as a binary string with length l_i, respectively. Here, each length is determined by its own range of $v_i \in [\underline{v}_i, \overline{v}_i]$, where the bounds $\underline{v}_i$ and $\overline{v}_i$ are preassigned. Clearly, the precision p_{v_i} under the above linearly mapped coding can be obtained by $p_{v_i} = \frac{\overline{v}_i - \underline{v}_i}{2^{l_i} - 1}$. By doing so, v is converted to a binary-based chromosome with length $\sum_{i=1}^{4} l_i$.

- (Population Initialization): Randomly generate N chromosomes as initial population.

- (Fitness Calculation): In order to solve the minimization problem (9.51), the fitness function is chosen as:

$$\mathbf{Fitness} = \frac{1}{\zeta(\theta_1, \theta_2)}.$$

Convert every individual in population to be real value and then obtain the corresponding fitness value via solving the LMI (9.8). If for a set of β_1, β_2, β_3, and γ, the LMI (9.8) is infeasible, then a sufficiently small value is assigned to the individual artificially.

- (Applying Genetic Operations): As per the fitness values obtained in Step 3, the next population is obtained via executing the sequence of genetic operations *Selection*, *Crossover* with probability p_m and *Mutation* with probability p_n as in [134].

- (Stop Criterion and Design Phase): The algorithm will repeat from step 3 to 6 until it reaches the maximum generation $G_{\max}$. Decode the best chromosome v into real value, and then obtain the optimized ESO (9.6) and the optimized periodic event-triggered TSMC law (9.20) with sampling period λ in (9.32).

Main Results 175

Based upon the developed GA-optimized ESO, the following implementation steps are formulated to design the proposed periodic event-triggered TSMC scheme (9.20):

1. Given the parameters β, p, and q with satisfying $\beta > 0$ and $1 < \frac{p}{q} < 2$. Construct the terminal sliding function by (9.14).

2. Substitute the known parameters ψ_f, n_p, J, and B_v into the PMSM speed regulation system (9.2) and TSMC law (9.20). The reference current $i_q^*(t)$ of the current loop $i_q(t)$ is obtained by integrating the control law $u(t)$, i.e., $i_q^*(t) = \int_0^t u(\tau)\mathrm{d}\tau$.

3. Given the reference speed $\omega_n(t)$, the load torque $T_L(t)$, and the event-triggered parameter α. Determine the searching ranges of the parameters β_1, β_2, β_3 and γ in GA.

4. Choose appropriate GA parameters N, p_m, p_n, $G_{\max}$, l_1, l_2, l_3, l_4 and the weighting parameters θ_1, θ_2 in (9.8).

5. Select the initial values $z_1(0)$, $z_2(0)$, $z_3(0)$ of the ESO (9.6) and the parameter L_0. Get the optimized parameters β_1, β_2, β_3, γ and matrix P by the proposed GA, then obtain the optimized $\bar{\gamma}$ according to (9.13).

6. Compute ϱ in (9.23). According to the known rated speed $\bar{\omega}$ and rated current $\bar{i}_q$ of the utilized PMSM, calculate μ by (9.30). Then, obtain h_0 according to (9.31).

7. According to Hölder continuous condition, select c and r according to $f(x_2(t)) = \frac{q\beta}{p}x_2^{2-\frac{p}{q}}(t)$.

8. Choose small positive parameters η and σ, compute the controller gain k and the scalar ρ by (9.36) and (9.34), respectively.

9. Obtain the upper bound of the sampling period λ^* by (9.33) and choose a sampling period $\lambda \in (0, \lambda^*]$.

10. Get the periodic event-triggered TSMC law (9.20) with the GA-optimized ESO (9.6) and apply it to the networked PMSM speed regulation system (9.2).

Remark 9.4 *This chapter proposes a novel periodic event-triggered TSMC strategy for the networked PMSM speed regulation system. The key ideas here are that the periodic event-triggered mechanism is employed to avoid the continuous sampling with reducing the communication burden and the GA-optimized ESO is introduced to reduce the chattering in the proposed control scheme. The design and theoretical analysis procedure of the proposed periodic event-triggered TSMC approach is shown in Fig. 9.3. First, for the networked PMSM, an ESO is constructed in (9.6) to estimate the unknown external disturbance and then an ESO-based TSMC is designed in (9.19). In order*

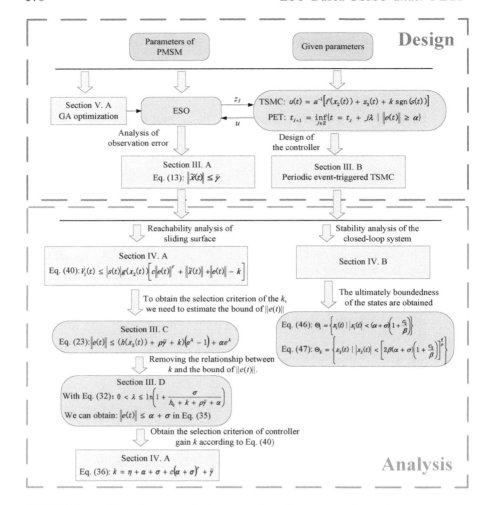

FIGURE 9.3: The design and theoretical analysis procedure of the proposed periodic event-triggered TSMC.

to reduce the communication burden, the periodic event-triggered protocol in (9.22) is introduced and then the actual triggering error is analyzed explicitly. Furthermore, by utilizing the known parameters of the PMSM, a proper selection criterion of the periodic sampling period is designed in (9.32), which removes the dependence of the control gain k and the periodic sampling period λ in the upper-bound of $\|e(t)\|$. Meanwhile, the reachability of the nonsingular terminal sliding surface and the stability of the sliding mode dynamics are analyzed. Finally, a GA is introduced to obtain the optimized ESO parameters.

9.3 Simulation and Experiment

9.3.1 Simulation

In this section, we consider a PMSM speed system (9.2) with the parameters specified in Table 9.1. The simulations are conducted firstly in MATLAB (R2018b) to illustrate the effectiveness of the proposed control scheme.

As per the design procedure developed in Section 9.2.6, a satisfied periodic event-triggered TSMC (9.20) is designed by the following step-by-step:

- **Step 1.** Select the parameters $\beta = 6$, $p = 11$, and $q = 9$. Then, the terminal sliding function is designed as:

$$s(t) = x_1(t) + \frac{1}{6}x_2^{\frac{11}{9}}(t). \tag{9.53}$$

- **Step 2.** By using the parameters given in Table 9.1, the PMSM speed regulation system (9.4) is obtained by

$$\begin{cases} \dot{x}_1(t) = x_2(t), \\ \dot{x}_2(t) = -54.39u(t_i) + d(t), \end{cases}$$

and the periodic event-triggered TSMC law (9.20) is designed as

$$u(t_i) = \frac{1}{54.39}\left[4.91x_2^{\frac{7}{9}}(t_i) + z_3(t_i) + k \cdot \mathrm{sgn}(s(t_i))\right]. \tag{9.54}$$

- **Step 3.** In the simulation and experiment, we assign $\omega_n(t) = 500 \text{ r/min}$, $T_L(t) = 0$ N, and $\alpha = 5$. In GA, the searching ranges of the parameters β_1, β_2, β_3, γ are determined by: $\underline{v_1} = \underline{v_2} = \underline{v_3} = 1, \overline{v_1} = \overline{v_2} = \overline{v_3} = 80, \underline{v_4} = 0.001$, and $\overline{v_4} = 10$.

- **Step 4.** The GA parameters are specified as: population size $N = 1600$; crossover probability $p_m = 0.8$; mutation probability $p_n = 0.05$; maximum of generations $G_{\max} = 300$; and the lengths of binary strings $l_1 = l_2 = l_3 = 17$,

TABLE 9.1: Parameters in the concerned PMSM.

Parameter	Symbol	Value
Rated current	i_q	7.1 A
Rated speed	$\bar{\omega}$	6000 rpm
Stator Resistance	R_s	0.72 Ω
Stator Inductances	$L_d = L_q$	0.2 mH
Rotor Flux Linkage	ψ_f	0.0064 wb
Number of Pole Pairs	n_p	4
Moment of Inertia	J	7.06×10^{-4}kg·m^2
Viscous Friction Coefficient	B_v	5.6×10^{-6}N·m·s·rad^{-1}

$l_4 = 14$. The weighting parameters in the optimization problem (9.51) are supposed as $\theta_1 = 0.2$, $\theta_2 = 0.8$.

- **Step 5.** In the practical applications, in order to ensure the initial estimation error of ESO possessing $\tilde{x}(0) = 0$, the initial states of the ESO are selected as $z_1(0) = x_1(0) = \omega_n(0) - \omega(0)$, $z_2(0) = x_2(0) = \dot{\omega}_n(0) - ai_q(0) + b\omega(0) + T_L(0)/J$, and $z_3(0) = x_3(0) = \ddot{\omega}_n(0) + b(\dot{\omega}_n(0) - x_2(0)) + \dot{T}_L(0)/J$. In this simulation example, it has $\omega_n(t) = 500$, $T_L(t) = 0$, $\omega(0) = 0$ and $i_q(0) = 0$. Thus, we select the initial states of ESO as $z_1(0) = 500, z_2(0) = z_3(0) = 0$. The parameter L_0 is supposed to be 10. By utilizing the above parameter settings, the proposed GA finds the following optimized solutions:

$$v = [4.086, 12.166, 7.173, 0.469],$$
$$P = \begin{bmatrix} 8.022 & -2.425 & 0.666 \\ -2.425 & 2.055 & -1.454 \\ 0.666 & -1.454 & 1.822 \end{bmatrix}. \quad (9.55)$$

Fig. 9.4 shows the best fitness values and average fitness values for each generation in running GA.

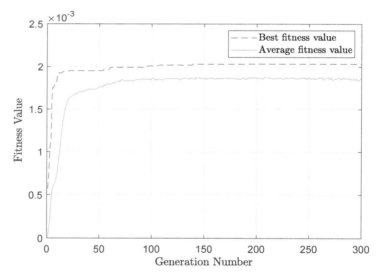

FIGURE 9.4: The best and average fitness values in each generation.

By means of the above obtained optimized solutions, we get $\bar{\gamma} = 85.705$ from (9.13).

- **Step 6.** We obtain $\varrho = \max\{1, \beta_3\} = 7.173$. In the concerned PMSM platform, the rated current and the rated speed are known as 7.1 A and 6000 rmp, respectively. Now, it gets from (9.30) and (9.31) that $\mu = 433.749$ and $h_0 = 702.355$, respectively.

- **Step 7.** As per $f(x_2(t)) = 4.91x_2^{\frac{7}{9}}(t)$, it finds $c = 5$ with $r = \frac{7}{9}$ since the Hölder continuous condition can be verified via

$$\lim_{\zeta \to 0} \frac{|f(\zeta) - f(0)|}{|\zeta - 0|^r} < 5.$$

- **Step 8.** In (9.34), we choose $\rho = 1327.117$ with $\sigma = 5$. Now, the controller gain in (9.36) can be chosen as $k = 500$ with $\eta = 374.321$.
- **Step 9.** It is computed from (9.33) that $\lambda^* = 0.0029\,\text{s}$. Here, we choose the sampling period as $\lambda = 0.0020\,\text{s}$ due to (9.32).
- **Step 10.** In the simulation, the sampling step size in MATLAB for the concerned PMSM speed system is set as 0.0001s and the simulation time is chosen as 15 s.

As shown in the Figs. 9.5–9.8, the following three groups of simulations are conducted:

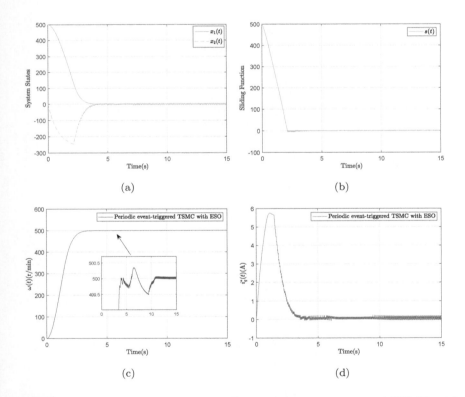

FIGURE 9.5: Simulation results for the periodic event-triggered TSMC with ESO under $T_L(t) = 0$: (a) Response of the state trajectories $x_1(t)$ and $x_2(t)$; (b) Response of the terminal sliding function $s(t)$; (c) Response of the speed $\omega(t)$; (d) Response of the reference q-axis current $i_q^*(t)$.

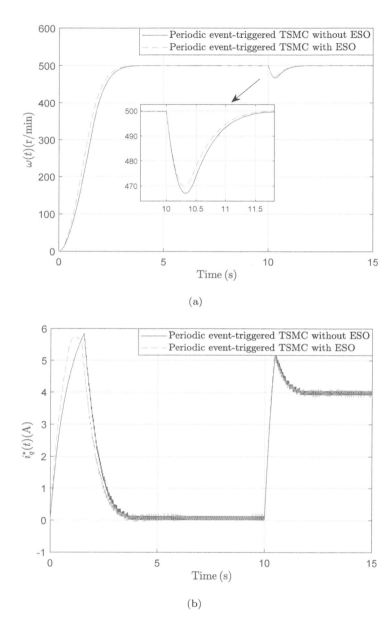

FIGURE 9.6: Simulation comparison of the periodic event-triggered TSMC with and without ESO: (a) Response of the speed $\omega(t)$; (b) Response of the reference q-axis current $i_q^*(t)$.

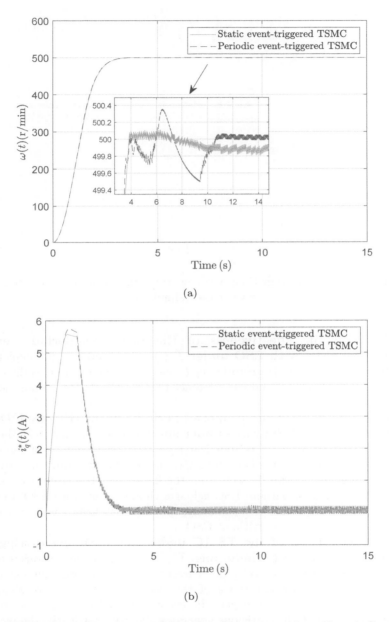

FIGURE 9.7: Simulation comparison between static and periodic event-triggered TSMC schemes: (a) Response of the speed $\omega(t)$; (b) Response of the reference q-axis current $i_q^*(t)$.

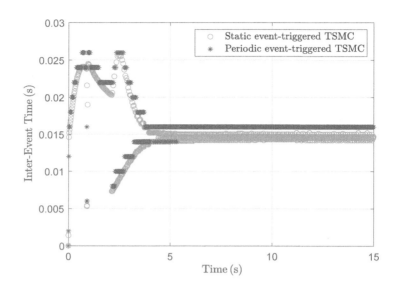

FIGURE 9.8: Inter-event time between static and periodic event-triggered TSMC cases (simulation)

i) **Verify the effectiveness of the proposed periodic event-triggered TSMC with ESO under** $T_L(t) = 0$ N · m. Fig. 9.5 depicts the responses of the system trajectories $x_1(t)$ and $x_2(t)$, the terminal sliding function $s(t)$, the speed $\omega(t)$ and the reference q-axis current $i_q^*(t)$ simultaneously in this case.

ii) **Comparison of the proposed periodic event-triggered TSMC with and without ESO cases under sudden load.** In this case, the load torque $T_L(t) = 0.15$ N · m is added at $t = 10$s. The responses of the speed $\omega(t)$ and the reference q-axis current $i_q^*(t)$ under the two different cases are shown in Fig. 9.6. It can be seen from the figure that the rising time of the speed $\omega(t)$ in the case without ESO is longer than the case with ESO. Besides, when the load torque is added suddenly, the case with ESO gives fewer speed fluctuations than the case without ESO.

iii) **Comparison of the TSMC under the static event-triggered mechanism and the periodic one.** Fig. 9.7 depicts the responses of the speed $\omega(t)$ and the reference q-axis current $i_q^*(t)$ in the two different cases. Meanwhile, the inter-event times of the two different cases are shown in Fig. 9.8. It is found that the periodic event-triggered mechanism gives 921 triggers in 15s while the static one generates 995 events. Clearly, the periodic event-triggered mechanism can save more communication resource than the static one at the cost of sacrificing some control performance as shown in Fig. 9.7(a).

Simulation and Experiment 183

Remark 9.5 *It is clear that the convergence time of the PMSM speed $\omega(t)$ is consistent with the state $x_1(t)$ due to $x_1(t) \triangleq \omega_n(t) - \omega(t)$. On the other hand, since the state trajectories will be driven to the sliding surface $s(t) = 0$ in finite time at first by the proposed periodic event-triggered TSMC, the convergence time of sliding function $s(t)$ is faster than the PMSM speed $\omega(t)$ as shown in Figs. 9.5(b) and 9.5(c).*

9.3.2 Experiment

To further demonstrate the applicability of the proposed control strategy, a real experiment is carried out for a PMSM platform with the parameters given in Table 9.1. The experimental test setup is shown in Fig. 9.9(a) and its configuration is depicted in Fig. 9.9(b). The motor is driven by DRV8305EVM, which consists of CSD1854Q5B Power Mosfet and DRV8305 Gate Driver. In the experiment, the sampling time of the speed is $0.002\,s$. The PI parameters of the two current loops are selected as $k_p = 0.4$ and $k_I = 0.8$.

By applying the above designed periodic event-triggered TSMC law to the PMSM platform, the following three groups of comparison experiments shown in Figs. 9.10–9.14 are executed:

i) **Comparison of the conventional time-triggered TSMC and the proposed periodic event-triggered TSMC under $T_L(t) = 0\ \mathrm{N} \cdot \mathrm{m}$.** Figs. 9.10–9.11 show the responses of the speed $\omega(t)$ and the reference q-axis current $i_q^*(t)$ in this comparison case, in which the inter-event times of the two different approaches are compared in Fig. 9.12. One can observe that the periodic event-triggered TSMC can reduce the communication burden significantly at the expense of some control performance.

ii) **Comparison of the time-triggered PI, periodic event-triggered TSMC and time-triggered TSMC under sudden load.** In this case, the responses of the speed $\omega(t)$ and the reference q-axis current $i_q^*(t)$ under the three different cases are depicted in Fig. 9.13. When the load torque $T_L(t) = 0.3\ \mathrm{N} \cdot \mathrm{m}$ is added suddenly, it is readily found that the time-triggered TSMC and the periodic event-triggered TSMC render fewer speed fluctuations than the time-triggered PI. However, it is worth stressing that the implementation of the time-triggered TSMC requires the continuous sampling of the PMSM speed $\omega(t)$ but not in the proposed periodic event-triggered TSMC strategy. Therefore, the proposed periodic event-triggered TSMC strategy may be the best choice from the actual application aspect.

iii) **Comparison of the proposed periodic event-triggered TSMC with and without ESO cases under sudden load.** While the load torque $T_L(t) = 0.3\ \mathrm{N} \cdot \mathrm{m}$ is added suddenly, the responses of the speed $\omega(t)$ and the reference q-axis current $i_q^*(t)$ under the two different cases are shown in Fig. 9.14. Clearly, similar to the simulation results presented in Fig. 9.6, the periodic event-triggered TSMC with ESO yields fewer speed fluctuations than the one without ESO.

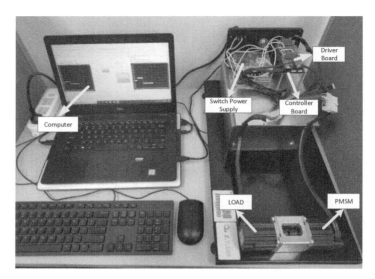

(a) Experimental test setup

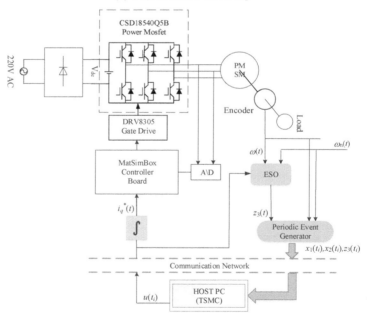

(b) The configuration of an experimental PMSM system used in this chapter

FIGURE 9.9: Schematic diagram of the considered PMSM speed regulation problem.

Simulation and Experiment

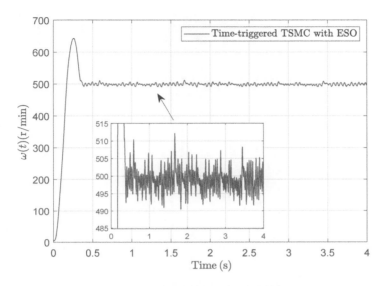

(a) Response of the speed $\omega(t)$ in time-triggered TSMC

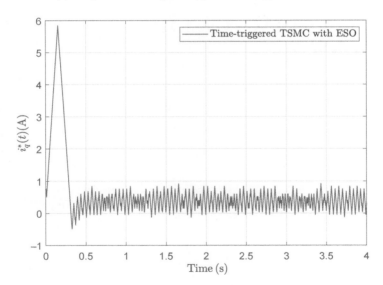

(b) Response of the reference q-axis current $i_q^*(t)$ in time-triggered TSMC

FIGURE 9.10: Experiment results in time-triggered TSMC.

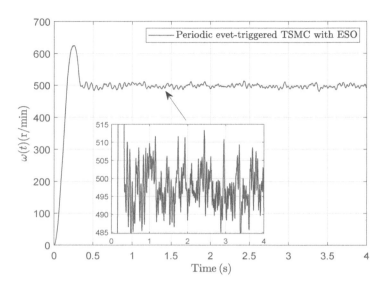

(a) Response of the speed $\omega(t)$ in periodic event-triggered TSMC

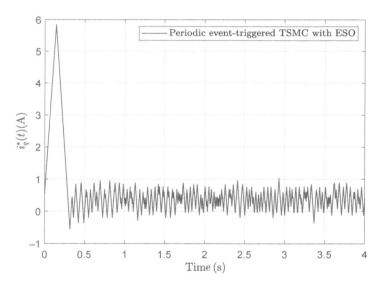

(b) Response of the reference q-axis current $i_q^*(t)$ in periodic event-triggered TSMC

FIGURE 9.11: Experiment results in periodic event-triggered TSMC.

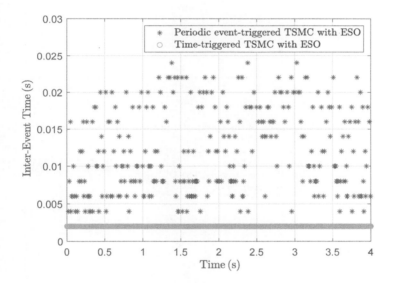

FIGURE 9.12: Inter-event time between periodic event-triggered and time-triggered TSMC cases (experiment)

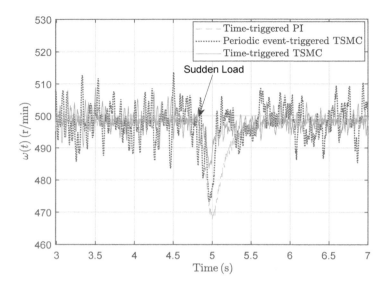

(a) Comparison of the speed $\omega(t)$ in three different cases

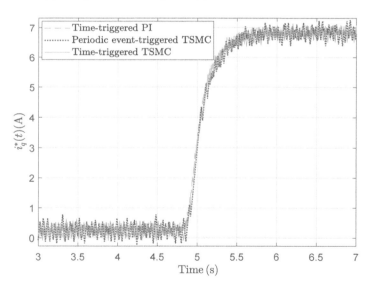

(b) Comparison of the reference q-axis current $i_q^*(t)$ in three different cases

FIGURE 9.13: Experimental comparison in three different cases when the load torque $T_L(t) = 0.3$ N · m is added suddenly.

Simulation and Experiment

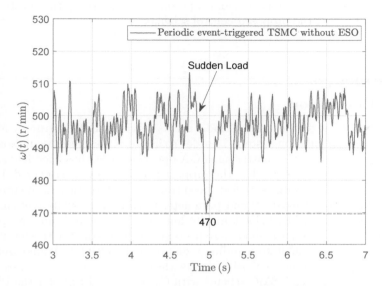

(a) Response of the speed $\omega(t)$ in periodic event-triggered TSMC without ESO

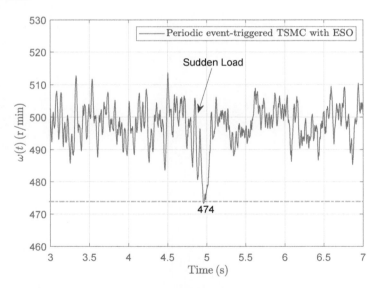

(b) Response of the speed $\omega(t)$ in periodic event-triggered TSMC with ESO

FIGURE 9.14: Experimental comparison of the proposed periodic event-triggered TSMC with and without ESO cases when the load torque $T_L(t) = 0.3$ N · m is added suddenly.

Overall, both the simulation and experiment results verify the effectiveness and advantage of the proposed control scheme.

9.4 Conclusion

This chapter has developed a periodic event-triggered TSMC strategy for the PMSM speed regulation system. In order to improve the speed regulation performance with reducing chattering, a GA-optimized ESO has been introduced in the design. It is shown that the proposed periodic event-triggered TSMC scheme can guarantee the satisfied speed tracking performance in the presence of the external disturbances and parameter uncertainties while reducing the communication burden. Selection criterions of the periodic sampling period and the control gain parameter were proposed for ensuring the reachability of terminal sliding variable to a sliding region and the ultimate boundedness of the tracking error simultaneously. Finally, the effectiveness of the proposed periodic event-triggered TSMC scheme with GA-optimized ESO for the PMSM speed regulation system has been verified in both of simulation and experiment.

PART II: 2-D System Case

10

2-D Sliding Mode Control Under Event-Triggered Protocol

Over the past few decades, a great number research interest has been attracted to two-dimensional (2-D) systems for their promising application insights in a variety of practical situations, such as repetitive processes, image data processing and transmission, thermal process, gas absorption, water stream heating and iterative learning control. Fundamentally, comparing with the traditional one-dimensional (1-D) systems, a key feature in 2-D systems is that their states propagate along two independent directions giving rise to more theoretical complexities and mathematical challenges for analyzing/synthesizing the 2-D systems. Some representative mathematical models for 2-D systems have been developed, of which Roesser model, Fornasini-Marchesini (FM) second model, and Attasi model are arguably the most popular ones. Till now, a large amount of works has been published on control/filtering issues for various 2-D model.

The main interesting of this chapter is focused on the co-design problem of 2-D event generators and 2-D sliding mode control (SMC) scheme for 2-D Roesser model subject to matched external disturbance. Both of the reachability of the specified horizontal and vertical sliding surfaces are guaranteed by the proposed event-based 2-D SMC law. By taking the scheduling effects from the designed 2-D event generators into account, the stability of the resultant sliding mode dynamics is analyzed as per the 2-D Lyapunov stability theory. *The main contributions of this chapter are outlined as follows. 1) In order to facilitate the analysis and synthesis for 2-D Roesser model, the event generating mechanisms in horizontal and vertical directions are considered to be independent of each other; 2) Based on the specified 2-D sliding functions, the event-based 2-D SMC laws are proposed for horizontal and vertical directions, respectively, while the desired 2-D event generators are designed to ensure the 2-D reaching law conditions; and 3) In dealing with the nonlinear stability condition for the sliding mode dynamics, a binary-based genetic algorithm (GA) is formulated based upon a trade-off-based multi-objective optimization problem so as to avoid introducing some further conservatism.*

DOI: 10.1201/9781003309499-10

10.1 Problem Formulation

In this chapter, we study the following uncertain 2-D system represented by the Roesser model:

$$x^h(i+1,j) = A_{11}x^h(i,j) + A_{12}x^v(i,j) + B_1\left(u^h(i,j) + f^h(i,j)\right), \quad (10.1)$$
$$x^v(i,j+1) = A_{21}x^h(i,j) + A_{22}x^v(i,j) + B_2\left(u^v(i,j) + f^v(i,j)\right), \quad (10.2)$$

where $x^h(i,j) \in \mathbb{R}^{n_h}$ and $x^v(i,j) \in \mathbb{R}^{n_v}$ represent, respectively, the horizontal and vertical states; $u(i,j) \in \mathbb{R}^m$ denotes the control input; $f^h(i,j) \in \mathbb{R}^m$ and $f^v(i,j) \in \mathbb{R}^m$ denote the matched external disturbances in horizontal and vertical directions, respectively. $A_{11} \in \mathbb{R}^{n_h \times n_h}$, $A_{12} \in \mathbb{R}^{n_h \times n_v}$, $A_{21} \in \mathbb{R}^{n_v \times n_h}$, $A_{22} \in \mathbb{R}^{n_v \times n_v}$, $B_1 \in \mathbb{R}^{n_h \times m}$, and $B_2 \in \mathbb{R}^{n_v \times m}$ are known constant matrices. Without loss of generality, the uncertain Roesser model (10.1)–(10.2) satisfies the following assumptions:

1) Both of the input matrices B_1 and B_2 have full column ranks with $\text{rank}(B_1) = \text{rank}(B_2) = 1$ (i.e. $m = 1$ in this chapter).

2) The matched disturbances $f^h(i,j)$ and $f^v(i,j)$ satisfy

$$\|f^h(i,j)\| \le \varepsilon_f^h, \ \|f^v(i,j)\| \le \varepsilon_f^v, \quad (10.3)$$

where $\varepsilon_f^h \ge 0$ and $\varepsilon_f^v \ge 0$ are known bounds.

3) The initial and boundary conditions are specified as

$$x^h(i,j) = \begin{cases} \psi^h(0,j); & 0 \le j \le z_1 \\ 0; & \forall\, z_1 < j \end{cases}, \quad x^v(i,j) = \begin{cases} \psi^v(i,0); & 0 \le i \le z_2 \\ 0; & \forall\, z_2 < i \end{cases} \quad (10.4)$$

where z_1 and z_2 are given positive scalars.

It should be noted that the external disturbances are not considered in the existing literatures on SMC of 2-D Roesser model [2, 5, 171]. Thus, the 2-D Roesser model (10.1)–(10.2) in this chapter is more general for SMC issues.

Definition 10.1 *[193] The nonlinear 2-D Roesser system (10.1)–(10.2) with $u(i,j) = f(i,j) = 0$ is said to be asymptotically stable if $\lim_{i+j\to\infty} \|x(i,j)\|^2 = 0$ holds for every initial and boundary conditions satisfying (10.4).*

Remark 10.1 *Many possible applications for 2-D control systems will start from continuous-time dynamics. For instance, Dymkov et. al. [39] considered a sorption process arising in waste water and sewage treatment, which can be modelled as a 2-D continuous-time Roesser model, also known as Goursat-type equations. However, in the framework of the networked control*

Main Results

195

system, the discrete-time system dynamics is necessarily considered for the digital signal transmission. By properly sampling the functions in partial differential equation with respect to certain spatial coordinates in an appropriate way [3, 167, 168, 171], the continuous-time Roesser model under consideration can be approximately converted into the discrete-time Roesser model (10.1)–(10.2), where the horizontal and vertical external disturbances are involved for reflecting the practical applications more closely.

10.2 Main Results

10.2.1 2-D sliding surface

This chapter employs the following 2-D sliding functions:

$$s^h(i,j) = B_1^{\mathrm{T}} X^h x^h(i,j), \tag{10.5}$$

$$s^v(i,j) = B_2^{\mathrm{T}} X^v x^v(i,j), \tag{10.6}$$

where the horizontal and vertical sliding matrices $X^h > 0$ and $X^v > 0$ will be determined later.

Considering the 2-D sliding functions (10.5)–(10.6) for the uncertain 2-D Roessor system (10.1)–(10.2), we have

$$\begin{aligned}
s^h(i+1,j) =& B_1^{\mathrm{T}} X^h A_{11} x^h(i,j) + B_1^{\mathrm{T}} X^h A_{12} x^v(i,j) + B_1^{\mathrm{T}} X^h B_1 u^h(i,j) \\
& + B_1^{\mathrm{T}} X^h B_1 f^h(i,j), \tag{10.7}
\end{aligned}$$

$$\begin{aligned}
s^v(i,j+1) =& B_2^{\mathrm{T}} X^v A_{21} x^h(i,j) + B_2^{\mathrm{T}} X^v A_{22} x^v(i,j) + B_2^{\mathrm{T}} X^v B_2 u^v(i,j) \\
& + B_2^{\mathrm{T}} X^v B_2 f^v(i,j). \tag{10.8}
\end{aligned}$$

According to the SMC theory, we have the following equivalent control laws from $s^h(i+1,j) = s^h(i,j) = 0$ and $s^v(i,j+1) = s^v(i,j) = 0$:

$$u_{eq}^h(i,j) = -(B_1^{\mathrm{T}} X^h B_1)^{-1} \Big(B_1^{\mathrm{T}} X^h A_{11} x^h(i,j) + B_1^{\mathrm{T}} X^h A_{12} x^v(i,j) \Big) - f^h(i,j), \tag{10.9}$$

$$u_{eq}^v(i,j) = -(B_2^{\mathrm{T}} X^v B_2)^{-1} \Big(B_2^{\mathrm{T}} X^v A_{21} x^h(i,j) + B_2^{\mathrm{T}} X^v A_{22} x^v(i,j) \Big) - f^v(i,j). \tag{10.10}$$

10.2.2 Design of 2-D event generator

According to the assumption in (10.3), one has

$$\|B_1^{\mathrm{T}} X^h B_1 f^h(i,j)\| \leq \varepsilon_f^h \|B_1^{\mathrm{T}} X^h B_1\|, \tag{10.11}$$

$$\|B_2^{\mathrm{T}} X^v B_2 f^v(i,j)\| \leq \varepsilon_f^v \|B_2^{\mathrm{T}} X^v B_2\|. \tag{10.12}$$

In this chapter, we introduce the following reaching law conditions in the horizontal and vertical directions, respectively:

$$s^h(i+1,j) - s^h(i,j) = -q_1 T s^h(i,j) - T(s^h(i,j))^2 \text{sgn}(s^h(i,j)), \quad (10.13)$$

$$s^v(i,j+1) - s^v(i,j) = -q_2 T s^v(i,j) - T(s^v(i,j))^2 \text{sgn}(s^v(i,j)), \quad (10.14)$$

where T is the sampling time, $q_1 > 0$ and $q_2 > 0$ are the approximation rates along the horizontal and vertical directions, respectively. A judicious choice of parameters $q_1 > 0$ and $q_2 > 0$ satisfying $0 < 1 - q_1 T < 1$ and $0 < 1 - q_2 T < 1$ can achieve desirable reaching mode responses for the predesigned sliding variables $s^h(i,j)$ and $s^v(i,j)$.

In order to enhance the utilization efficiency of the communication resource between the plant and the controller, we shall design a 2-D event-triggering mechanism, which are *independent* with each other in horizontal and vertical directions [165]. Denote the current state of 2-D Roessor system (10.1)–(10.2) as $x^h(i,j)$, $x^v(i,j)$, and the latest transmitted state as $x^h(i_k,j)$, $x^v(i,j_l)$, $(k,l \in \mathbb{N}_0^+, i_0 = j_0 = 0)$. Then, the transmission errors can be defined by

$$e^h(i,j) \triangleq x^h(i,j) - x^h(i_k,j), \quad (10.15)$$

$$e^v(i,j) \triangleq x^v(i,j) - x^v(i,j_l). \quad (10.16)$$

It is worth stressing that under the event-triggering mechanism, the event-based SMC laws $u^h(i,j)$ and $u^v(i,j)$ can *only* employ the latest transmitted state $x^h(i_k,j)$, $x^v(i,j_l)$, but not the current state $x^h(i,j)$, $x^v(i,j)$ or the transmission errors $e^h(i,j)$, $e^v(i,j)$, for any $(i,j) \in [i_k, i_{k+1}) \times [j_l, j_{l+1})$, where the next controller update "instants" i_{k+1} and j_{l+1} are determined by a desired 2-D event generator. In what follows, we derive a 2-D event generator based on the reaching law conditions (10.13)–(10.14).

Theorem 10.1 *Consider the horizontal and vertical sliding functions (10.5)–(10.6) and the 2-D reaching law conditions (10.13)–(10.14). If, for given scalars $\delta^h > 0$ and $\delta^v > 0$, the following inequalities hold:*

$$\|B_1^T X^h A_{11} + B_1^T X^h\| \cdot \|e^h(i,j)\| + \|B_1^T X^h A_{12}\| \cdot \|e^v(i,j)\| \le \delta^h \|x^h(i_k,j)\|, \quad (10.17)$$

$$\|B_2^T X^v A_{21}\| \cdot \|e^h(i,j)\| + \|B_2^T X^v A_{22} + B_2^T X^v\| \cdot \|e^v(i,j)\| \le \delta^v \|x^v(i,j_l)\|, \quad (10.18)$$

then, the horizontal and vertical state trajectories of the resultant SMC system can be forced, respectively, to the sliding surfaces $s^h(i,j) = 0$ and $s^v(i,j) = 0$

Main Results 197

in finite iterations by the event-based 2-D SMC laws as follows:

$$u^h(i,j) = (B_1^T X^h B_1)^{-1}\left[(1 - q_1 T)s^h(i_k, j) - B_1^T X^h A_{11} x^h(i_k, j)\right.$$
$$\left. - B_1^T X^h A_{12} x^v(i, j_l) - \vartheta^h(i_k, j)\mathrm{sgn}(s^h(i_k, j))\right], \qquad (10.19)$$

$$u^v(i,j) = (B_2^T X^v B_2)^{-1}\left[(1 - q_2 T)s^v(i, j_l) - B_2^T X^v A_{21} x^h(i_k, j)\right.$$
$$\left. - B_2^T X^v A_{22} x^v(i, j_l) - \vartheta^v(i, j_l)\mathrm{sgn}(s^v(i, j_l))\right], \qquad (10.20)$$

where the robust terms are given as

$$\vartheta^h(i_k, j) \triangleq T(s^h(i_k, j))^2 + \varepsilon_f^h \|B_1^T X^h B_1\| + \delta^h \|x^h(i_k, j)\|, \qquad (10.21)$$
$$\vartheta^v(i, j_l) \triangleq T(s^v(i, j_l))^2 + \varepsilon_f^v \|B_2^T X^v B_2\| + \delta^v \|x^v(i, j_l)\|. \qquad (10.22)$$

Proof *Firstly, we consider the horizontal direction. Combining the sliding function (10.5) and the event-based SMC law (10.19), one gets*

$$\Delta s^h(i,j)$$
$$\triangleq s^h(i+1, j) - s^h(i, j)$$
$$= B_1^T X^h A_{11} x^h(i, j) + B_1^T X^h A_{12} x^v(i, j) + B_1^T X^h B_1 u^h(i_k, j_l)$$
$$\quad + B_1^T X^h B_1 f^h(i, j) - s^h(i, j)$$
$$= - q_1 T s^h(i_k, j) - T(s^h(i_k, j))^2 \mathrm{sgn}(s^h(i_k, j)) + B_1^T X^h B_1 f^h(i, j)$$
$$\quad - \varepsilon_f^h \|B_1^T X^h B_1\|\mathrm{sgn}(s^h(i_k, j)) + B_1^T X^h A_{11} e^h(i, j) + B_1^T X^h A_{12} e^v(i, j)$$
$$\quad + B_1^T X^h e^h(i, j) - \delta^h \|x^h(i_k, j)\|\mathrm{sgn}(s^h(i_k, j)). \qquad (10.23)$$

With the similar lines to the ones in [51, 142], it is easily shown from the conditions (10.11) and (10.17) that if $s^h(i_k, j) > 0$, then it yields $\Delta s^h(i,j) \leq -q_1 T s^h(i_k, j) - T(s^h(i_k, j))^2 \mathrm{sgn}(s^h(i_k, j)) < 0$; and if $s^h(i_k, j) < 0$, then it has $\Delta s^h(i,j) \geq -q_1 T s^h(i_k, j) - T(s^h(i_k, j))^2 \mathrm{sgn}(s^h(i_k, j)) > 0$. Therefore, the event-based SMC law (10.19) and the condition (10.17) ensure the reaching law condition (10.13) for discrete-time sliding mode $s^h(i,j) = 0$ in the horizontal direction.

Similarly, it is also obtained that in the vertical direction, the reaching law condition (10.14) for discrete-time sliding mode $s^v(i,j) = 0$ can be guaranteed by the event-based SMC law (10.20) and the condition (10.18). This completes the proof.

Now, according to Theorem 10.1, in order to ensure the reaching law conditions (10.13)–(10.14), we design the event generators in horizontal and vertical directions as follows:

$$i_{k+1} \triangleq \min_{i \in \mathbb{N}_0^+} \left\{ i > i_k \,\middle|\, (\|B_1^T X^h A_{11} + B_1^T X^h\| + \gamma^h) \cdot \|e^h(i, j)\| \right.$$
$$\left. + \|B_1^T X^h A_{12}\| \cdot \|e^v(i, j)\| > \delta^h \|x^h(i_k, j)\| \right\}, \qquad (10.24)$$

198 *2-D SMC Under ETP*

and

$$j_{l+1} \triangleq \min_{j \in \mathbb{N}_0^+} \Big\{ j > j_l \ \Big| \ (\|B_2^{\mathrm{T}} X^v A_{22} + B_2^{\mathrm{T}} X^v\| + \gamma^v) \cdot \|e^v(i,j)\|$$

$$+ \|B_2^{\mathrm{T}} X^v A_{21}\| \cdot \|e^h(i,j)\| > \delta^v \|x^v(i,j_l)\| \Big\}. \qquad (10.25)$$

where $\gamma^h \geq 0$ and $\gamma^v \geq 0$ are the assigned small scalars.

Remark 10.2 *In order to ensure the feasibility of the designed 2-D event generators (10.24)–(10.25), two ancillary parameters γ^h and γ^v are introduced. Specifically, as shown in (10.36)–(10.37) later, we choose $\gamma^h > 0$ if and only if $\|B_1^{\mathrm{T}} X^h A_{11} + B_1^{\mathrm{T}} X^h\| = 0$, $\gamma^h = 0$ if and only if $\|B_1^{\mathrm{T}} X^h A_{11} + B_1^{\mathrm{T}} X^h\| > 0$, and the same to the vertical direction. Besides, due to the discrete characteristics of the controlled system in this chapter, the Zeno execution of 2-D SMC input can be effectively avoided.*

Remark 10.3 *As well known, FM model is another widely investigated 2-D system model, see [14] and the references therein for more details on stabilization and estimation problems. It should be noted that the 2-D event generators (10.24)–(10.25) is designed by using the special characteristic of Roesser model, that is, the state dynamics of Roesser model are partitioned into independent horizontal and vertical state sub-vectors [165]. However, FM model just utilizes a single state vector, which means that the state sub-vectors in horizontal and vertical directions are dependent on each other. Hence, the proposed 2-D event-triggered SMC method for Roesser model may not be extended to the FM model readily. This open problem is served as one of future researches.*

10.2.3 Stability of sliding mode dynamics

Under the event generators (10.24)–(10.25), the equivalent control laws $u_{eq}^h(i,j)$ and $u_{eq}^v(i,j)$ in (10.9)–(10.10) become

$$u_{eq}^h(i,j) = -(B_1^{\mathrm{T}} X^h B_1)^{-1} \Big(B_1^{\mathrm{T}} X^h A_{11} x^h(i_k,j) + B_1^{\mathrm{T}} X^h A_{12} x^v(i,j_l) \Big) - f^h(i,j), \qquad (10.26)$$

$$u_{eq}^v(i,j) = -(B_2^{\mathrm{T}} X^v B_2)^{-1} \Big(B_2^{\mathrm{T}} X^v A_{21} x^h(i_k,j) + B_2^{\mathrm{T}} X^v A_{22} x^v(i,j_l) \Big) - f^v(i,j), \qquad (10.27)$$

where $i \in [i_k, i_{k+1})$ and $j \in [j_l, j_{l+1})$.

Substituting (10.26)–(10.27) into (10.1)–(10.2), the following sliding mode dynamics is obtained:

$$x^h(i+1,j) = G^h A_{11} x^h(i,j) + \tilde{X}^h A_{11} e^h(i,j) + G^h A_{12} x^v(i,j) + \tilde{X}^h A_{12} e^v(i,j), \qquad (10.28)$$

$$x^v(i,j+1) = G^v A_{21} x^h(i,j) + \tilde{X}^v A_{21} e^h(i,j) + G^v A_{22} x^v(i,j) + \tilde{X}^v A_{22} e^v(i,j), \qquad (10.29)$$

Main Results　199

with $\tilde{X}^h \triangleq B_1(B_1^T X^h B_1)^{-1}B_1^T X^h$, $\tilde{X}^v \triangleq B_2(B_2^T X^v B_2)^{-1}B_2^T X^v$, $G^h \triangleq I - \tilde{X}^h$ and $G^v \triangleq I - \tilde{X}^v$.

In the following theorem, the stability of the sliding mode dynamics (10.28)–(10.29) is analyzed by employing the 2-D Lyapunov stability theory.

Theorem 10.2 *Consider the 2-D Roesser model (10.1)–(10.2) with the sliding surfaces (10.5)–(10.6) and the event generators (10.24)–(10.25). If there exist matrices $X^h > 0$, $X^v > 0$, $P^h > 0$, $P^v > 0$, and scalars $\mu > 0$, $\varsigma > 0$ such that the following matrix inequality holds:*

$$\tilde{\Psi} \triangleq \begin{bmatrix} \tilde{\Theta} & \Gamma P^h & \Lambda P^v \\ \star & -P^h & 0 \\ \star & \star & -P^v \end{bmatrix} < 0, \tag{10.30}$$

where

$$\tilde{\Theta} \triangleq - \operatorname{diag}\{P^h, P^v, \mu I, \varsigma I\},$$
$$\Gamma \triangleq \begin{bmatrix} G^h A_{11} & G^h A_{12} & \tilde{X}^h A_{11} & \tilde{X}^h A_{12} \end{bmatrix}^T,$$
$$\Lambda \triangleq \begin{bmatrix} G^v A_{21} & G^v A_{22} & \tilde{X}^v A_{21} & \tilde{X}^v A_{22} \end{bmatrix}^T,$$

then the state trajectories of the sliding mode dynamics (10.28)–(10.29) will be forced into the following horizontal and vertical neighborhoods $\tilde{\Xi}^h$, $\tilde{\Xi}^v$ of the origins by the designed 2-D event generators (10.24)–(10.25):

$$\tilde{\Xi}^h \triangleq \{x^h(i,j) \mid \|x^h(i,j)\| \le \tilde{\varrho}(i,j)\}, \tag{10.31}$$
$$\tilde{\Xi}^v \triangleq \{x^v(i,j) \mid \|x^v(i,j)\| \le \tilde{\varrho}(i,j)\}, \tag{10.32}$$

where

$$\tilde{\varrho}(i,j) \triangleq \sqrt{\frac{\varrho(i,j)}{\lambda_{\min}(-\tilde{\Psi})}}, \tag{10.33}$$

$$\varrho(i,j) \triangleq \mu \left(\frac{\delta^h \|x^h(i_k,j)\|}{\|B_1^T X^h A_{11} + B_1^T X^h\| + \gamma^h} \right)^2$$
$$+ \varsigma \left(\frac{\delta^v \|x^v(i,j_l)\|}{\|B_2^T X^v A_{22} + B_2^T X^v\| + \gamma^v} \right)^2. \tag{10.34}$$

Proof *Consider the 2-D Lyapunov function* $V(x^h(i,j), x^v(i,j)) \triangleq x^{hT}(i,j)P^h x^h(i,j) + x^{vT}(i,j)P^v x^v(i,j)$. *Along the trajectories of the sliding mode dynamics (10.28) and (10.29), it yields*

$$\Delta V(x^h(i,j), x^v(i,j))$$
$$= x^{hT}(i+1,j)P^h x^h(i+1,j) - x^{hT}(i,j)P^h x^h(i,j) + x^{vT}(i,j+1)P^v x^v(i,j+1)$$
$$- x^{vT}(i,j)P^v x^v(i,j)$$
$$= \xi^T(i,j)\Psi\xi(i,j), \tag{10.35}$$

where

$$\xi(i,j) \triangleq \left[\begin{array}{cccc} x^{h\mathrm{T}}(i,j) & x^{v\mathrm{T}}(i,j) & e^{h\mathrm{T}}(i,j) & e^{v\mathrm{T}}(i,j) \end{array}\right]^{\mathrm{T}},$$

$$\Psi \triangleq \begin{bmatrix} \Theta & \Gamma P^h & \Lambda P^v \\ \star & -P^h & 0 \\ \star & \star & -P^v \end{bmatrix}, \Theta \triangleq -\mathrm{diag}\{P^h, P^v, 0, 0\}.$$

It is noted that by applying the 2-D event generators (10.24)–(10.25), we have

$$\|e^h(i,j)\| \leq \frac{\delta^h\|x^h(i_k,j)\|}{\|B_1^{\mathrm{T}}X^h A_{11} + B_1^{\mathrm{T}}X^h\| + \gamma^h}, \tag{10.36}$$

$$\|e^v(i,j)\| \leq \frac{\delta^v\|x^v(i,j_l)\|}{\|B_2^{\mathrm{T}}X^v A_{22} + B_2^{\mathrm{T}}X^v\| + \gamma^v}, \tag{10.37}$$

for any $(i,j) \in [i_k, i_{k+1}) \times [j_l, j_{l+1})$.

Thus, combining (10.36)–(10.37) and (10.35), for any $\mu > 0$ *and* $\varsigma > 0$, *it renders*

$$\Delta V(x^h(i,j), x^v(i,j))$$

$$\leq \xi^{\mathrm{T}}(i,j)\Psi\xi(i,j) + \mu\left[\left(\frac{\delta^h\|x^h(i_k,j)\|}{\|B_1^{\mathrm{T}}X^h A_{11} + B_1^{\mathrm{T}}X^h\| + \gamma^h}\right)^2 - \|e^h(i,j)\|^2\right]$$

$$+ \varsigma\left[\left(\frac{\delta^v\|x^v(i,j_l)\|}{\|B_2^{\mathrm{T}}X^v A_{22} + B_2^{\mathrm{T}}X^v\| + \gamma^v}\right)^2 - \|e^v(i,j)\|^2\right]$$

$$= \xi^{\mathrm{T}}(i,j)\tilde{\Psi}\xi(i,j) + \varrho(i,j). \tag{10.38}$$

If the condition (10.30) is feasible, then one has from (10.38) that

$$\Delta V(x^h(i,j), x^v(i,j)) \leq -\lambda_{\min}(-\tilde{\Psi})\left[\|\xi(i,j)\|^2 - \frac{\varrho(i,j)}{\lambda_{\min}(-\tilde{\Psi})}\right]. \tag{10.39}$$

Clearly, when the state vector $\xi(i,j)$ *lies outside the domain*

$$\Xi \triangleq \left\{\xi(i,j) \mid \|\xi(i,j)\| > \tilde{\varrho}(i,j)\right\},$$

one has $\Delta V(x^h(i,j), x^v(i,j)) < 0$. *Thus, both of the horizontal state* $x^h(i,j)$ *and the vertical state* $x^v(i,j)$ *of the ideal sliding mode dynamics (10.28)–(10.29) will be derived into the regions* $\tilde{\Xi}^h$ *and* $\tilde{\Xi}^v$ *defined in (10.31)–(10.32), respectively, by the designed 2-D event generators (10.24)–(10.25). The proof is now complete.*

10.2.4 Solving algorithm

It is seen from Theorems 10.1–10.2 that a key issue in designing 2-D event-based SMC laws (10.19)–(10.20) and 2-D event generators (10.24)–(10.25) is

Main Results

to determine the sliding matrices $X^h > 0$ and $X^v > 0$ in the sliding functions (10.5)–(10.6). However, the matrix inequality (10.30) is *nonlinear* for the decision matrices $X^h > 0$, $X^v > 0$, $P^h > 0$, and $P^v > 0$ and thus cannot be solved readily by using convex algorithms. By letting $X^h = P^h$ and $X^v = P^v$, it would be possible to employ the matrix inequality techniques as in [62,196] to overcome the above obstacle in spite of inducing much conservatism inevitably. Fortunately, if the sliding matrices X^h and X^v are specified in the matrix inequality (10.30), then it will reduce to be a LMI condition for P^h, P^v, μ, and ς, which can be solved directly by MATLAB LMI toolbox. This fact motivates us to solve the proposed co-design problem by combining GA and LMI technique.

As a typical evolutionary algorithm, GA has been successfully applied to controller synthesis problems as in [17,37,134]. The key idea of GA is to search the global optimization solution stochastically by simulating human trial-and-error procedure with Darwinian principle of "survival of the fittest". To this end, a meaningful optimization objective needs to be developed firstly for the proposed co-design problem in this chapter.

Remark 10.4 *It is well known that nature-inspired evolutionary algorithms can be used to solve the nonlinear optimization problem effectively via directed random researches. Actually, the SMC problem of multi-input system just involves a nonlinear constraint for ensuring the existence of the sliding surface. For example, the sliding gain matrices X^h and X^v in (10.5)–(10.6) should be determined properly such that both of the horizontal and vertical sliding mode dynamics (10.28)–(10.29) are stable. In this chapter, we solve the above nonlinear constrained problem by using GA. Obviously, the introduce of the existing evolutionary algorithms (e.g., GA, particle swarm optimization (PSO) and simulated annealing (SA)) into the design procedure will help us to produce a "smarter" SMC strategy [197]. This chapter represents the first attempt to this interesting research direction in 2-D systems.*

It is observed from (10.31)–(10.32) that if $\tilde{\varrho}(i,j)$ is more smaller, then the convergence of the sliding mode dynamics (10.28)–(10.29) is faster. However, a smaller $\tilde{\varrho}(i,j)$ may also imply that more events would be triggered as per the 2-D event generators (10.24)–(10.25), that is, more data packet $(x^h(i_k,j)$ and $x^v(i,j_l))$ would be sent to the controller via the communication network. Hence, the trade-off between the convergence of the sliding mode dynamics and the scheduling performance of the designed 2-D event generators (10.24)–(10.25) may be considered while solving sliding matrices X^h and X^v.

By applying the inequality $\sqrt{x^2 + y^2} \le x + y$ $(x, y \in \mathbb{R}^+)$ to $\tilde{\varrho}(i,j)$ in (10.33), we introduce the following multi-objectives minimization problem:

$$\min_{X^h > 0, X^v > 0} \quad g(X^h, X^v) \tag{10.40}$$

subject to: $P^h > 0, P^v > 0, \mu > 0, \varsigma > 0$ and LMI (10.30), where the objective function $g(X^h, X^v)$ is defined as

$$g(X^h, X^v) \triangleq \alpha_1 \left[\frac{1}{\sqrt{\lambda_{\min}(-\tilde{\Psi})}} \left(\frac{\delta^h \sqrt{\mu}}{\|B_1^T X^h A_{11} + B_1^T X^h\| + \gamma^h} \right. \right.$$

$$\left. \left. + \frac{\delta^v \sqrt{\varsigma}}{\|B_2^T X^v A_{22} + B_2^T X^v\| + \gamma^v} \right) \right] + \alpha_2 \frac{\|B_1^T X^h A_{11} + B_1^T X^h\|}{\delta^h}$$

$$+ \alpha_3 \frac{\|B_2^T X^v A_{22} + B_2^T X^v\|}{\delta^v}, \tag{10.41}$$

with weighting parameters α_b ($b = 1, 2, 3$) satisfying $\alpha_b \geq 0$ and $\sum_{b=1}^{3} \alpha_b = 1$. In fact, each weighting parameter α_b exhibits its own physical meaning in the optimizing co-design, that is, α_1 denotes the actual requirement for the control performance, α_2 presents the performance of the horizontal event generator, and α_3 reflects the performance of the vertical event generator.

In what follows, a binary-based GA is formulated to solve the constructed multi-objectives optimization problem (10.40) for a set of prespecified weighting parameters $\{\alpha_1, \alpha_2, \alpha_3\}$:

• **Step 1: Parameter encoding.** Due to $X^h = [\varphi_{t_1 r_1}]_{n^h \times n^h} > 0$ and $X^v = [\phi_{t_2 r_2}]_{n^v \times n^v} > 0$, so there are $\frac{n^h(n^h+1)}{2}$ independent variables $\varphi_{t_1 r_1}$ in X^h and $\frac{n^v(n^v+1)}{2}$ independent variables $\phi_{t_2 r_2}$ in X^v. Now, we express the phenotype in the search space as the following row vector $w \in \mathbb{R}^{1 \times \frac{n^h(n^h+1) + n^v(n^v+1)}{2}}$:

$$[X^h, X^v] \to w$$
$$\triangleq \begin{bmatrix} \varphi_{11} & \varphi_{12} & \cdots & \varphi_{1n^h} & \varphi_{22} & \varphi_{23} & \cdots & \varphi_{2n^h} & \varphi_{33} & \cdots & \varphi_{n^h n^h} \\ \phi_{11} & \phi_{12} & \cdots & \phi_{1n^v} & \phi_{22} & \phi_{23} & \cdots & \phi_{2n^v} & \phi_{33} & \cdots & \phi_{n^v n^v} \end{bmatrix}.$$

Each element $\varphi_{t_1 r_1}$ and $\phi_{t_2 r_2}$ in w are coded as a binary string of length $\ell_{\varphi_{t_1 r_1}}$ and $\ell_{\phi_{t_2 r_2}}$, respectively, which are determined by their own ranges of $\varphi_{t_1 r_1} \in [\underline{\varphi}_{t_1 r_1}, \overline{\varphi}_{t_1 r_1}]$ and $\phi_{t_2 r_2} \in [\underline{\phi}_{t_2 r_2}, \overline{\phi}_{t_2 r_2}]$. Specifically, due to $X^h > 0$ and $X^v > 0$, one has $\underline{\varphi}_{t_1 r_1} > 0$ for $t_1 = r_1 \in \{1, \cdots, n^h\}$ and $\underline{\phi}_{t_2 r_2} > 0$ for $t_2 = r_2 \in \{1, \cdots, n^v\}$. Hence, the precisions $q_{\varphi_{t_1 r_1}}$ and $q_{\phi_{t_2 r_2}}$ under the linearly-mapped coding can be obtained by

$$q_{\varphi_{t_1 r_1}} = \frac{\overline{\varphi}_{t_1 r_1} - \underline{\varphi}_{t_1 r_1}}{2^{\ell_{\varphi_{t_1 r_1}}} - 1}, \quad q_{\phi_{t_2 r_2}} = \frac{\overline{\phi}_{t_2 r_2} - \underline{\phi}_{t_2 r_2}}{2^{\ell_{\phi_{t_2 r_2}}} - 1}.$$

By doing so, the following binary-based chromosome w (an individual) of length $\sum \ell_{\varphi_{t_1 r_1}} + \sum \ell_{\phi_{t_2 r_2}}$ is produced by concatenating all the strings (genotype):

$$w \to \overbrace{\underbrace{001010}_{\varphi_{11}} \cdots \underbrace{011100}_{\varphi_{n^h n^h}} \underbrace{101101}_{\phi_{11}} \cdots \underbrace{010101}_{\phi_{n^v n^v}}}^{\sum \ell_{\varphi_{t_1 r_1}} + \sum \ell_{\phi_{t_2 r_2}} \text{ bits}}.$$

Main Results

- **Step 2: Population initialization.** Randomly generate an initial population of N chromosomes w_c, $c = 1, 2, \ldots, N$.
- **Step 3: Fitness function and assignment.** Decode the initial population produced in Step 2 into a real values for every sliding matrices X_c^h and X_c^v, $c = 1, 2, \ldots, N$. In this chapter, we define the fitness function as:

$$\mathbf{F}(g(X^h, X^v)) \triangleq \frac{1}{g(X^h, X^v)}.$$

Obviously, minimizing the objective function $g(X^h, X^v)$ is equivalent to getting a maximum fitness value in the genetic search. Next, compute the fitness function $g(X_c^h, X_c^v)$ for every X_c^h and X_c^v by solving LMI (10.30). If for a set of X_c^h and X_c^v, LMI (10.30) is infeasible or $X_c^h \leq 0$ or $X_c^v \leq 0$, then the fitness function $\mathbf{F}(g(X_c^h, X_c^v))$ will be artificially assigned a sufficiently small value for reducing its opportunity to survive in the next generation.

- **Step 4: Performing genetic operations.** According to the assigned fitness in Step 3, the next population will be obtained by executing the sequence of genetic operations step-by-step as follows:

1. *Selection.* In order to guarantee the survival of the strongest elements in the population, the classic "Roulette Wheel" selection method is employed here to select some individuals w_c for populating the next generation via recombination. Meanwhile, the elitist reinsertion strategy is implemented to guarantee that the best chromosomes in the present population always survives and is retained in the next generation.

2. *Crossover.* The single-point crossover strategy is applied to a newly selected chromosomes (parents). In each pair of chromosomes, the crossover bit position is probabilistically determined via the crossover probability p_c. And then, produce pairs of new chromosomes (offspring) by pairing together.

3. *Mutation.* After crossover, each chromosome is subjected to a single bit mutation with a small mutation probability p_m, that is, change a bit from 1 to 0 and vice versa.

- **Step 5: Stop criterion.** The evolution process will be repeated from Step 3 to 4 in each generation until the maximum generations $T_{\max}$ is reached or the prespecified accuracy $\epsilon > 0$ is converged. When the evolution process stops, the *best* chromosome w_c is decoded into real values to produce the sliding matrices X^h and X^v.
- **Step 6: Co-design of 2-D event generators and SMC laws.** Produce the 2-D event generators (10.24)–(10.25) and the SMC law (10.19)–(10.20) by using the sliding matrices X^h and X^v obtained in Step 5, and then apply them to stabilize the 2-D Roesser model (10.1)–(10.2).

204　　2-D SMC Under ETP

Remark 10.5 *It is worth emphasizing that the proposed co-design approach via GA in this chapter exhibits the following two distinct features: 1) the designed 2-D event generators (10.24)–(10.25) not only ensure the sliding motion under the event-based SMC laws (10.19)–(10.20) but also guarantee the ultimate boundedness of the 2-D sliding mode dynamics (10.28)–(10.29); and 2) in order to avoid introducing conservatism in solving the sliding matrices, the obtained nonlinear matrix inequality condition is dealt with by a combination of GA and LMI, where a trade-off-based multi-objective performance is considered for facilitating different engineering requirements.*

10.3　Example

Consider the 2-D Roesser model in (10.1)–(10.2) with the following parameters:

$$A_{11} = \begin{bmatrix} 1 & 0.35 \\ -0.28 & 0.8 \end{bmatrix}, \ A_{12} = \begin{bmatrix} 0.25 & -0.30 \\ -0.30 & 0.15 \end{bmatrix}, \ A_{21} = \begin{bmatrix} 0.45 & 0.20 \\ 0.25 & -0.30 \end{bmatrix},$$

$$A_{22} = \begin{bmatrix} 0.6 & 0.25 \\ -0.75 & -0.41 \end{bmatrix}, \ B_1 = \begin{bmatrix} 0 \\ 2 \end{bmatrix}, B_2 = \begin{bmatrix} 1 \\ 1 \end{bmatrix},$$

$$f^h(i,j) = 0.1\sin(0.01(i+j)), \ f^v(i,j) = 0.1\sin(|30(i+j-1)|).$$

Thus, one can choose $\varepsilon_f^h = \varepsilon_f^v = 0.1$ in (10.3). The initial and boundary conditions in (10.4) are given as $\psi^h(0,l) = 0.3$, $\psi^v(k,0) = -0.4$, $z_1 = 21$ and $z_2 = 12$. The open-loop case of the 2-D Roesser model is shown in Fig. 10.1.

Select the design parameters in the 2-D event generators (10.24)–(10.25) as $\delta^h = \delta^v = 0.3$ and $\gamma^h = \gamma^v = 0$. We are now in position to solve the sliding matrices X^h and X^v in (10.5)–(10.6) by using the proposed GA in Section 10.2.4, where the weighting parameters α_b ($b = 1,2,3$) in the objective function $g(X^h, X^v)$ are set as $\alpha_1 = 0.2$ and $\alpha_2 = \alpha_3 = 0.4$. The GA parameters employed here are: population size $N = 80$; maximum of generations $T_{\max} = 300$; crossover probability $p_c = 0.8$; mutation probability $p_m = 0.05$; bounds of the elements $\overline{\varphi}_{11} = \overline{\varphi}_{22} = \overline{\phi}_{11} = \overline{\phi}_{22} = 1$, $\underline{\varphi}_{11} = \underline{\varphi}_{22} = \underline{\phi}_{11} = \underline{\phi}_{22} = 0.0001$, $\overline{\varphi}_{12} = \overline{\phi}_{12} = 1$, and $\underline{\varphi}_{12} = \underline{\phi}_{12} = -1$; and lengths of binary strings $\ell_{\varphi_{11}} = \ell_{\varphi_{22}} = \ell_{\phi_{11}} = \ell_{\phi_{22}} = 10$ and $\ell_{\varphi_{12}} = \ell_{\phi_{12}} = 11$.

Under the above parameters, the proposed GA products a set of optimized sliding matrices as follows:

$$X^h = \begin{bmatrix} 0.5543 & 0.1099 \\ 0.1099 & 0.2982 \end{bmatrix}, \ X^v = \begin{bmatrix} 0.9257 & -0.6063 \\ -0.6063 & 0.9345 \end{bmatrix}.$$

Fig. 10.2 depicts the best and average fitness values of each generation in the computation process of GA.

Example

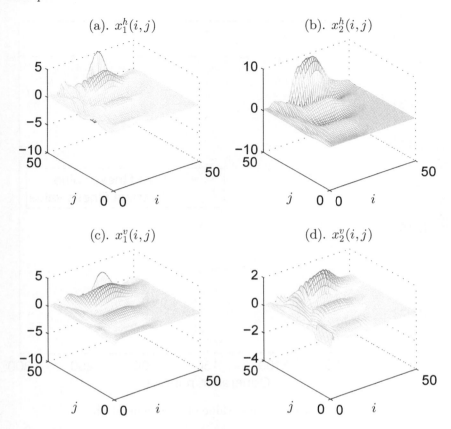

FIGURE 10.1: State trajectories $x^h(i,j)$ and $x^v(i,j)$ in open-loop case.

By substituting the obtained sliding matrices into (10.24)–(10.25), the following 2-D event generators are developed:

$$i_{k+1} \triangleq \min_{i \in \mathbb{N}_0^+} \left\{ i > i_k \mid 1.1824 \|e^h(i,j)\| + 0.1262 \|e^v(i,j)\| > 0.3 \|x^h(i_k,j)\| \right\}, \tag{10.42}$$

$$j_{l+1} \triangleq \min_{j \in \mathbb{N}_0^+} \left\{ j > j_l \mid 0.3832 \|e^v(i,j)\| + 0.2285 \|e^h(i,j)\| > 0.3 \|x^v(i,j_l)\| \right\}. \tag{10.43}$$

Furthermore, for $q_1 = q_2 = 0.3$ and $T = 0.2$, the event-based SMC laws (10.19)–(10.20) are computed as

$$u^h(i,j) = 0.8383 \times \Big[0.94 s^h(i_k,j) - \begin{bmatrix} 0.0528 & 0.5541 \end{bmatrix} x^h(i_k,j)$$
$$- \begin{bmatrix} -0.1240 & 0.0235 \end{bmatrix} x^v(i,j_l) - \vartheta^h(i_k,j) \operatorname{sgn}(s^h(i_k,j)) \Big], \tag{10.44}$$

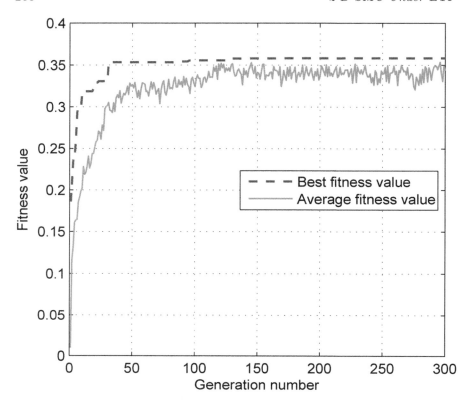

FIGURE 10.2: Fitness value of each generation.

$$u^v(i,j) = 1.5439 \times \Big[0.94 s^v(i,j_l) - \begin{bmatrix} 0.2258 & -0.0346 \end{bmatrix} x^h(i_k, j)$$
$$- \begin{bmatrix} -0.0545 & -0.0514 \end{bmatrix} x^v(i, j_l) - \vartheta^v(i, j_l) \mathrm{sgn}(s^v(i, j_l)) \Big], \tag{10.45}$$

where the robust terms are $\vartheta^h(i_k, j) = 0.2(s^h(i_k, j))^2 + 0.1193 + 0.3\|x^h(i_k, j)\|$ and $\vartheta^v(i, j_l) = 0.2(s^v(i, j_l))^2 + 0.0648 + 0.3\|x^v(i, j_l)\|$.

Figs. 10.3–10.6 show the simulation results. It is observed from Fig. 10.3 that under the obtained 2-D SMC laws (10.44)–(10.45) with the designed 2-D event generators (10.42)–(10.43), the 2-D Roesser system can achieve asymptotic stability. The evolutions of the sliding variables $s^h(i,j)$, $s^v(i,j)$, and the event-based SMC schemes $u^h(i,j)$, $u^v(i,j)$ are shown in Figs. 10.4–10.5, respectively. Fig. 10.6 depicts the horizontal and vertical scheduling instants under the designed 2-D event generators (10.42)–(10.43), respectively, where "1" means the event is triggered and "0" means not. One can observe that the communication burden can be reduced much by the designed 2-D event generators.

Example

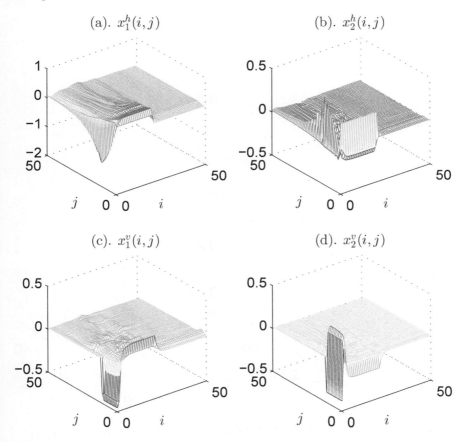

FIGURE 10.3: State trajectories $x^h(i,j)$ and $x^v(i,j)$ in closed-loop case.

Furthermore, we are interesting to illustrate the ultimate boundedness conditions (10.31)–(10.32). To this end, we introduce the following auxiliary functions:

$$AF^h(i,j) = \begin{cases} 1, & \text{if } \|x^h(i,j)\| \leq \tilde{\varrho}(i,j) \\ 0, & \text{if } \|x^h(i,j)\| > \tilde{\varrho}(i,j) \end{cases},$$

$$AF^v(i,j) = \begin{cases} 1, & \text{if } \|x^v(i,j)\| \leq \tilde{\varrho}(i,j) \\ 0, & \text{if } \|x^v(i,j)\| > \tilde{\varrho}(i,j) \end{cases}.$$

Fig. 10.7 plots the evolutions of the above auxiliary functions $AF^h(i,j)$ and $AF^v(i,j)$, from which one can observe that the ultimate boundedness conditions (10.31)–(10.32) can be achieved indeed in both of the horizontal and vertical directions by the obtained 2-D event generators (10.42)–(10.43) and 2-D SMC laws (10.44)–(10.45).

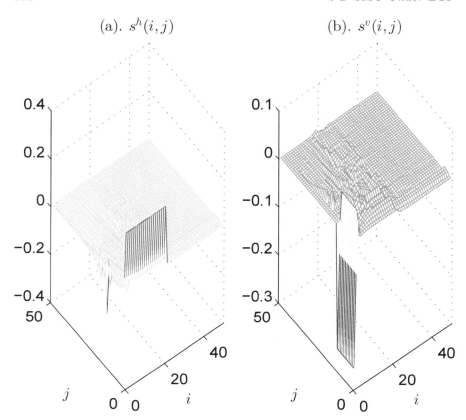

FIGURE 10.4: Evolutions of sliding variables $s^h(i,j)$ and $s^v(i,j)$.

Finally, we are in a position to explore the effects of the event-triggered parameters δ^h and δ^v to the 2-D SMC performance and the trigger times generated by the 2-D event generators (10.42)–(10.43). Five cases (i.e., $\delta^h = \delta^v = 0$, $\delta^h = \delta^v = 0.1$, $\delta^h = \delta^v = 0.2$, $\delta^h = \delta^v = 0.3$ and $\delta^h = \delta^v = 0.4$) are compared within the sampling region $(i,j) \in [0,50] \times [0,50]$, that is, there are 2500 sampling points in each direction. Besides, we compute the value of $SUM = \sum_{i=0}^{50} \sum_{j=0}^{50} (\|x^h(i,j)\| + \|x^v(i,j)\|)$ to reflect the 2-D SMC performance. Clearly, the smaller value SUM implies the better 2-D SMC performance over the sampling region $(i,j) \in [0,50] \times [0,50]$. Let TT^h and TT^v represent the triggering times in horizontal and vertical direction, respectively. Table 10.1 shows the comparing results, from which we find that the larger event-triggered parameters δ^h and δ^v will render the less triggering times at the cost of the worse 2-D SMC performance reflecting in the larger value SUM. All of simulation results have verified the effectiveness of the proposed co-design approach for the 2-D Roesser model via GA.

Example   209

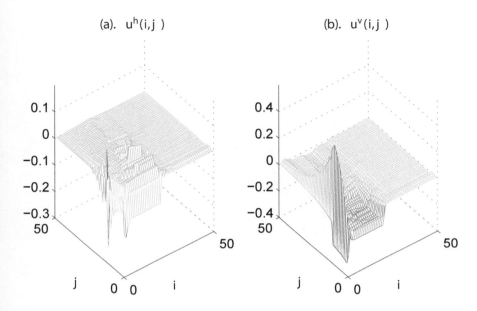

FIGURE 10.5: Evolutions of event-based SMC inputs $u^h(i,j)$ and $u^v(i,j)$.

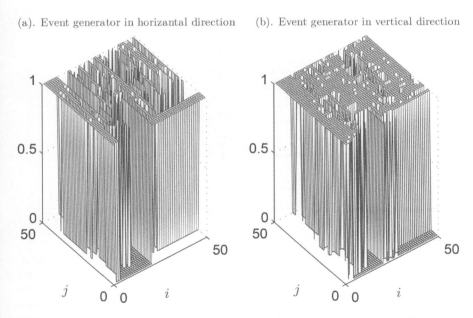

FIGURE 10.6: Horizontal and vertical scheduling instants under 2-D event generators.

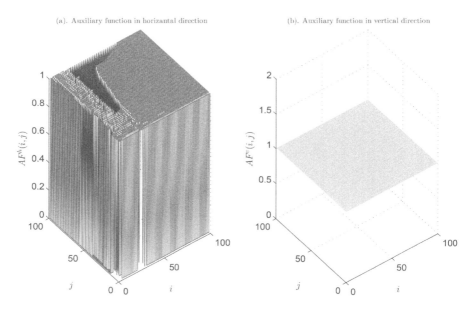

FIGURE 10.7: Auxiliary functions in horizontal and vertical directions.

TABLE 10.1: Comparing results for five sets of δ^h and δ^v.

Triggering parameters	TT^h	TT^v	SUM
$\delta^h = \delta^v = 0$	2500	2500	2.0855×10^2
$\delta^h = \delta^v = 0.1$	1953	2228	3.6874×10^2
$\delta^h = \delta^v = 0.2$	1398	1865	4.3590×10^2
$\delta^h = \delta^v = 0.3$	1050	1779	4.5018×10^2
$\delta^h = \delta^v = 0.4$	722	1165	5.9759×10^2

10.4 Conclusion

This chapter has investigated the event-triggering SMC problem for the 2-D Roesser model by using GA. Based on the 2-D reaching law conditions, the event-based 2-D SMC schemes have been derived with giving 2-D event generators for the horizontal and vertical directions independently. By resorting to the 2-D Lyapunov stability theory, the stability of the sliding mode dynamics has been analyzed. In order to solve the sliding matrices without introducing conservatism, a binary-based GA has been formulated, where the

Conclusion 211

multi-objective optimization problem was constructed based on a trade-off between the convergence of the sliding mode dynamics and the scheduling performance of 2-D event generators. The effectiveness of the proposed co-design approach with GA has been demonstrated via an numerical example.

11

2-D Sliding Mode Control Under Round-Robin Protocol

Over the past few decades, two-dimensional (2-D) systems have gained considerable attention since a large number of practical systems can be modelled as 2-D systems such as those in repetitive processes, image data processing and transmission, thermal process, gas absorption, water stream heating and iterative learning control. It is well known that 2-D systems can be represented by different models such as the Roesser model, Fornasini-Marchesini (FM) model, and Attasi model. So far, the stability analysis of 2-D systems described by these models has attracted a great deal of interest and some significant results have been obtained.

In this chapter, we endeavour to develop a framework of designing the 2-D first- and second-order sliding mode control (SMC) laws for an uncertain Roesser model under the Round-Robin protocol. Based on the notion of "time instant" for the 2-D switched systems, we introduce a Round-Robin protocol for the 2-D systems, in which only one actuator is allowed to obtain access to the communication network at each "time instant" and the Round-Robin scheduling signal is employed to determine which actuator should be given the access to the network. Besides, a set of zero-order holders (ZOHs) is exploited to keep the other actuator nodes unchanged until the next renewed control signal arrives. By taking into account the impacts from the Round-Robin scheduling signal and the ZOHs, token-dependent 2-D first- and second-order SMC laws are constructed, respectively, based on a new 2-D common sliding function. Sufficient conditions guaranteeing the ultimate boundedness and the reachability of the resultant 2-D closed-loop systems are obtained by employing some token-dependent Lyapunov-like functions. *The main contributions of this chapter are highlighted as follows: 1) With the aid of the notion of "time instant" for 2-D switched systems, the Round-Robin protocol with the ZOHs strategy is introduced to the 2-D NCSs, in which the access token among actuators is regulated orderly so as to reduce the communication burden in controller-to-actuator (C/A) network; 2) By constructing a novel 2-D common sliding function, token-dependent 2-D first- and second-order SMC laws are designed, respectively, with the consideration of the impacts from the Round-Robin scheduling signal and the ZOHs. It is shown that the existence condition of the proposed 2-D common sliding surface is more relaxed that the existing 2-D directional sliding surface; 3) Sufficient conditions guaranteeing*

DOI: 10.1201/9781003309499-11

both the ultimate boundedness of the state trajectories and the sliding variable are obtained by exploiting some token-dependent Lyapunov-like functions; and 4) The solving algorithms for acquiring the optimized controller parameters are established via two optimization problems, which have been illustrated clearly in two examples.

11.1 Problem Formulation

Define $A \triangleq \begin{bmatrix} A_{11} & A_{12} \\ A_{21} & A_{22} \end{bmatrix}$, $B \triangleq \begin{bmatrix} B_1 \\ B_2 \end{bmatrix}$, $x(k,l) \triangleq \begin{bmatrix} x^h(k,l) \\ x^v(k,l) \end{bmatrix}$, and $x^+(k,l) \triangleq \begin{bmatrix} x^h(k+1,l) \\ x^v(k,l+1) \end{bmatrix}$. In this chapter, we study the following uncertain 2-D system represented by the Roesser model:

$$x^+(k,l) = (A + \Delta A(k,l)) x(k,l) + B\left(u(k,l) + f(x(k,l))\right), \qquad (11.1)$$

where $x^h(k,l) \in \mathbb{R}^{n_h}$ and $x^v(k,l) \in \mathbb{R}^{n_v}$ represent, respectively, the horizontal and vertical states. $u(k,l) \in \mathbb{R}^m$ denotes the actuators signal. $A_{11} \in \mathbb{R}^{n_h \times n_h}$, $A_{12} \in \mathbb{R}^{n_h \times n_v}$, $A_{21} \in \mathbb{R}^{n_v \times n_h}$, $A_{22} \in \mathbb{R}^{n_v \times n_v}$, $B_1 \in \mathbb{R}^{n_h \times m}$, and $B_2 \in \mathbb{R}^{n_v \times m}$ are known constant matrices. The time-varying uncertainty $\Delta A(k,l)$ is assumed to be norm-bounded, that is, $\Delta A(k,l) = MY(k,l)N$, where M and N are known constant matrices, and the unknown time-varying matrix $Y(k,l)$ satisfies $Y^T(k,l)Y(k,l) \leq I$. Without loss of generality, we make the following assumptions on the uncertain Roesser model (11.1):

1) The input matrix B has full column rank, that is, rank$(B) = m$.

2) The nonlinear function $f(x(k,l))$ satisfies

$$\|f(x(k,l))\| \leq \varepsilon \|x(k,l)\| + d, \qquad (11.2)$$

where $\varepsilon \geq 0$ and $d \geq 0$ are two known parameters.

3) The initial and boundary conditions are specified as

$$x^h(k,l) = \begin{cases} \phi^h(0,l); & 0 \leq l \leq z_1 \\ 0; & \forall z_1 < l \end{cases}$$
$$x^v(k,l) = \begin{cases} \phi^v(k,0); & 0 \leq k \leq z_2 \\ 0; & \forall z_2 < k \end{cases} \qquad (11.3)$$

where z_1 and z_2 are given positive scalars.

It should be noted that the effects of the communication delays may be also coped within the parameter uncertainty $\Delta A(k,l)x(k,l)$ and the external disturbance $f(x(k,l))$. In existing literature on SMC problem of 2-D systems

Problem Formulation

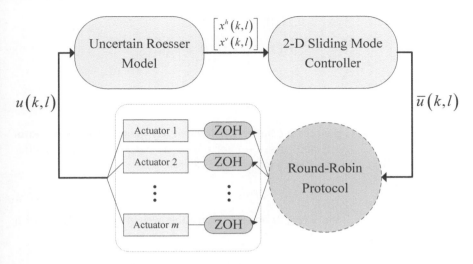

FIGURE 11.1: 2-D SMC subject to Round-Robin scheduling protocol

[2, 5, 171], neither the parameter uncertainty nor the external disturbance has been considered, which means that the uncertain Roesser model (11.1) in this chapter is more general than those investigated in the literature.

In this chapter, the controller signal $\bar{u}(k,l)$ is transmitted to the actuators $u(k,l)$ via a shared communication channel subject to the Round-Robin communication protocol. As shown in Fig. 11.1, a set of ZOHs is employed to store the received values at the actuators' side. For technical analysis, we denote

$$\bar{u}(k,l) \triangleq \begin{bmatrix} \bar{u}_1^{\mathrm{T}}(k,l) & \bar{u}_2^{\mathrm{T}}(k,l) & \cdots & \bar{u}_m^{\mathrm{T}}(k,l) \end{bmatrix}^{\mathrm{T}} \in \mathbb{R}^m, \qquad (11.4)$$

$$u(k,l) \triangleq \begin{bmatrix} u_1^{\mathrm{T}}(k,l) & u_2^{\mathrm{T}}(k,l) & \cdots & u_m^{\mathrm{T}}(k,l) \end{bmatrix}^{\mathrm{T}} \in \mathbb{R}^m, \qquad (11.5)$$

where $\bar{u}_i(k,l)$ is the ith controller signal and $u_i(k,l)$ is the received signal by the ith actuator.

It is noted from the evolution of the 2-D system (11.1) that the local state results from a global state assignment $x^h(k+1,l)$ and $x^v(k,l+1)$, which implies that the 2-D system causality imposes an increment depending on $\xi \triangleq k+l$. Similar to the works in [10, 46, 173], ξ is regarded as the "time instant" in 2-D systems (11.1). Now, by resorting to the notion ξ, we can extend the Round-Robin protocol for 1-D systems ([153, 219]) to the 2-D case as follows:

- At a certain "time instant" ξ, **only one actuator** obtains the token to access the C/A network;

- Let $\sigma(\xi) \in \mathcal{M} \triangleq \{1, 2, \ldots, m\}$ denote the scheduling signal at the "time instant" ξ. For the ith actuator:

216   *2-D SMC Under RRP*

- IF $\mathrm{mod}(\xi - i, m) = 0$, THEN

$$\sigma(\xi) = i, \ u_i(k, l) = \bar{u}_i(k, l), \tag{11.6}$$

- ELSE

$$u_i(k, l) = v_i(k, l), \tag{11.7}$$

where $v_i(k, l)$ denotes a compensation value. By means of the **previous** values $u_i(k-1, l)$ and $u_i(k, l-1)$ of the ith actuator stored in the buffer of a ZOH, this chapter chooses $v_i(k, l)$ as follows:

$$v_i(k, l) \triangleq \alpha_1 u_i(k-1, l) + \alpha_2 u_i(k, l-1), \tag{11.8}$$

where $\alpha_1 \geq 0$ and $\alpha_2 \geq 0$ are two adjustable weighting parameters with satisfying $\alpha_1 + \alpha_2 = 1$.

- END.

According to the above 2-D Round-Robin communication protocol (11.6)–(11.8), we obtain the updating rule for the ith actuator $u_i(k, l)$ ($i \in \mathcal{M}$) as follows:

$$u_i(k, l) = \begin{cases} \bar{u}_i(k, l), & \text{if } \sigma(\xi) = i, \\ v_i(k, l), & \text{otherwise.} \end{cases} \tag{11.9}$$

Obviously, the communication burden could be significantly reduced at every "time instant" ξ, thereby relieving the pressure on the communication channel. Besides, it is easily seen that the scheduling signal $\sigma(\xi)$ satisfies $\sigma(\xi + m) = \sigma(\xi)$ and can be calculated by

$$\sigma(\xi) = \mathrm{mod}(\xi - 1, m) + 1, \ \forall \xi \in \mathbb{N}^+. \tag{11.10}$$

By exploiting the updating rule (11.9) for each actuator, we formulate the *actual* control input $u(k, l)$ at the plant side as

$$u(k, l) = \Phi_{\sigma(\xi)} \bar{u}(k, l) + \left(I - \Phi_{\sigma(\xi)}\right) v(k, l), \xi \in \mathbb{N}^+, \tag{11.11}$$

where the update matrix is defined as $\Phi_{\sigma(\xi)} \triangleq \mathrm{diag}\{\tilde{\delta}^1_{\sigma(\xi)}, \tilde{\delta}^2_{\sigma(\xi)}, \cdots, \tilde{\delta}^m_{\sigma(\xi)}\}$ in which $\tilde{\delta}^b_a \triangleq \delta(a - b)$ with $\delta(\cdot)$ as the Kronecker delta function. Here, we assume that the input components $u(k, l-1)$ and $u(k-1, l)$ have the initial and boundary conditions as $u(k, -1) = u(k, 0), \forall k \geq 0$ and $u(-1, l) = u(0, l), \forall l \geq 0$

Remark 11.1 *As shown in the proposed 2-D Round-Robin protocol (11.6)–(11.8), a set of ZOHs is employed to keep the received values unchanged in the actuators end until the next renewed controller signal arrives. It is readily seen that the proposed 2-D SMC approach in this chapter can be reduced to the case with the zero-input strategy, that is, if the ith actuator does not obtain*

Main Results   217

the access token at the "time instant" $\xi = k + l$, _then the value of $\bar{u}_i(k,l)$
is generated as zero. Thus, the updating rule for the ith actuator $u_i(k,l)$ under the 2-D Round-Robin protocol with zero-input strategy can be formulated
as_ $u_i(k,l) = \begin{cases} \bar{u}_i(k,l), & \text{if } \sigma(\xi) = i, \\ 0, & \text{otherwise.} \end{cases}$ _Accordingly, the actual control input_
$u(k,l)$ _in (11.11) becomes_ $u(k,l) = \Phi_{\sigma(\xi)}\bar{u}(k,l), \ \xi \in \mathbb{N}^+$.

Substituting (11.11) into (11.1) yields the following closed-loop system:

$$x^+(k,l) = (A + \Delta A(k,l)) x(k,l) + Bf(x(k,l))$$
$$+ B \left(I - \Phi_{\sigma(\xi)}\right) v(k,l) + B\Phi_{\sigma(\xi)}\bar{u}(k,l). \tag{11.12}$$

The objective in this chapter is to design a 2-D sliding mode controller
$\bar{u}(k,l)$ such that the closed-loop system (11.12) is ultimately bounded subject
to the matched disturbance $f(x(k,l))$.

11.2 Main Results

11.2.1 A novel 2-D common sliding function

In this chapter, we employ _a 2-D common sliding function_ as follows:

$$s(k,l) = B^{\mathrm{T}}X \left[\begin{array}{c} x^h(k,l) \\ x^v(k,l) \end{array} \right] - B^{\mathrm{T}}XA \left[\begin{array}{c} x^h(k-1,l) \\ x^v(k,l-1) \end{array} \right], \tag{11.13}$$

where $X > 0$ is a preassigned matrix. Clearly, the nonsingularity of $B^{\mathrm{T}}XB$ is
ensured due to the full column rank of B.

Remark 11.2 _In this chapter, the state trajectories along the horizontal and
the vertical directions possess the same sliding surface (11.13), and this is
the reason why we call the sliding surface (11.13) as a 2-D common sliding
surface. Notice that the existence condition for the proposed common sliding
surface (11.13) is more relaxed than the ones for the directional sliding surfaces employed in [2, 171, 192]. To be specific, to ensure the feasibility of the
directional sliding surfaces, the following assumptions have been made on the
input matrix B:_

- _Denoting the input matrix $B \triangleq \begin{bmatrix} B_1^{\mathrm{T}} & B_2^{\mathrm{T}} \end{bmatrix}^{\mathrm{T}}$ in [171, 192], both of the
  matrices B_1 and B_2 should be full column rank (see Assumption 3 therein);_

- _In [2], the input matrix $B \in \mathbb{R}^{(n_h+n_v) \times m}$ should satisfy $m \geq 2$ (see Eq. (1)
  therein)._

218　　　　　　　　　　　　　　　　　　　　　　　　　　*2-D SMC Under RRP*

Clearly, the key precondition for the 2-D common sliding surface (11.13) is easier to be attained in practical applications since it just requires the full column rank of the input matrix B with $m \geq 1$. On the other hand, by converting the 2-D systems to the 1-D form, an interesting SMC design scheme has been reported in [5] for the first FM model. Unfortunately, it is required in [5] that one of the distinct variables of 2-D systems should be finite, which may not be satisfied in practical applications.

Remark 11.3 *It is worth emphasizing that the equivalent control law approach proposed in [2, 171] is no longer applicable in this chapter. In fact, from (11.1), (11.11), and (11.13), one obtains*

$$s^+(k,l) \triangleq B^{\mathrm{T}} x^+(k,l) - B^{\mathrm{T}} X A x(k,l)$$
$$= B^{\mathrm{T}} X B \Phi_{\sigma(\xi)} \bar{u}(k,l) + D(k,l) + B^{\mathrm{T}} X B \left(I - \Phi_{\sigma(\xi)}\right) v(k,l), \quad (11.14)$$

where $D(k,l) \triangleq B^{\mathrm{T}} X \Delta A(k,l) x(k,l) + B^{\mathrm{T}} X B f(x(k,l))$. Notice that the matrix $B^{\mathrm{T}} X B \Phi_{\sigma(\xi)}$ is not an invertible matrix due to the update matrix $\Phi_{\sigma(\xi)}$, which also reflects that the utilization of Round-Robin protocol contributes much complexity to the stability analysis of the addressed 2-D SMC problem as compared the one in [2, 171].

To stabilize the uncertain closed-loop system (11.12), two design frameworks of the SMC laws $\bar{u}(k,l)$ with first- and second-order sliding mode will be developed, respectively, by utilizing the novel 2-D common sliding function $s(k,l)$ in (11.13). It is found from (11.12) that the introduction of Round-Robin scheduling mechanism (11.6)–(11.7) will bring two kinds of complexities to 2-D SMC design, that is, the scheduling signal $\sigma(\xi)$ and the input component $v(k,l)$ caused by ZOHs. Thus, one should tackle these complexities properly in designing SMC law.

11.2.2　First-order sliding mode case

By applying the 2-D common sliding function (11.13), the following token-dependent 2-D SMC law $\bar{u}(k,l)$ with first-order sliding mode is firstly designed for the closed-loop system (11.12):

$$\bar{u}(k,l) = - \left(B^{\mathrm{T}} X B\right)^{-1} \left(K_{\sigma(\xi)} x(k,l) + \delta(k,l) \cdot \mathrm{sgn}(s(k,l))\right), \quad (11.15)$$

where $\delta(k,l) \triangleq \bar{\varepsilon} \|x(k,l)\| + d \|B^{\mathrm{T}} X B\|$ with $\bar{\varepsilon} \triangleq \|B^{\mathrm{T}} X M\| \cdot \|N\| + \varepsilon \|B^{\mathrm{T}} X B\|$ and the matrices $K_{\sigma(\xi)} \in \mathbb{R}^{m \times (n_h + n_v)}$ ($\sigma(\xi) \in \mathcal{M}$) are to be designed. It should be pointed out that the effect of Round-Robin protocol from one actuator to another is reflected in SMC law $\bar{u}(k,l)$ through the token-dependent matrices $K_{\sigma(\xi)}$ that will be the solutions to a set of coupled matrix inequalities concerning the periodic scheduler $\sigma(\xi)$ (as will be seen in Theorem 11.3 later).

Main Results 219

Combining (11.11) and (11.15), we have the actual control input $u(k,l)$ as follows:

$$u(k,l) = - \Phi_{\sigma(\xi)}(B^{\mathrm{T}}XB)^{-1}K_{\sigma(\xi)}x(k,l)$$
$$- \Phi_{\sigma(\xi)}(B^{\mathrm{T}}XB)^{-1}\tilde{\delta}(k,l) + (I - \Phi_{\sigma(\xi)})v(k,l), \qquad (11.16)$$

where $\tilde{\delta}(k,l) \triangleq \delta(k,l) \cdot \mathrm{sgn}(s(k,l))$.

Substituting (11.16) into (11.1) gives rise to the following closed-loop system (11.12):

$$x^+(k,l) = \tilde{A}_{\sigma(\xi)}(k,l)x(k,l) + B\left(I - \Phi_{\sigma(\xi)}\right)f(x(k,l))$$
$$+ B(I - \Phi_{\sigma(\xi)})v(k,l)$$
$$+ B\Phi_{\sigma(\xi)}(B^{\mathrm{T}}XB)^{-1}\left[D(k,l) - \tilde{\delta}(k,l)\right], \qquad (11.17)$$

where $\tilde{A}_{\sigma(\xi)}(k,l) \triangleq A - B\Phi_{\sigma(\xi)}(B^{\mathrm{T}}XB)^{-1}K_{\sigma(\xi)} + \tilde{X}_{\sigma(\xi)}\Delta A(k,l)$ and $\tilde{X}_{\sigma(\xi)} \triangleq I - B\Phi_{\sigma(\xi)}(B^{\mathrm{T}}XB)^{-1}B^{\mathrm{T}}X$.

Furthermore, combining (11.14) and (11.15) yields

$$s^+(k,l) = - \Phi_{\sigma(\xi)}K_{\sigma(\xi)}x(k,l) + D(k,l) - \Phi_{\sigma(\xi)}\tilde{\delta}(k,l)$$
$$+ B^{\mathrm{T}}XB\left(I - \Phi_{\sigma(\xi)}\right)v(k,l). \qquad (11.18)$$

It is clear that

$$\|\tilde{\delta}(k,l)\|^2 \leq 2m\left(\bar{\varepsilon}^2\|x(k,l)\|^2 + d^2\|B^{\mathrm{T}}XB\|^2\right), \qquad (11.19)$$
$$\|D(k,l) - \tilde{\delta}(k,l)\| \leq (1 + \sqrt{m})\delta(k,l). \qquad (11.20)$$

In the sequel, by resorting to a token-dependent Lyapunov-like function, a sufficient condition will be derived to ensure the state trajectories of the closed-loop system (11.17) entering into a 2-D sliding region Ω around the specified 2-D sliding surface $s(k,l) = 0$.

Theorem 11.1 *Consider the 2-D system (11.1) with the Round-Robin protocol (11.6)–(11.7) and the token-dependent 2-D first-order SMC law (11.15). For preassigned parameters $\alpha_1 \geq 0$, $\alpha_2 \geq 0$, a matrix $X > 0$, and any $\sigma(\xi) \in \mathcal{M}$ satisfying the condition (11.10), assume that there exist matrices $K_{\sigma(\xi)} \in \mathbb{R}^{m \times (n_h + n_v)}$, $P^h_{\sigma(\xi)} > 0$, $P^v_{\sigma(\xi)} > 0$, $P^h_{\sigma(\xi+1)} > 0$, $P^v_{\sigma(\xi+1)} > 0$, $Q^h_{\sigma(\xi)} > 0$, $Q^v_{\sigma(\xi)} > 0$, $Q^h_{\sigma(\xi+1)} > 0$, $Q^v_{\sigma(\xi+1)} > 0$, $W_{\sigma(\xi)} > 0$, $W_{\sigma(\xi+1)} > 0$, and scalars $\rho > 0$, $\omega_{\sigma(\xi),\sigma(\xi+1)} > 0$, $\varpi_{\sigma(\xi),\sigma(\xi+1)} > 0$, $\theta_{\sigma(\xi),\sigma(\xi+1)} > 0$, $\zeta_{\sigma(\xi+1)} > 0$*

such that the following matrix inequalities hold:

$$(I - \Phi_{\sigma(\xi)})B^{\mathrm{T}}P_{\sigma(\xi+1)}B(I - \Phi_{\sigma(\xi)}) \leq \omega_{\sigma(\xi),\sigma(\xi+1)}I, \qquad (11.21)$$

$$(B^{\mathrm{T}}XB)^{-1}\Phi_{\sigma(\xi)}B^{\mathrm{T}}P_{\sigma(\xi+1)}B\Phi_{\sigma(\xi)}(B^{\mathrm{T}}XB)^{-1} \leq \varpi_{\sigma(\xi),\sigma(\xi+1)}I, \qquad (11.22)$$

$$\left(B^{\mathrm{T}}XB\right)^{-1}\Phi_{\sigma(\xi)}Q_{\sigma(\xi+1)}\Phi_{\sigma(\xi)}\left(B^{\mathrm{T}}XB\right)^{-1} \leq \theta_{\sigma(\xi),\sigma(\xi+1)}I, \qquad (11.23)$$

$$W_{\sigma(\xi+1)} \leq \zeta_{\sigma(\xi+1)}I, \qquad (11.24)$$

$$\tilde{\Pi}_{\sigma(\xi),\sigma(\xi+1)} \triangleq \begin{bmatrix} \tilde{\Pi}^{(11)}_{\sigma(\xi),\sigma(\xi+1)} & \alpha_1\mathbf{K}_{\sigma(\xi),\sigma(\xi+1)} & \alpha_2\mathbf{K}_{\sigma(\xi),\sigma(\xi+1)} \\ \star & \tilde{\Pi}^{(22)}_{\sigma(\xi),\sigma(\xi+1)} & \tilde{\Pi}^{(23)}_{\sigma(\xi),\sigma(\xi+1)} \\ \star & \star & \tilde{\Pi}^{(33)}_{\sigma(\xi),\sigma(\xi+1)} \end{bmatrix} < 0,$$

$$(11.25)$$

where $P_{\sigma(\xi)} \triangleq \mathrm{diag}\left\{P^h_{\sigma(\xi)}, P^v_{\sigma(\xi)}\right\}$, $Q_{\sigma(\xi)} \triangleq Q^h_{\sigma(\xi)} + Q^v_{\sigma(\xi)}$, *and*

$$\tilde{\Pi}^{(11)}_{\sigma(\xi),\sigma(\xi+1)} \triangleq -P_{\sigma(\xi)} + 3\tilde{A}^{\mathrm{T}}_{\sigma(\xi)}(k,l)P_{\sigma(\xi+1)}\tilde{A}_{\sigma(\xi)}(k,l) + 8\varepsilon^2\omega_{\sigma(\xi),\sigma(\xi+1)}I$$

$$+ 8\bar{\varepsilon}^2\left(1 + \sqrt{m}\right)^2\varpi_{\sigma(\xi),\sigma(\xi+1)}I + 2K^{\mathrm{T}}_{\sigma(\xi)}\left(B^{\mathrm{T}}XB\right)^{-1}\Phi_{\sigma(\xi)}Q_{\sigma(\xi+1)}$$

$$\times \Phi_{\sigma(\xi)}\left(B^{\mathrm{T}}XB\right)^{-1}K_{\sigma(\xi)} + 6m\bar{\varepsilon}^2\theta_{\sigma(\xi),\sigma(\xi+1)}I$$

$$+ 2K^{\mathrm{T}}_{\sigma(\xi)}\Phi_{\sigma(\xi)}W_{\sigma(\xi+1)}\Phi_{\sigma(\xi)}K_{\sigma(\xi)} + 6\bar{\varepsilon}^2(1 + \sqrt{m})^2\zeta_{\sigma(\xi+1)}I + \rho I,$$

$$\mathbf{K}_{\sigma(\xi),\sigma(\xi+1)} \triangleq \tilde{A}^{\mathrm{T}}_{\sigma(\xi)}(k,l)P_{\sigma(\xi+1)}B(I - \Phi_{\sigma(\xi)})$$

$$- K^{\mathrm{T}}_{\sigma(\xi)}\left(B^{\mathrm{T}}XB\right)^{-1}\Phi_{\sigma(\xi)}Q_{\sigma(\xi+1)}\left(I - \Phi_{\sigma(\xi)}\right)$$

$$- K^{\mathrm{T}}_{\sigma(\xi)}\Phi_{\sigma(\xi)}W_{\sigma(\xi+1)}B^{\mathrm{T}}XB\left(I - \Phi_{\sigma(\xi)}\right),$$

$$\mathbf{W}_{\sigma(\xi),\sigma(\xi+1)} \triangleq 3(I - \Phi_{\sigma(\xi)})B^{\mathrm{T}}P_{\sigma(\xi+1)}B(I - \Phi_{\sigma(\xi)})$$

$$+ 2\left(I - \Phi_{\sigma(\xi)}\right)Q_{\sigma(\xi+1)}\left(I - \Phi_{\sigma(\xi)}\right)$$

$$+ 2\left(I - \Phi_{\sigma(\xi)}\right)B^{\mathrm{T}}XBW_{\sigma(\xi+1)}B^{\mathrm{T}}XB\left(I - \Phi_{\sigma(\xi)}\right),$$

$$\tilde{\Pi}^{(22)}_{\sigma(\xi),\sigma(\xi+1)} \triangleq \alpha_1^2\mathbf{W}_{\sigma(\xi),\sigma(\xi+1)} - Q^h_{\sigma(\xi)},$$

$$\tilde{\Pi}^{(23)}_{\sigma(\xi),\sigma(\xi+1)} \triangleq \alpha_1\alpha_2\mathbf{W}_{\sigma(\xi),\sigma(\xi+1)},$$

$$\tilde{\Pi}^{(33)}_{\sigma(\xi),\sigma(\xi+1)} \triangleq \alpha_2^2\mathbf{W}_{\sigma(\xi),\sigma(\xi+1)} - Q^v_{\sigma(\xi)}.$$

Then, both the state trajectories of the closed-loop system (11.17) and the sliding function of (11.18) are driven into the following domain Ω *around the origin:*

$$\Omega \triangleq \left\{\vec{x}(k,l) \in \mathbb{R}^{n_h+n_v+m} \mid \|\vec{x}(k,l)\| \leq \hat{\rho}\right\}. \qquad (11.26)$$

where $\vec{x}(k,l) \triangleq \begin{bmatrix} x^{\mathrm{T}}(k,l) & s^{\mathrm{T}}(k,l) \end{bmatrix}^{\mathrm{T}}$, $\hat{\rho} \triangleq \max_{\sigma(\xi),\sigma(\xi+1)\in\mathcal{M}}\left\{\sqrt{\frac{\tilde{d}_{\sigma(\xi),\sigma(\xi+1)}}{\tilde{\rho}_{\sigma(\xi)}}}\right\}$,

$$\tilde{d}_{\sigma(\xi),\sigma(\xi+1)} \triangleq 8d^2\left(1 + \sqrt{m}\right)^2\|B^{\mathrm{T}}XB\|^2\varpi_{\sigma(\xi),\sigma(\xi+1)} + 8d^2\omega_{\sigma(\xi),\sigma(\xi+1)} +$$

Main Results 221

$6md^2\theta_{\sigma(\xi),\sigma(\xi+1)}\|B^{\mathrm{T}}XB\|^2 + 6d^2(1 + \sqrt{m})^2\zeta_{\sigma(\xi+1)}\|B^{\mathrm{T}}XB\|^2$ *and* $\tilde{\rho}_{\sigma(\xi)} \triangleq$ $\min\{\rho, \lambda_{\min}(W_{\sigma(\xi)})\}$.

Define the token-dependent Lyapunov-like function as $\tilde{V}(k,l,\sigma(\xi)) \triangleq$ $x^{h\mathrm{T}}(k,l)P_{\sigma(\xi)}^h x^h(k,l) + x^{v\mathrm{T}}(k,l)P_{\sigma(\xi)}^v x^v(k,l) + u^{\mathrm{T}}(k-1,l)Q_{\sigma(\xi)}^h u(k-1,l) +$ $u^{\mathrm{T}}(k,l-1)Q_{\sigma(\xi)}^v u(k,l-1) + s^{\mathrm{T}}(k,l)W_{\sigma(\xi)}s(k,l)$. We further have

$$\begin{aligned}
&\Delta\tilde{V}(k,l,\sigma(\xi))\\
=&x^{+\mathrm{T}}(k,l)P_{\sigma(\xi+1)}x^+(k,l) - x^{\mathrm{T}}(k,l)P_{\sigma(\xi)}x(k,l)\\
&+ u^{\mathrm{T}}(k,l)Q_{\sigma(\xi+1)}u(k,l) - u^{\mathrm{T}}(k-1,l)Q_{\sigma(\xi)}^h u(k-1,l)\\
&- u^{\mathrm{T}}(k,l-1)Q_{\sigma(\xi)}^v u(k,l-1) + s^{+\mathrm{T}}(k,l)W_{\sigma(\xi+1)}s^+(k,l)\\
&- s^{\mathrm{T}}(k,l)W_{\sigma(\xi)}s(k,l).
\end{aligned} \tag{11.27}$$

Along the trajectories of the closed-loop system (11.17), we obtain

$$\begin{aligned}
&x^{+\mathrm{T}}(k,l)P_{\sigma(\xi+1)}x^+(k,l)\\
=&x^{\mathrm{T}}(k,l)\tilde{A}_{\sigma(\xi)}^{\mathrm{T}}(k,l)P_{\sigma(\xi+1)}\tilde{A}_{\sigma(\xi)}(k,l)x(k,l)\\
&+ v^{\mathrm{T}}(k,l)(I - \Phi_{\sigma(\xi)})B^{\mathrm{T}}P_{\sigma(\xi+1)}B(I - \Phi_{\sigma(\xi)})v(k,l)\\
&+ f^{\mathrm{T}}(x(k,l))(I - \Phi_{\sigma(\xi)})B^{\mathrm{T}}P_{\sigma(\xi+1)}B(I - \Phi_{\sigma(\xi)})f(x(k,l))\\
&+ \left[D(k,l) - \tilde{\delta}(k,l)\right]^{\mathrm{T}}(B^{\mathrm{T}}XB)^{-1}\Phi_{\sigma(\xi)}B^{\mathrm{T}}P_{\sigma(\xi+1)}\\
&\times B\Phi_{\sigma(\xi)}(B^{\mathrm{T}}XB)^{-1}\left[D(k,l) - \tilde{\delta}(k,l)\right]\\
&+ 2x^{\mathrm{T}}(k,l)\tilde{A}_{\sigma(\xi)}^{\mathrm{T}}(k,l)P_{\sigma(\xi+1)}B(I - \Phi_{\sigma(\xi)})v(k,l)\\
&+ 2x^{\mathrm{T}}(k,l)\tilde{A}_{\sigma(\xi)}^{\mathrm{T}}(k,l)P_{\sigma(\xi+1)}B(I - \Phi_{\sigma(\xi)})f(x(k,l))\\
&+ 2x^{\mathrm{T}}(k,l)\tilde{A}_{\sigma(\xi)}^{\mathrm{T}}(k,l)P_{\sigma(\xi+1)}B^{\mathrm{T}}P_{\sigma(\xi+1)}\\
&\times B\Phi_{\sigma(\xi)}(B^{\mathrm{T}}XB)^{-1}\left[D(k,l) - \tilde{\delta}(k,l)\right]\\
&+ 2v^{\mathrm{T}}(k,l)(I - \Phi_{\sigma(\xi)})B^{\mathrm{T}}P_{\sigma(\xi+1)}B(I - \Phi_{\sigma(\xi)})f(x(k,l))\\
&+ 2v^{\mathrm{T}}(k,l)(I - \Phi_{\sigma(\xi)})B^{\mathrm{T}}P_{\sigma(\xi+1)}B\Phi_{\sigma(\xi)}\\
&\times (B^{\mathrm{T}}XB)^{-1}\left[D(k,l) - \tilde{\delta}(k,l)\right]\\
&+ 2f^{\mathrm{T}}(x(k,l))(I - \Phi_{\sigma(\xi)})B^{\mathrm{T}}P_{\sigma(\xi+1)}B^{\mathrm{T}}P_{\sigma(\xi+1)}\\
&\times B\Phi_{\sigma(\xi)}(B^{\mathrm{T}}XB)^{-1}\left[D(k,l) - \tilde{\delta}(k,l)\right].
\end{aligned} \tag{11.28}$$

By means of the relationship (11.20) and the conditions (11.21)–(11.22), one has

$$f^{\mathrm{T}}(x(k,l))(I - \Phi_{\sigma(\xi)})B^{\mathrm{T}}P_{\sigma(\xi+1)}B(I - \Phi_{\sigma(\xi)})f(x(k,l))$$
$$\leq 2\omega_{\sigma(\xi),\sigma(\xi+1)}\left(\varepsilon^2\|x(k,l)\|^2 + d^2\right), \tag{11.29}$$

$$\left[D(k,l) - \tilde{\delta}(k,l)\right]^{\mathrm{T}}(B^{\mathrm{T}}XB)^{-1}\Phi_{\sigma(\xi)}B^{\mathrm{T}}P_{\sigma(\xi+1)}$$
$$\times B\Phi_{\sigma(\xi)}(B^{\mathrm{T}}XB)^{-1}\left[D(k,l) - \tilde{\delta}(k,l)\right]$$
$$\leq 2\left(1 + \sqrt{m}\right)^2\varpi_{\sigma(\xi),\sigma(\xi+1)}\left(\bar{\varepsilon}^2\|x(k,l)\|^2 + d^2\|B^{\mathrm{T}}XB\|^2\right), \tag{11.30}$$

$$2x^{\mathrm{T}}(k,l)\tilde{A}_{\sigma(\xi)}^{\mathrm{T}}(k,l)P_{\sigma(\xi+1)}B(I - \Phi_{\sigma(\xi)})f(x(k,l))$$
$$\leq x^{\mathrm{T}}(k,l)\tilde{A}_{\sigma(\xi)}^{\mathrm{T}}(k,l)P_{\sigma(\xi+1)}\tilde{A}_{\sigma(\xi)}(k,l)x(k,l)$$
$$+ 2\omega_{\sigma(\xi),\sigma(\xi+1)}\left(\varepsilon^2\|x(k,l)\|^2 + d^2\right), \tag{11.31}$$

$$2x^{\mathrm{T}}(k,l)\tilde{A}_{\sigma(\xi)}^{\mathrm{T}}(k,l)P_{\sigma(\xi+1)}B^{\mathrm{T}}P_{\sigma(\xi+1)}$$
$$\times B\Phi_{\sigma(\xi)}(B^{\mathrm{T}}XB)^{-1}\left[D(k,l) - \tilde{\delta}(k,l)\right]$$
$$\leq x^{\mathrm{T}}(k,l)\tilde{A}_{\sigma(\xi)}^{\mathrm{T}}(k,l)P_{\sigma(\xi+1)}\tilde{A}_{\sigma(\xi)}(k,l)x(k,l)$$
$$+ 2\left(1 + \sqrt{m}\right)^2\varpi_{\sigma(\xi),\sigma(\xi+1)}\left(\bar{\varepsilon}^2\|x(k,l)\|^2 + d^2\|B^{\mathrm{T}}XB\|^2\right), \tag{11.32}$$

$$2v^{\mathrm{T}}(k,l)(I - \Phi_{\sigma(\xi)})B^{\mathrm{T}}P_{\sigma(\xi+1)}B(I - \Phi_{\sigma(\xi)})f(x(k,l))$$
$$\leq v^{\mathrm{T}}(k,l)(I - \Phi_{\sigma(\xi)})B^{\mathrm{T}}P_{\sigma(\xi+1)}B(I - \Phi_{\sigma(\xi)})v(k,l)$$
$$+ 2\omega_{\sigma(\xi),\sigma(\xi+1)}\left(\varepsilon^2\|x(k,l)\|^2 + d^2\right), \tag{11.33}$$

$$2v^{\mathrm{T}}(k,l)(I - \Phi_{\sigma(\xi)})B^{\mathrm{T}}P_{\sigma(\xi+1)}$$
$$\times B\Phi_{\sigma(\xi)}(B^{\mathrm{T}}XB)^{-1}\left[D(k,l) - \tilde{\delta}(k,l)\right]$$
$$\leq v^{\mathrm{T}}(k,l)(I - \Phi_{\sigma(\xi)})B^{\mathrm{T}}P_{\sigma(\xi+1)}B(I - \Phi_{\sigma(\xi)})v(k,l)$$
$$+ 2\left(1 + \sqrt{m}\right)^2\varpi_{\sigma(\xi),\sigma(\xi+1)}\left(\bar{\varepsilon}^2\|x(k,l)\|^2 + d^2\|B^{\mathrm{T}}XB\|^2\right), \tag{11.34}$$

$$2f^{\mathrm{T}}(x(k,l))(I - \Phi_{\sigma(\xi)})B^{\mathrm{T}}P_{\sigma(\xi+1)}B^{\mathrm{T}}P_{\sigma(\xi+1)}$$
$$\times B\Phi_{\sigma(\xi)}(B^{\mathrm{T}}XB)^{-1}\left[D(k,l) - \tilde{\delta}(k,l)\right]$$
$$\leq 2\omega_{\sigma(\xi),\sigma(\xi+1)}\left(\varepsilon^2\|x(k,l)\|^2 + d^2\right) + 2\left(1 + \sqrt{m}\right)^2\varpi_{\sigma(\xi),\sigma(\xi+1)}$$
$$\times \left(\bar{\varepsilon}^2\|x(k,l)\|^2 + d^2\|B^{\mathrm{T}}XB\|^2\right). \tag{11.35}$$

By means of the relationships in (11.29)–(11.35), one has from (11.28) that

$$x^{+\mathrm{T}}(k,l)P_{\sigma(\xi+1)}x^+(k,l)$$
$$\leq x^{\mathrm{T}}(k,l)\left[3\tilde{A}_{\sigma(\xi)}^{\mathrm{T}}(k,l)P_{\sigma(\xi+1)}\tilde{A}_{\sigma(\xi)}(k,l) + 8\varepsilon^2\omega_{\sigma(\xi),\sigma(\xi+1)}I\right.$$
$$\left. + 8\bar{\varepsilon}^2\left(1 + \sqrt{m}\right)^2\varpi_{\sigma(\xi),\sigma(\xi+1)}I\right]x(k,l)$$
$$+ 3v^{\mathrm{T}}(k,l)(I - \Phi_{\sigma(\xi)})B^{\mathrm{T}}P_{\sigma(\xi+1)}B(I - \Phi_{\sigma(\xi)})v(k,l)$$

Main Results

$$+ 2x^{\mathrm{T}}(k,l)\tilde{A}_{\sigma(\xi)}^{\mathrm{T}}(k,l)P_{\sigma(\xi+1)}B(I - \Phi_{\sigma(\xi)})v(k,l)$$

$$+ 8d^2\left(1+\sqrt{m}\right)^2\|B^{\mathrm{T}}XB\|^2\varpi_{\sigma(\xi),\sigma(\xi+1)} + 8d^2\omega_{\sigma(\xi),\sigma(\xi+1)}. \tag{11.36}$$

Next, let us tackle the term $u^{\mathrm{T}}(k,l)Q_{\sigma(\xi+1)}u(k,l)$ in (11.27). For this purpose, it follows from (11.16) that

$$
\begin{aligned}
&u^{\mathrm{T}}(k,l)Q_{\sigma(\xi+1)}u(k,l)\\
&= x^{\mathrm{T}}(k,l)K_{\sigma(\xi)}^{\mathrm{T}}\left(B^{\mathrm{T}}XB\right)^{-1}\Phi_{\sigma(\xi)}Q_{\sigma(\xi+1)}\\
&\quad \times \Phi_{\sigma(\xi)}\left(B^{\mathrm{T}}XB\right)^{-1}K_{\sigma(\xi)}x(k,l)\\
&\quad + \tilde{\delta}^{\mathrm{T}}(k,l)\left(B^{\mathrm{T}}XB\right)^{-1}\Phi_{\sigma(\xi)}Q_{\sigma(\xi+1)}\Phi_{\sigma(\xi)}\left(B^{\mathrm{T}}XB\right)^{-1}\tilde{\delta}(k,l)\\
&\quad + v^{\mathrm{T}}(k,l)\left(I - \Phi_{\sigma(\xi)}\right)Q_{\sigma(\xi+1)}\left(I - \Phi_{\sigma(\xi)}\right)v(k,l)\\
&\quad + 2x^{\mathrm{T}}(k,l)K_{\sigma(\xi)}^{\mathrm{T}}\left(B^{\mathrm{T}}XB\right)^{-1}\Phi_{\sigma(\xi)}Q_{\sigma(\xi+1)}\\
&\quad \times \Phi_{\sigma(\xi)}\left(B^{\mathrm{T}}XB\right)^{-1}\tilde{\delta}(k,l)\\
&\quad - 2x^{\mathrm{T}}(k,l)K_{\sigma(\xi)}^{\mathrm{T}}\left(B^{\mathrm{T}}XB\right)^{-1}\Phi_{\sigma(\xi)}Q_{\sigma(\xi+1)}\left(I - \Phi_{\sigma(\xi)}\right)v(k,l)\\
&\quad - 2\tilde{\delta}^{\mathrm{T}}(k,l)\left(B^{\mathrm{T}}XB\right)^{-1}\Phi_{\sigma(\xi)}Q_{\sigma(\xi+1)}\left(I - \Phi_{\sigma(\xi)}\right)v(k,l). \tag{11.37}
\end{aligned}
$$

By using inequalities (11.19) and (11.23), one has

$$
\begin{aligned}
&\tilde{\delta}^{\mathrm{T}}(k,l)\left(B^{\mathrm{T}}XB\right)^{-1}\Phi_{\sigma(\xi)}Q_{\sigma(\xi+1)}\Phi_{\sigma(\xi)}\left(B^{\mathrm{T}}XB\right)^{-1}\tilde{\delta}(k,l)\\
&\leq 2m\theta_{\sigma(\xi),\sigma(\xi+1)}\left(\bar{\varepsilon}^2\|x(k,l)\|^2 + d^2\|B^{\mathrm{T}}XB\|^2\right), \tag{11.38}
\end{aligned}
$$

$$
\begin{aligned}
&2x^{\mathrm{T}}(k,l)K_{\sigma(\xi)}^{\mathrm{T}}\left(B^{\mathrm{T}}XB\right)^{-1}\Phi_{\sigma(\xi)}Q_{\sigma(\xi+1)}\\
&\quad \times \Phi_{\sigma(\xi)}\left(B^{\mathrm{T}}XB\right)^{-1}\tilde{\delta}(k,l)\\
&\leq x^{\mathrm{T}}(k,l)K_{\sigma(\xi)}^{\mathrm{T}}\left(B^{\mathrm{T}}XB\right)^{-1}\Phi_{\sigma(\xi)}Q_{\sigma(\xi+1)}\\
&\quad \times \Phi_{\sigma(\xi)}\left(B^{\mathrm{T}}XB\right)^{-1}K_{\sigma(\xi)}x(k,l)\\
&\quad + 2m\theta_{\sigma(\xi),\sigma(\xi+1)}\left(\bar{\varepsilon}^2\|x(k,l)\|^2 + d^2\|B^{\mathrm{T}}XB\|^2\right), \tag{11.39}
\end{aligned}
$$

$$
\begin{aligned}
&- 2\tilde{\delta}^{\mathrm{T}}(k,l)\left(B^{\mathrm{T}}XB\right)^{-1}\Phi_{\sigma(\xi)}Q_{\sigma(\xi+1)}\left(I - \Phi_{\sigma(\xi)}\right)v(k,l)\\
&\leq 2m\theta_{\sigma(\xi),\sigma(\xi+1)}\left(\bar{\varepsilon}^2\|x(k,l)\|^2 + d^2\|B^{\mathrm{T}}XB\|^2\right)\\
&\quad + v^{\mathrm{T}}(k,l)\left(I - \Phi_{\sigma(\xi)}\right)Q_{\sigma(\xi+1)}\left(I - \Phi_{\sigma(\xi)}\right)v(k,l). \tag{11.40}
\end{aligned}
$$

Keeping (11.35)–(11.40) in mind, we further obtain from (11.37) that

$$
\begin{aligned}
&u^{\mathrm{T}}(k,l)Q_{\sigma(\xi+1)}u(k,l)\\
&\leq x^{\mathrm{T}}(k,l)\Big[2K_{\sigma(\xi)}^{\mathrm{T}}\left(B^{\mathrm{T}}XB\right)^{-1}\Phi_{\sigma(\xi)}Q_{\sigma(\xi+1)}\\
&\quad \times \Phi_{\sigma(\xi)}\left(B^{\mathrm{T}}XB\right)^{-1}K_{\sigma(\xi)} + 6m\bar{\varepsilon}^2\theta_{\sigma(\xi),\sigma(\xi+1)}I\Big]x(k,l)\\
&\quad + 2v^{\mathrm{T}}(k,l)\left(I - \Phi_{\sigma(\xi)}\right)Q_{\sigma(\xi+1)}\left(I - \Phi_{\sigma(\xi)}\right)v(k,l)
\end{aligned}
$$

$$- 2x^{\mathrm{T}}(k,l)K_{\sigma(\xi)}^{\mathrm{T}}\left(B^{\mathrm{T}}XB\right)^{-1}\Phi_{\sigma(\xi)}Q_{\sigma(\xi+1)}$$
$$\times\left(I-\Phi_{\sigma(\xi)}\right)v(k,l)+6md^2\theta_{\sigma(\xi),\sigma(\xi+1)}\|B^{\mathrm{T}}XB\|^2. \tag{11.41}$$

Furthermore, it follows from (11.18) that

$$s^{+\mathrm{T}}(k,l)W_{\sigma(\xi+1)}s^{+}(k,l)$$
$$=x^{\mathrm{T}}(k,l)K_{\sigma(\xi)}^{\mathrm{T}}\Phi_{\sigma(\xi)}W_{\sigma(\xi+1)}\Phi_{\sigma(\xi)}K_{\sigma(\xi)}x(k,l)$$
$$+v^{\mathrm{T}}(k,l)\left(I-\Phi_{\sigma(\xi)}\right)B^{\mathrm{T}}XBW_{\sigma(\xi+1)}B^{\mathrm{T}}XB\left(I-\Phi_{\sigma(\xi)}\right)v(k,l)$$
$$+\left[D(k,l)-\Phi_{\sigma(\xi)}\tilde{\delta}(k,l)\right]^{\mathrm{T}}W_{\sigma(\xi+1)}\left[D(k,l)-\Phi_{\sigma(\xi)}\tilde{\delta}(k,l)\right]$$
$$-2x^{\mathrm{T}}(k,l)K_{\sigma(\xi)}^{\mathrm{T}}\Phi_{\sigma(\xi)}W_{\sigma(\xi+1)}B^{\mathrm{T}}XB\left(I-\Phi_{\sigma(\xi)}\right)v(k,l)$$
$$-2x^{\mathrm{T}}(k,l)K_{\sigma(\xi)}^{\mathrm{T}}\Phi_{\sigma(\xi)}W_{\sigma(\xi+1)}\left[D(k,l)-\Phi_{\sigma(\xi)}\tilde{\delta}(k,l)\right]$$
$$+2v^{\mathrm{T}}(k,l)\left(I-\Phi_{\sigma(\xi)}\right)B^{\mathrm{T}}XBW_{\sigma(\xi+1)}\left[D(k,l)-\Phi_{\sigma(\xi)}\tilde{\delta}(k,l)\right]. \tag{11.42}$$

It is noted from (11.19) and (11.24) that

$$\left[D(k,l)-\Phi_{\sigma(\xi)}\tilde{\delta}(k,l)\right]^{\mathrm{T}}W_{\sigma(\xi+1)}\left[D(k,l)-\Phi_{\sigma(\xi)}\tilde{\delta}(k,l)\right]$$
$$\leq 2(1+\sqrt{m})^2\zeta_{\sigma(\xi+1)}\left(\bar{\varepsilon}^2\|x(k,l)\|^2+d^2\|B^{\mathrm{T}}XB\|^2\right), \tag{11.43}$$
$$-2x^{\mathrm{T}}(k,l)K_{\sigma(\xi)}^{\mathrm{T}}\Phi_{\sigma(\xi)}W_{\sigma(\xi+1)}\left[D(k,l)-\Phi_{\sigma(\xi)}\tilde{\delta}(k,l)\right]$$
$$\leq x^{\mathrm{T}}(k,l)K_{\sigma(\xi)}^{\mathrm{T}}\Phi_{\sigma(\xi)}W_{\sigma(\xi+1)}\Phi_{\sigma(\xi)}K_{\sigma(\xi)}x(k,l)$$
$$+2(1+\sqrt{m})^2\zeta_{\sigma(\xi+1)}\left(\bar{\varepsilon}^2\|x(k,l)\|^2+d^2\|B^{\mathrm{T}}XB\|^2\right), \tag{11.44}$$
$$2v^{\mathrm{T}}(k,l)\left(I-\Phi_{\sigma(\xi)}\right)B^{\mathrm{T}}XBW_{\sigma(\xi+1)}\left[D(k,l)-\Phi_{\sigma(\xi)}\tilde{\delta}(k,l)\right]$$
$$\leq v^{\mathrm{T}}(k,l)\left(I-\Phi_{\sigma(\xi)}\right)B^{\mathrm{T}}XBW_{\sigma(\xi+1)}B^{\mathrm{T}}XB\left(I-\Phi_{\sigma(\xi)}\right)v(k,l)$$
$$+2(1+\sqrt{m})^2\zeta_{\sigma(\xi+1)}\left(\bar{\varepsilon}^2\|x(k,l)\|^2+d^2\|B^{\mathrm{T}}XB\|^2\right). \tag{11.45}$$

Thus, we have from (11.42)–(11.45) that

$$s^{+\mathrm{T}}(k,l)W_{\sigma(\xi+1)}s^{+}(k,l)$$
$$\leq x^{\mathrm{T}}(k,l)\left[2K_{\sigma(\xi)}^{\mathrm{T}}\Phi_{\sigma(\xi)}W_{\sigma(\xi+1)}\Phi_{\sigma(\xi)}K_{\sigma(\xi)}+6\bar{\varepsilon}^2(1+\sqrt{m})^2\zeta_{\sigma(\xi+1)}I\right]x(k,l)$$
$$+2v^{\mathrm{T}}(k,l)\left(I-\Phi_{\sigma(\xi)}\right)B^{\mathrm{T}}XBW_{\sigma(\xi+1)}$$
$$\times B^{\mathrm{T}}XB\left(I-\Phi_{\sigma(\xi)}\right)v(k,l)-2x^{\mathrm{T}}(k,l)K_{\sigma(\xi)}^{\mathrm{T}}\Phi_{\sigma(\xi)}W_{\sigma(\xi+1)}$$
$$\times B^{\mathrm{T}}XB\left(I-\Phi_{\sigma(\xi)}\right)v(k,l)-s^{\mathrm{T}}(k,l)W_{\sigma(\xi)}s(k,l)$$
$$+6d^2(1+\sqrt{m})^2\zeta_{\sigma(\xi+1)}\|B^{\mathrm{T}}XB\|^2. \tag{11.46}$$

Main Results 225

Combining (11.36), (11.41), and (11.46) results in

$$\Delta \tilde{V}(k,l,\sigma(\xi)) \leq \begin{bmatrix} x(k,l) \\ u(k-1,l) \\ u(k,l-1) \end{bmatrix}^{\mathrm{T}} \tilde{\Pi}_{\sigma(\xi),\sigma(\xi+1)} \begin{bmatrix} x(k,l) \\ u(k-1,l) \\ u(k,l-1) \end{bmatrix}$$
$$- \rho \|x(k,l)\|^2 - s^{\mathrm{T}}(k,l) W_{\sigma(\xi)} s(k,l) + \tilde{d}_{\sigma(\xi),\sigma(\xi+1)}$$
$$\leq \begin{bmatrix} x(k,l) \\ u(k-1,l) \\ u(k,l-1) \end{bmatrix}^{\mathrm{T}} \tilde{\Pi}_{\sigma(\xi),\sigma(\xi+1)} \begin{bmatrix} x(k,l) \\ u(k-1,l) \\ u(k,l-1) \end{bmatrix}$$
$$- \left[\tilde{\rho}_{\sigma(\xi)} \|\vec{x}(k,l)\|^2 - \tilde{d}_{\sigma(\xi),\sigma(\xi+1)} \right], \tag{11.47}$$

where $\vec{x}(k,l) \triangleq \begin{bmatrix} x^{\mathrm{T}}(k,l) & s^{\mathrm{T}}(k,l) \end{bmatrix}^{\mathrm{T}}$.

For the state trajectories outside the domain Ω, we have

$$\|\vec{x}(k,l)\| > \max_{\sigma(\xi),\sigma(\xi+1)\in\mathcal{M}} \left\{ \sqrt{\frac{\tilde{d}_{\sigma(\xi),\sigma(\xi+1)}}{\tilde{\rho}_{\sigma(\xi)}}} \right\} \geq \sqrt{\frac{\tilde{d}_{\sigma(\xi),\sigma(\xi+1)}}{\tilde{\rho}_{\sigma(\xi)}}},$$

which means that the second term in (11.47) is great than or equal to zero. Then, the conditions (11.25) and (11.47) render

$$\Delta \tilde{V}(k,l,\sigma(\xi)) \leq \begin{bmatrix} x(k,l) \\ u(k-1,l) \\ u(k,l-1) \end{bmatrix}^{\mathrm{T}} \tilde{\Pi}_{\sigma(\xi),\sigma(\xi+1)} \begin{bmatrix} x(k,l) \\ u(k-1,l) \\ u(k,l-1) \end{bmatrix}$$
$$\leq - \lambda_{\min} \left(-\tilde{\Pi}_{\sigma(\xi),\sigma(\xi+1)} \right) \left(\|x(k,l)\|^2 \right.$$
$$\left. + \|u(k-1,l)\|^2 + \|u(k,l-1)\|^2 \right). \tag{11.48}$$

According to Theorem 6.2 in [114], the state trajectories of the closed-loop system (11.17) are exponentially deceasing outside the domain Ω defined in (11.26). Thus, the extended state $\vec{x}(k,l)$ will be driven into the sliding domain Ω within a finite region. The proof is now complete.

In the following theorem, we shall establish a sufficient condition to solve gain matrices $K_{\sigma(\xi)}$ in the 2-D first-order SMC law (11.15).

Theorem 11.2 *Consider the 2-D system (11.1) with the Round-Robin protocol (11.11) and the token-dependent 2-D first-order SMC law (11.15). For preassigned parameters $\alpha_1 \geq 0$, $\alpha_2 \geq 0$, a matrix $X > 0$ and any $j \in \mathcal{M}$, assume that there exist matrices $\vec{K}_j \in \mathbb{R}^{m \times (n_h + n_v)}$, $\vec{P}_j^h > 0$, $\vec{P}_j^v > 0$, $\vec{Q}_j^h > 0$, $\vec{Q}_j^v > 0$, $\vec{W}_j > 0$, and scalars $\vec{\rho} > 0$, $\nu_j > 0$, $\mu_j > 0$, $\vec{\omega}_{j,j+1} > 0$, $\vec{\varpi}_{j,j+1} > 0$,*

$\vec{\theta}_{j,j+1} > 0$, $\vec{\zeta}_{j+1} > 0$ such that the following coupled LMIs hold:

$$
\begin{bmatrix}
-\vec{\omega}_{j,j+1}I & (I - \Phi_j)B^{\mathrm{T}}\vec{\omega}_{j,j+1} \\
\star & -\vec{P}_{j+1}
\end{bmatrix} \leq 0, \tag{11.49}
$$

$$
\begin{bmatrix}
-\vec{\varpi}_{j,j+1}I & \left(B^{\mathrm{T}}XB\right)^{-1}\Phi_j B^{\mathrm{T}}\vec{\varpi}_{j,j+1} \\
\star & -\vec{P}_{j+1}
\end{bmatrix} \leq 0, \tag{11.50}
$$

$$
\begin{bmatrix}
-\vec{\theta}_{j,j+1}I & \left(B^{\mathrm{T}}XB\right)^{-1}\Phi_j\vec{\theta}_{j,j+1} & 0 & 0 \\
\star & -2\mu_{j+1}I & \mu_{j+1}I & \mu_{j+1}I \\
\star & \star & -\vec{Q}_{j+1}^h & 0 \\
\star & \star & \star & -\vec{Q}_{j+1}^v
\end{bmatrix} \leq 0, \tag{11.51}
$$

$$
\begin{bmatrix}
-\vec{\zeta}_{j+1}I & \vec{\zeta}_{j+1}I \\
\star & -\vec{W}_{j+1}
\end{bmatrix} \leq 0, \tag{11.52}
$$

$$
\vec{\Pi}_{j,j+1} \triangleq \begin{bmatrix}
\vec{\Pi}_{j,j+1}^{(11)} & \vec{\Pi}_{j,j+1}^{(12)} \\
\star & \vec{\Pi}_{j,j+1}^{(22)}
\end{bmatrix} < 0, \tag{11.53}
$$

where $\vec{P}_j \triangleq \mathrm{diag}\left\{\vec{P}_j^h, \vec{P}_j^v\right\}$ ($j \in \mathcal{M}$), and

$$
\vec{\Pi}_{j,j+1}^{(11)} \triangleq \begin{bmatrix}
-\mathrm{diag}\{\vec{P}_j, \vec{Q}_j^h, \vec{Q}_j^v\} & \mathbb{K}_j \\
\star & \bar{\Sigma}_{j+1}
\end{bmatrix},
$$

$$
\bar{\Sigma}_{j+1} \triangleq -\mathrm{diag}\{\vec{P}_{j+1}, \vec{P}_{j+1}, \vec{P}_{j+1}, 2\mu_{j+1}I, 2\mu_{j+1}I, 2\mu_{j+1}I, \vec{\rho}I\},
$$

$$
\mathbb{K}_j \triangleq \begin{bmatrix}
\vec{A}_j^{\mathrm{T}} & \sqrt{2}\vec{A}_j^{\mathrm{T}} & 0 \\
\alpha_1\vec{Q}_j^h(I - \Phi_j)B^{\mathrm{T}} & 0 & \sqrt{2}\alpha_1\vec{Q}_j^h(I - \Phi_j)B^{\mathrm{T}} \\
\alpha_2\vec{Q}_j^v(I - \Phi_j)B^{\mathrm{T}} & 0 & \sqrt{2}\alpha_2\vec{Q}_j^v(I - \Phi_j)B^{\mathrm{T}}
\end{bmatrix}
$$

$$
\begin{matrix}
-\vec{K}_j^{\mathrm{T}}(B^{\mathrm{T}}XB)^{-1}\Phi_j & -\vec{K}_j^{\mathrm{T}}(B^{\mathrm{T}}XB)^{-1}\Phi_j & 0 & \vec{P}_j \\
\alpha_1\vec{Q}_j^h(I-\Phi_j) & 0 & \alpha_1\vec{Q}_j^h(I-\Phi_j) & 0 \\
\alpha_2\vec{Q}_j^v(I-\Phi_j) & 0 & \alpha_2\vec{Q}_j^v(I-\Phi_j) & 0
\end{matrix} \Bigg],
$$

$$
\vec{A}_j \triangleq A\vec{P}_j - B\Phi_j(B^{\mathrm{T}}XB)^{-1}\vec{K}_j, \quad \vec{\Pi}_{j,j+1}^{(12)} \triangleq \begin{bmatrix}
\mathbb{X}_j & \mathbb{N}_j & 0 \\
0 & \mathbb{M}_j & 0 \\
0 & 0 & \mathbb{U}_{j+1}
\end{bmatrix},
$$

$$
\mathbb{X}_j \triangleq \begin{bmatrix}
-\vec{K}_j^{\mathrm{T}}\Phi_j & -\vec{K}_j^{\mathrm{T}}\Phi_j & 0 \\
\alpha_1\vec{Q}_j^h(I-\Phi_j)B^{\mathrm{T}}XB & 0 & \alpha_1\vec{Q}_j^h(I-\Phi_j)B^{\mathrm{T}}XB \\
\alpha_2\vec{Q}_j^v(I-\Phi_j)B^{\mathrm{T}}XB & 0 & \alpha_2\vec{Q}_j^v(I-\Phi_j)B^{\mathrm{T}}XB
\end{bmatrix}
$$

$$
\begin{matrix}
2\sqrt{2}\varepsilon\vec{P}_j & 2\sqrt{2}\bar{\varepsilon}(1+\sqrt{m})\vec{P}_j & \sqrt{6m}\bar{\varepsilon}\vec{P}_j & \sqrt{6}\bar{\varepsilon}(1+\sqrt{m})\vec{P}_j \\
0 & 0 & 0 & 0 \\
0 & 0 & 0 & 0
\end{matrix} \Bigg],
$$

$$
\mathbb{N}_j \triangleq \begin{bmatrix}
\vec{P}_jN^{\mathrm{T}} & 0 \\
0 & 0 \\
0 & 0
\end{bmatrix}, \quad \mathbb{M}_j \triangleq \begin{bmatrix}
0 & \tilde{X}_jM\nu_j \\
0 & \sqrt{2}\hat{X}_jM\nu_j \\
0 & 0
\end{bmatrix},
$$

Main Results

$$\mathbb{U}_{j+1} \triangleq \begin{bmatrix} \mu_{j+1}I & \mu_{j+1}I & 0 & 0 & 0 & 0 \\ 0 & 0 & \mu_{j+1}I & \mu_{j+1}I & 0 & 0 \\ 0 & 0 & 0 & 0 & \mu_{j+1}I & \mu_{j+1}I \\ 0 & 0 & 0 & 0 & 0 & 0 \end{bmatrix},$$

$$\Pi_{j,j+1}^{(22)} \triangleq -\mathrm{diag}\{\vec{W}_{j+1}, \vec{W}_{j+1}, \vec{W}_{j+1}, \vec{\omega}_{j,j+1}I, \vec{\varpi}_{j,j+1}I, \vec{\theta}_{j,j+1}I,$$
$$\vec{\zeta}_{j,j+1}I, \nu_j I, \nu_j I, \vec{Q}_{j+1}^h, \vec{Q}_{j+1}^v, \vec{Q}_{j+1}^h, \vec{Q}_{j+1}^v, \vec{Q}_{j+1}^h, \vec{Q}_{j+1}^v\},$$

with $\vec{P}_{m+1}^h = \vec{P}_1^h$, $\vec{P}_{m+1}^v = \vec{P}_1^v$, $\vec{Q}_{m+1}^h = \vec{Q}_1^h$, $\vec{Q}_{m+1}^v = \vec{Q}_1^v$, $\vec{W}_{m+1} = \vec{W}_1$, $\mu_{m+1} = \mu_1$, $\vec{\omega}_{m,m+1} = \vec{\omega}_{m,1}$, $\vec{\varpi}_{m,m+1} = \vec{\varpi}_{m,1}$, $\vec{\theta}_{m,m+1} = \vec{\theta}_{m,1}$, $\vec{\zeta}_{m+1} = \vec{\zeta}_1$. *Then, both the state trajectories of the closed-loop system (11.17) and the sliding function of (11.18) are driven into the following domain Ω around the origin by the 2-D first-order SMC law $\bar{u}(k,l)$ in (11.15) with the gain matrices* $K_j = \vec{K}_j \cdot \mathrm{diag}\left\{\vec{P}_j^h, \vec{P}_j^v\right\}^{-1}$, $j \in \mathcal{M}$.

Proof Denote $\vec{P}_{\sigma(\xi)} = P_{\sigma(\xi)}^{-1}$, $\vec{Q}_{\sigma(\xi)}^h = (Q_{\sigma(\xi)}^h)^{-1}$, $\vec{Q}_{\sigma(\xi)}^v = (Q_{\sigma(\xi)}^v)^{-1}$, $\vec{W}_{\sigma(\xi)} = W_{\sigma(\xi)}^{-1}$, $\vec{\rho} = \rho^{-1}$, $\vec{\zeta}_{\sigma(\xi),\sigma(\xi+1)} = \zeta_{\sigma(\xi),\sigma(\xi+1)}^{-1}$, $\vec{\omega}_{\sigma(\xi),\sigma(\xi+1)} = \omega_{\sigma(\xi),\sigma(\xi+1)}^{-1}$, $\vec{\varpi}_{\sigma(\xi),\sigma(\xi+1)} = \varpi_{\sigma(\xi),\sigma(\xi+1)}^{-1}$, and $\vec{\theta}_{\sigma(\xi),\sigma(\xi+1)} = \theta_{\sigma(\xi),\sigma(\xi+1)}^{-1}$. Letting $\sigma(\xi) = j \in \mathcal{M}$ and noticing that

$$\begin{cases} \sigma(\xi+1) - \sigma(\xi) = 1, & \text{if } k \neq rm, \\ \sigma(\xi+1) = 1 \text{ and } \sigma(\xi) = m, & \text{if } k = rm, \ r \in \mathbb{N}^+, \end{cases} \tag{11.54}$$

the LMIs (11.49)–(11.53) follow immediately from Schur complement when considering the parameter uncertainty $\Delta A(k,l)$ and using the following matrix inequality:

$$-Q_{j+1}^{-1} \leq -2\mu_{j+1}I + \mu_{j+1}^2(Q_{j+1}^h + Q_{j+1}^v). \tag{11.55}$$

The proof is now complete.

11.2.3 Second-order sliding mode case

In this subsection, we develop a 2-D *second-order* SMC scheme for the uncertain Roesser model (11.1) under the Round-Robin protocol (11.6)–(11.7). A 2-D second-order sliding mode is employed to enhance the robustness of the proposed 2-D SMC scheme, which is motivated from the 1-D super-twisting-like algorithm as in [123, 124]. The novel *2-D super-twisting-like algorithm* for uncertain Roesser model (11.1) is designed as follows:

$$\bar{u}(k,l) = -\left(B^{\mathrm{T}}XB\right)^{-1}\left(K_{\sigma(\xi)}x(k,l) - \hbar(k,l)\right), \tag{11.56}$$

where

$$\hbar(k,l) = -\gamma_{1\sigma(\xi)}\|s(k,l)\|^{\frac{1}{2}} \cdot \mathrm{sgn}(s(k,l)) + \lambda(k,l), \tag{11.57}$$

with

$$\lambda(k,l) \triangleq \beta_1 \lambda^h(k,l) + \beta_2 \lambda^v(k,l), \tag{11.58}$$

$$\lambda^h(k+1,l) = \lambda^h(k,l) - T^h \gamma^h_{2\sigma(\xi)} \cdot \mathrm{sgn}(s(k,l)), \tag{11.59}$$

$$\lambda^v(k,l+1) = \lambda^v(k,l) - T^v \gamma^v_{2\sigma(\xi)} \cdot \mathrm{sgn}(s(k,l)). \tag{11.60}$$

Here, $\beta_1 \geq 0$ and $\beta_2 \geq 0$ are the given weighting parameters with satisfying $\beta_1 + \beta_2 = 1$; the scalars $T^h > 0$ and $T^v > 0$ represent the sampling period in the horizontal and vertical directions, respectively; $\gamma_{1\sigma(\xi)} > 0$, $\gamma^h_{2\sigma(\xi)} > 0$ and $\gamma^v_{2\sigma(\xi)} > 0$ are the preassigned design parameters; and the token-dependent matrices $K_{\sigma(\xi)} \in \mathbb{R}^{m \times (n_h + n_v)}$ ($\sigma(\xi) \in \mathcal{M}$) are to be determined later. Furthermore, it yields from (11.58)–(11.60) that

$$\lambda^+(k,l) = \lambda(k,l) - \mathbf{T}_{\sigma(\xi)} \cdot \mathrm{sgn}(s(k,l)), \tag{11.61}$$

where $\mathbf{T}_{\sigma(\xi)} \triangleq \beta_1 T^h \gamma^h_{2\sigma(\xi)} + \beta_2 T^v \gamma^v_{2\sigma(\xi)}$.

Remark 11.4 *As one of important and famous second-order SMC approaches, the continuous-time super-twisting algorithm was first developed in [73] for reducing the chattering and providing the finite-time convergence of the sliding motion. Recently, the discrete-time 1-D super-twisting-like SMC scheme has been studied in [123,124] by applying the Euler discretization to the continuous-time super-twisting algorithm. It has been shown in [123,124] that the discrete-time 1-D super-twisting-like algorithm renders more robustness than the 1-D first-order SMC law since it may reduce* quasi-*sliding mode band. Following a similar idea of the above fact, this chapter makes the first attempt to develop a novel 2-D super-twisting-like algorithm (11.56) with (11.57)–(11.60) for uncertain Roesser model (11.1) under the purpose of enhancing the robustness of the proposed 2-D SMC scheme.*

Substituting (11.56) into (11.11), the actual control input with the second-order sliding mode is obtained as follows:

$$\begin{aligned} u(k,l) = &- \Phi_{\sigma(\xi)} \left(B^{\mathrm{T}} X B\right)^{-1} K_{\sigma(\xi)} x(k,l) \\ &+ \Phi_{\sigma(\xi)} \left(B^{\mathrm{T}} X B\right)^{-1} \hbar(k,l) + \left(I - \Phi_{\sigma(\xi)}\right) v(k,l). \end{aligned} \tag{11.62}$$

Now, one can has the following resulted closed-loop system and controlled sliding variable from (11.1), (11.14), and (11.62) that:

$$\begin{aligned} x^+(k,l) = &\tilde{A}_{\sigma(\xi)}(k,l) x(k,l) + B \left(I - \Phi_{\sigma(\xi)}\right) f(x(k,l)) + B(I - \Phi_{\sigma(\xi)}) v(k,l) \\ &- \gamma_{1\sigma(\xi)} \|s(k,l)\|^{\frac{1}{2}} \cdot B \Phi_{\sigma(\xi)} (B^{\mathrm{T}} X B)^{-1} \mathrm{sgn}(s(k,l)) \\ &+ B \Phi_{\sigma(\xi)} (B^{\mathrm{T}} X B)^{-1} \lambda(k,l) + B \Phi_{\sigma(\xi)} (B^{\mathrm{T}} X B)^{-1} D(k,l), \end{aligned} \tag{11.63}$$

and

$$\begin{aligned} s^+(k,l) = &- \Phi_{\sigma(\xi)} K_{\sigma(\xi)} x(k,l) + B^{\mathrm{T}} X B \left(I - \Phi_{\sigma(\xi)}\right) v(k,l) \\ &- \gamma_{1\sigma(\xi)} \|s(k,l)\|^{\frac{1}{2}} \cdot \Phi_{\sigma(\xi)} \mathrm{sgn}(s(k,l)) + \Phi_{\sigma(\xi)} \lambda(k,l) + D(k,l). \end{aligned} \tag{11.64}$$

Main Results 229

Define an extended state variable $\chi(k,l) \triangleq \left[\begin{array}{ccc} x^{\mathrm{T}}(k,l) & s^{\mathrm{T}}(k,l) & \lambda^{\mathrm{T}}(k,l) \end{array}\right]^{\mathrm{T}}$. Combining (11.61), (11.63), and (11.64) gives

$$
\begin{aligned}
\chi^+(k,l) =& \mathbf{A}_{\sigma(\xi)}(k,l)\chi(k,l) + \bar{\mathbf{B}}_{\sigma(\xi)}f(x(k,l)) + \tilde{\mathbf{B}}_{\sigma(\xi)}D(k,l) \\
& + \hat{\mathbf{B}}_{\sigma(\xi)}v(k,l) + \mathbf{F}_{\sigma(\xi)}(k,l)\mathrm{sgn}(s(k,l)),
\end{aligned} \qquad (11.65)
$$

where

$$
\mathbf{A}_{\sigma(\xi)}(k,l) \triangleq \left[\begin{array}{ccc} \tilde{A}_{\sigma(\xi)}(k,l) & 0 & B\Phi_{\sigma(\xi)}(B^{\mathrm{T}}XB)^{-1} \\ -\Phi_{\sigma(\xi)}K_{\sigma(\xi)} & 0 & \Phi_{\sigma(\xi)} \\ 0 & 0 & I \end{array}\right],
$$

$$
\bar{\mathbf{B}}_{\sigma(\xi)} \triangleq \left[\begin{array}{c} B\left(I - \Phi_{\sigma(\xi)}\right) \\ 0 \\ 0 \end{array}\right], \tilde{\mathbf{B}}_{\sigma(\xi)} \triangleq \left[\begin{array}{c} B\Phi_{\sigma(\xi)}(B^{\mathrm{T}}XB)^{-1} \\ I \\ 0 \end{array}\right],
$$

$$
\hat{\mathbf{B}}_{\sigma(\xi)} \triangleq \left[\begin{array}{c} B(I - \Phi_{\sigma(\xi)}) \\ B^{\mathrm{T}}XB\left(I - \Phi_{\sigma(\xi)}\right) \\ 0 \end{array}\right],
$$

$$
\mathbf{F}_{\sigma(\xi)}(k,l) \triangleq \left[\begin{array}{c} -\gamma_{1\sigma(\xi)}B\Phi_{\sigma(\xi)}(B^{\mathrm{T}}XB)^{-1} \cdot \|s(k,l)\|^{\frac{1}{2}} \\ -\gamma_{1\sigma(\xi)}\Phi_{\sigma(\xi)} \cdot \|s(k,l)\|^{\frac{1}{2}} \\ -\mathbf{T}_{\sigma(\xi)} \end{array}\right].
$$

According to the definition of Euclidean vector norm, it gives

$$
\|\mathbf{F}_{\sigma(\xi)}(k,l)\|^2 \leq \tilde{\gamma}_{\sigma(\xi)}\|s(k,l)\| + \mathbf{T}_{\sigma(\xi)}^2, \qquad (11.66)
$$

where $\tilde{\gamma}_{\sigma(\xi)} \triangleq \gamma_{1\sigma(\xi)}^2 \left(\|B\Phi_{\sigma(\xi)}(B^{\mathrm{T}}XB)^{-1}\|^2 + 1\right)$.

The following theorem proposes a sufficient condition to guarantee the ultimate boundedness of the controlled state trajectories and the common sliding variable under the token-dependent 2-D second-order SMC law (11.56) with (11.57)–(11.60).

Theorem 11.3 *Consider the 2-D system (11.1) with the Round-Robin protocol (11.6)–(11.7) and the token-dependent 2-D second-order SMC law (11.56) with (11.57)–(11.60). For preassigned parameters $\alpha_1 \geq 0$, $\alpha_2 \geq 0$, a matrix $X > 0$ and any $\sigma(\xi) \in \mathcal{M}$ satisfying the condition (11.10), assume that there exist matrices $K_{\sigma(\xi)} \in \mathbb{R}^{m \times (n_h + n_v)}$, $P_{\sigma(\xi)}^h > 0$, $P_{\sigma(\xi)}^v > 0$, $P_{\sigma(\xi+1)}^h > 0$, $P_{\sigma(\xi+1)}^v > 0$, $Q_{\sigma(\xi)}^h > 0$, $Q_{\sigma(\xi)}^v > 0$, $Q_{\sigma(\xi+1)}^h > 0$, $Q_{\sigma(\xi+1)}^v > 0$, $W_{\sigma(\xi)} > 0$,*

$W_{\sigma(\xi+1)} > 0$, $Z_{\sigma(\xi)} > 0$, $Z_{\sigma(\xi+1)} > 0$, and scalars $\kappa > 0$, $\pi_{\sigma(\xi),\sigma(\xi+1)} > 0$, $\psi_{\sigma(\xi),\sigma(\xi+1)} > 0$ such that the following matrix inequalities hold:

$$\bar{\mathbf{B}}_{\sigma(\xi)}^{\mathrm{T}}\mathbf{P}_{\sigma(\xi+1)}\bar{\mathbf{B}}_{\sigma(\xi)} \leq \pi_{\sigma(\xi),\sigma(\xi+1)}I, \tag{11.67}$$

$$\tilde{\mathbf{B}}_{\sigma(\xi)}^{\mathrm{T}}\mathbf{P}_{\sigma(\xi+1)}\tilde{\mathbf{B}}_{\sigma(\xi)} \leq \psi_{\sigma(\xi),\sigma(\xi+1)}I, \tag{11.68}$$

$$\tilde{\mathbf{X}}_{\sigma(\xi),\sigma(\xi+1)}(k,l) \triangleq \begin{bmatrix} \mathbf{X}_{\sigma(\xi),\sigma(\xi+1)}(k,l) & \Upsilon_{\sigma(\xi),\sigma(\xi+1)}(k,l) & \bar{\Upsilon}_{\sigma(\xi),\sigma(\xi+1)}(k,l) \\ \star & \alpha_1^2\mathbf{Y}_{\sigma(\xi),\sigma(\xi+1)} - Q_{\sigma(\xi)}^h & \alpha_1\alpha_2\mathbf{Y}_{\sigma(\xi),\sigma(\xi+1)} \\ \star & \star & \alpha_2^2\mathbf{Y}_{\sigma(\xi),\sigma(\xi+1)} - Q_{\sigma(\xi)}^v \end{bmatrix} < 0, \tag{11.69}$$

where $\mathbf{P}_{\sigma(\xi)} \triangleq \mathrm{diag}\{P_{\sigma(\xi)}, W_{\sigma(\xi)}, Z_{\sigma(\xi)}\}$, $P_{\sigma(\xi)} \triangleq \mathrm{diag}\left\{P_{\sigma(\xi)}^h, P_{\sigma(\xi)}^v\right\}$, $Q_{\sigma(\xi)} \triangleq Q_{\sigma(\xi)}^h + Q_{\sigma(\xi)}^v$, and

$$\mathbf{X}_{\sigma(\xi),\sigma(\xi+1)}(k,l) \triangleq 4\mathbf{A}_{\sigma(\xi)}^{\mathrm{T}}(k,l)\mathbf{P}_{\sigma(\xi+1)}\mathbf{A}_{\sigma(\xi)}(k,l) + 10\varepsilon^2\pi_{\sigma(\xi),\sigma(\xi+1)}\check{\mathbf{I}}$$
$$+ 10\bar{\varepsilon}^2\psi_{\sigma(\xi),\sigma(\xi+1)}\check{\mathbf{I}} + \bar{\mathbf{X}}_{\sigma(\xi),\sigma(\xi+1)} + \kappa I - \mathbf{P}_{\sigma(\xi)},$$
$$\bar{\mathbf{X}}_{\sigma(\xi),\sigma(\xi+1)} \triangleq \mathrm{diag}\{\hat{\mathbf{X}}_{\sigma(\xi),\sigma(\xi+1)}, 0, \vec{\mathbf{X}}_{\sigma(\xi),\sigma(\xi+1)}\},$$
$$\hat{\mathbf{X}}_{\sigma(\xi),\sigma(\xi+1)} \triangleq 2K_{\sigma(\xi)}^{\mathrm{T}}(B^{\mathrm{T}}XB)^{-1}\Phi_{\sigma(\xi)}Q_{\sigma(\xi)}\Phi_{\sigma(\xi)}(B^{\mathrm{T}}XB)^{-1}K_{\sigma(\xi)},$$
$$\vec{\mathbf{X}}_{\sigma(\xi),\sigma(\xi+1)} \triangleq 6\left(B^{\mathrm{T}}XB\right)^{-1}\Phi_{\sigma(\xi)}Q_{\sigma(\xi)}\Phi_{\sigma(\xi)}\left(B^{\mathrm{T}}XB\right)^{-1},$$
$$\Upsilon_{\sigma(\xi),\sigma(\xi+1)}(k,l) \triangleq \alpha_1\left(\mathbf{A}_{\sigma(\xi)}^{\mathrm{T}}(k,l)\mathbf{P}_{\sigma(\xi+1)}\hat{\mathbf{B}}_{\sigma(\xi)} - \mathbf{Q}_{\sigma(\xi),\sigma(\xi+1)}\right),$$
$$\bar{\Upsilon}_{\sigma(\xi),\sigma(\xi+1)}(k,l) \triangleq \alpha_2\left(\mathbf{A}_{\sigma(\xi)}^{\mathrm{T}}(k,l)\mathbf{P}_{\sigma(\xi+1)}\hat{\mathbf{B}}_{\sigma(\xi)} - \mathbf{Q}_{\sigma(\xi),\sigma(\xi+1)}\right),$$
$$\mathbf{Q}_{\sigma(k),\sigma(\xi+1)} \triangleq \bar{\mathbf{I}}^{\mathrm{T}}K_{\sigma(\xi)}^{\mathrm{T}}(B^{\mathrm{T}}XB)^{-1}\Phi_{\sigma(\xi)}Q_{\sigma(\xi)}\left(I - \Phi_{\sigma(\xi)}\right),$$
$$\bar{\mathbf{I}} \triangleq \begin{bmatrix} I & 0 & 0 \end{bmatrix}, \quad \check{\mathbf{I}} \triangleq \mathrm{diag}\{I, 0, 0\},$$
$$\mathbf{Y}_{\sigma(\xi),\sigma(\xi+1)} \triangleq 4\hat{\mathbf{B}}_{\sigma(\xi)}^{\mathrm{T}}\mathbf{P}_{\sigma(\xi+1)}\hat{\mathbf{B}}_{\sigma(\xi)} + 2\left(I - \Phi_{\sigma(\xi)}\right)Q_{\sigma(\xi)}\left(I - \Phi_{\sigma(\xi)}\right).$$

Then, both the state trajectories of the closed-loop system (11.63) and the sliding function of (11.64) are driven into the following domain $\mathbf{O}$ around the origin:

$$\mathbf{O} \triangleq \left\{\chi(k,l) \in \mathbb{R}^{n_h+n_v+2m} \mid \|\chi(k,l)\| \leq \bar{\kappa}\right\}, \tag{11.70}$$

where $\chi(k,l) \triangleq \begin{bmatrix} x^{\mathrm{T}}(k,l) & s^{\mathrm{T}}(k,l) & \lambda^{\mathrm{T}}(k,l) \end{bmatrix}^{\mathrm{T}}$, $\bar{\kappa} \triangleq \max_{\sigma(\xi),\sigma(\xi+1)\in\mathcal{M}}\{\bar{\kappa}_{\sigma(\xi),\sigma(\xi+1)}\}$, $\bar{\kappa}_{\sigma(\xi),\sigma(\xi+1)} = \frac{\vec{\gamma}_{\sigma(\xi),\sigma(\xi+1)}+\sqrt{\vec{\gamma}_{\sigma(\xi),\sigma(\xi+1)}^2+4\kappa\hat{\mathbf{T}}_{\sigma(\xi)}}}{2\kappa}$, $\vec{\gamma}_{\sigma(\xi),\sigma(\xi+1)} \triangleq 5m\|\mathbf{P}_{\sigma(\xi+1)}\|\tilde{\gamma}_{\sigma(\xi)} + 6m\gamma_{1\sigma(\xi)}^2\|\left(B^{\mathrm{T}}XB\right)^{-1}\Phi_{\sigma(\xi)}Q_{\sigma(\xi)} \times \Phi_{\sigma(\xi)}\left(B^{\mathrm{T}}XB\right)^{-1}\|$ and $\hat{\mathbf{T}}_{\sigma(\xi)} \triangleq 5m\|\mathbf{P}_{\sigma(\xi+1)}\|\cdot\mathbf{T}_{\sigma(\xi)}^2 + 10\left(\pi_{\sigma(\xi),\sigma(\xi+1)}+\psi_{\sigma(\xi),\sigma(\xi+1)}\|B^{\mathrm{T}}XB\|^2\right)d^2$.

Main Results 231

Proof *Choose the token-dependent Lyapunov-like function as* $\hat{V}(k,l,\sigma(\xi))$
$\triangleq \chi^{\mathrm{T}}(k,l)\mathbf{P}_{\sigma(\xi)}\chi(k,l)+u^{\mathrm{T}}(k-1,l)Q^h_{\sigma(\xi)}u(k-1,l)+u^{\mathrm{T}}(k,l-1)Q^v_{\sigma(\xi)}u(k,l-1)$,
where $\mathbf{P}_{\sigma(\xi)} \triangleq \mathrm{diag}\{P_{\sigma(\xi)},W_{\sigma(\xi)},Z_{\sigma(\xi)}\}$. *It further has*

$$
\begin{aligned}
\Delta\hat{V}&(k,l,\sigma(\xi))\\
=&\chi^{+\mathrm{T}}(k,l)\mathbf{P}_{\sigma(\xi+1)}\chi^+(k,l) - \chi^{\mathrm{T}}(k,l)\mathbf{P}_{\sigma(\xi)}\chi(k,l)\\
&+ u^{\mathrm{T}}(k,l)Q_{\sigma(\xi)}u(k,l) - u^{\mathrm{T}}(k-1,l)Q^h_{\sigma(\xi)}u(k-1,l)\\
&- u^{\mathrm{T}}(k,l-1)Q^v_{\sigma(\xi)}u(k,l-1).
\end{aligned}
\tag{11.71}
$$

Along the trajectories of the extended system dynamics (11.65), we have

$$
\begin{aligned}
\chi^{+\mathrm{T}}&(k,l)\mathbf{P}_{\sigma(\xi+1)}\chi^+(k,l)\\
=&\chi^{\mathrm{T}}(k,l)\mathbf{A}^{\mathrm{T}}_{\sigma(\xi)}(k,l)\mathbf{P}_{\sigma(\xi+1)}\mathbf{A}_{\sigma(\xi)}(k,l)\chi(k,l)\\
&+ f^{\mathrm{T}}(x(k,l))\bar{\mathbf{B}}^{\mathrm{T}}_{\sigma(\xi)}\mathbf{P}_{\sigma(\xi+1)}\bar{\mathbf{B}}_{\sigma(\xi)}f(x(k,l))\\
&+ D^{\mathrm{T}}(k,l)\tilde{\mathbf{B}}^{\mathrm{T}}_{\sigma(\xi)}\mathbf{P}_{\sigma(\xi+1)}\tilde{\mathbf{B}}_{\sigma(\xi)}D(k,l)\\
&+ v^{\mathrm{T}}(k,l)\hat{\mathbf{B}}^{\mathrm{T}}_{\sigma(\xi)}\mathbf{P}_{\sigma(\xi+1)}\hat{\mathbf{B}}_{\sigma(\xi)}v(k,l)\\
&+ \mathrm{sgn}^{\mathrm{T}}(s(k,l))\mathbf{F}^{\mathrm{T}}_{\sigma(\xi)}(k,l)\mathbf{P}_{\sigma(\xi+1)}\mathbf{F}_{\sigma(\xi)}(k,l)\mathrm{sgn}(s(k,l))\\
&+ 2\chi^{\mathrm{T}}(k,l)\mathbf{A}^{\mathrm{T}}_{\sigma(\xi)}(k,l)\mathbf{P}_{\sigma(\xi+1)}\bar{\mathbf{B}}_{\sigma(\xi)}f(x(k,l))\\
&+ 2\chi^{\mathrm{T}}(k,l)\mathbf{A}^{\mathrm{T}}_{\sigma(\xi)}(k,l)\mathbf{P}_{\sigma(\xi+1)}\tilde{\mathbf{B}}_{\sigma(\xi)}D(k,l)\\
&+ 2\chi^{\mathrm{T}}(k,l)\mathbf{A}^{\mathrm{T}}_{\sigma(\xi)}(k,l)\mathbf{P}_{\sigma(\xi+1)}\hat{\mathbf{B}}_{\sigma(\xi)}v(k,l)\\
&+ 2\chi^{\mathrm{T}}(k,l)\mathbf{A}^{\mathrm{T}}_{\sigma(\xi)}(k,l)\mathbf{P}_{\sigma(\xi+1)}\mathbf{F}_{\sigma(\xi)}(k,l)\mathrm{sgn}(s(k,l))\\
&+ 2f^{\mathrm{T}}(x(k,l))\bar{\mathbf{B}}^{\mathrm{T}}_{\sigma(\xi)}\mathbf{P}_{\sigma(\xi+1)}\tilde{\mathbf{B}}_{\sigma(\xi)}D(k,l)\\
&+ 2f^{\mathrm{T}}(x(k,l))\bar{\mathbf{B}}^{\mathrm{T}}_{\sigma(\xi)}\mathbf{P}_{\sigma(\xi+1)}\hat{\mathbf{B}}_{\sigma(\xi)}v(k,l)\\
&+ 2f^{\mathrm{T}}(x(k,l))\bar{\mathbf{B}}^{\mathrm{T}}_{\sigma(\xi)}\mathbf{P}_{\sigma(\xi+1)}\mathbf{F}_{\sigma(\xi)}(k,l)\mathrm{sgn}(s(k,l))\\
&+ 2D^{\mathrm{T}}(k,l)\tilde{\mathbf{B}}^{\mathrm{T}}_{\sigma(\xi)}\mathbf{P}_{\sigma(\xi+1)}\hat{\mathbf{B}}_{\sigma(\xi)}v(k,l)\\
&+ 2D^{\mathrm{T}}(k,l)\tilde{\mathbf{B}}^{\mathrm{T}}_{\sigma(\xi)}\mathbf{P}_{\sigma(\xi+1)}\mathbf{F}_{\sigma(\xi)}(k,l)\mathrm{sgn}(s(k,l))\\
&+ 2v^{\mathrm{T}}(k,l)\hat{\mathbf{B}}^{\mathrm{T}}_{\sigma(\xi)}\mathbf{P}_{\sigma(\xi+1)}\mathbf{F}_{\sigma(\xi)}(k,l)\mathrm{sgn}(s(k,l)).
\end{aligned}
\tag{11.72}
$$

The conditions (11.67) and (11.68) ensure the following relationships:

$$
\begin{aligned}
&f^{\mathrm{T}}(x(k,l))\bar{\mathbf{B}}^{\mathrm{T}}_{\sigma(\xi)}\mathbf{P}_{\sigma(\xi+1)}\bar{\mathbf{B}}_{\sigma(\xi)}f(x(k,l))\\
\leq&2\pi_{\sigma(\xi),\sigma(\xi+1)}(\varepsilon^2\|x(k,l)\|^2+d^2),
\end{aligned}
\tag{11.73}
$$

$$
\begin{aligned}
&D^{\mathrm{T}}(k,l)\tilde{\mathbf{B}}^{\mathrm{T}}_{\sigma(\xi)}\mathbf{P}_{\sigma(\xi+1)}\tilde{\mathbf{B}}_{\sigma(\xi)}D(k,l)\\
\leq&2\psi_{\sigma(\xi),\sigma(\xi+1)}(\bar{\varepsilon}^2\|x(k,l)\|^2+d^2\|B^{\mathrm{T}}XB\|^2),
\end{aligned}
\tag{11.74}
$$

$$
\begin{aligned}
&2\chi^{\mathrm{T}}(k,l)\mathbf{A}^{\mathrm{T}}_{\sigma(\xi)}(k,l)\mathbf{P}_{\sigma(\xi+1)}\bar{\mathbf{B}}_{\sigma(\xi)}f(x(k,l))\\
\leq&\chi^{\mathrm{T}}(k,l)\mathbf{A}^{\mathrm{T}}_{\sigma(\xi)}(k,l)\mathbf{P}_{\sigma(\xi+1)}\mathbf{A}_{\sigma(\xi)}(k,l)\chi(k,l)
\end{aligned}
$$

$$+ 2\pi_{\sigma(\xi),\sigma(\xi+1)}(\varepsilon^2\|x(k,l)\|^2 + d^2), \tag{11.75}$$

$$2\chi^{\mathrm{T}}(k,l)\mathbf{A}_{\sigma(\xi)}^{\mathrm{T}}(k,l)\mathbf{P}_{\sigma(\xi+1)}\tilde{\mathbf{B}}_{\sigma(\xi)}D(k,l)$$
$$\leq\chi^{\mathrm{T}}(k,l)\mathbf{A}_{\sigma(\xi)}^{\mathrm{T}}(k,l)\mathbf{P}_{\sigma(\xi+1)}\mathbf{A}_{\sigma(\xi)}(k,l)\chi(k,l)$$
$$+ 2\psi_{\sigma(\xi),\sigma(\xi+1)}(\bar{\varepsilon}^2\|x(k,l)\|^2 + d^2\|B^{\mathrm{T}}XB\|^2), \tag{11.76}$$

$$2\chi^{\mathrm{T}}(k,l)\mathbf{A}_{\sigma(\xi)}^{\mathrm{T}}(k,l)\mathbf{P}_{\sigma(\xi+1)}\mathbf{F}_{\sigma(\xi)}(k,l)\mathrm{sgn}(s(k,l))$$
$$\leq\chi^{\mathrm{T}}(k,l)\mathbf{A}_{\sigma(\xi)}^{\mathrm{T}}(k,l)\mathbf{P}_{\sigma(\xi+1)}\mathbf{A}_{\sigma(\xi)}(k,l)\chi(k,l)$$
$$+ \mathrm{sgn}^{\mathrm{T}}(s(k,l))\mathbf{F}_{\sigma(\xi)}^{\mathrm{T}}(k,l)\mathbf{P}_{\sigma(\xi+1)}\mathbf{F}_{\sigma(\xi)}(k,l)\mathrm{sgn}(s(k,l)), \tag{11.77}$$

$$2f^{\mathrm{T}}(x(k,l))\bar{\mathbf{B}}_{\sigma(\xi)}^{\mathrm{T}}\mathbf{P}_{\sigma(\xi+1)}\tilde{\mathbf{B}}_{\sigma(\xi)}D(k,l)$$
$$\leq 2\left(\varepsilon^2\pi_{\sigma(\xi),\sigma(\xi+1)} + \bar{\varepsilon}^2\psi_{\sigma(\xi),\sigma(\xi+1)}\right)\|x(k,l)\|^2$$
$$+ 2\left(\pi_{\sigma(\xi),\sigma(\xi+1)} + \psi_{\sigma(\xi),\sigma(\xi+1)}\|B^{\mathrm{T}}XB\|^2\right)d^2, \tag{11.78}$$

$$2f^{\mathrm{T}}(x(k,l))\bar{\mathbf{B}}_{\sigma(\xi)}^{\mathrm{T}}\mathbf{P}_{\sigma(\xi+1)}\hat{\mathbf{B}}_{\sigma(\xi)}v(k,l)$$
$$\leq v^{\mathrm{T}}(k,l)\hat{\mathbf{B}}_{\sigma(\xi)}^{\mathrm{T}}\mathbf{P}_{\sigma(\xi+1)}\hat{\mathbf{B}}_{\sigma(\xi)}v(k,l)$$
$$+ 2\pi_{\sigma(\xi),\sigma(\xi+1)}(\varepsilon^2\|x(k,l)\|^2 + d^2), \tag{11.79}$$

$$2f^{\mathrm{T}}(x(k,l))\bar{\mathbf{B}}_{\sigma(\xi)}^{\mathrm{T}}\mathbf{P}_{\sigma(\xi+1)}\mathbf{F}_{\sigma(\xi)}(k,l)\mathrm{sgn}(s(k,l))$$
$$\leq\mathrm{sgn}^{\mathrm{T}}(s(k,l))\mathbf{F}_{\sigma(\xi)}^{\mathrm{T}}(k,l)\mathbf{P}_{\sigma(\xi+1)}\mathbf{F}_{\sigma(\xi)}(k,l)\mathrm{sgn}(s(k,l))$$
$$+ 2\pi_{\sigma(\xi),\sigma(\xi+1)}(\varepsilon^2\|x(k,l)\|^2 + d^2), \tag{11.80}$$

$$2D^{\mathrm{T}}(k,l)\tilde{\mathbf{B}}_{\sigma(\xi)}^{\mathrm{T}}\mathbf{P}_{\sigma(\xi+1)}\hat{\mathbf{B}}_{\sigma(\xi)}v(k,l)$$
$$\leq v^{\mathrm{T}}(k,l)\hat{\mathbf{B}}_{\sigma(\xi)}^{\mathrm{T}}\mathbf{P}_{\sigma(\xi+1)}\hat{\mathbf{B}}_{\sigma(\xi)}v(k,l)$$
$$+ 2\psi_{\sigma(\xi),\sigma(\xi+1)}(\bar{\varepsilon}^2\|x(k,l)\|^2 + d^2\|B^{\mathrm{T}}XB\|^2), \tag{11.81}$$

$$2D^{\mathrm{T}}(k,l)\tilde{\mathbf{B}}_{\sigma(\xi)}^{\mathrm{T}}\mathbf{P}_{\sigma(\xi+1)}\mathbf{F}_{\sigma(\xi)}(k,l)\mathrm{sgn}(s(k,l))$$
$$\leq\mathrm{sgn}^{\mathrm{T}}(s(k,l))\mathbf{F}_{\sigma(\xi)}^{\mathrm{T}}(k,l)\mathbf{P}_{\sigma(\xi+1)}\mathbf{F}_{\sigma(\xi)}(k,l)\mathrm{sgn}(s(k,l))$$
$$+ 2\psi_{\sigma(\xi),\sigma(\xi+1)}(\bar{\varepsilon}^2\|x(k,l)\|^2 + d^2\|B^{\mathrm{T}}XB\|^2), \tag{11.82}$$

$$2v^{\mathrm{T}}(k,l)\hat{\mathbf{B}}_{\sigma(\xi)}^{\mathrm{T}}\mathbf{P}_{\sigma(\xi+1)}\mathbf{F}_{\sigma(\xi)}(k,l)\mathrm{sgn}(s(k,l))$$
$$\leq\mathrm{sgn}^{\mathrm{T}}(s(k,l))\mathbf{F}_{\sigma(\xi)}^{\mathrm{T}}(k,l)\mathbf{P}_{\sigma(\xi+1)}\mathbf{F}_{\sigma(\xi)}(k,l)\mathrm{sgn}(s(k,l))$$
$$+ v^{\mathrm{T}}(k,l)\hat{\mathbf{B}}_{\sigma(\xi)}^{\mathrm{T}}\mathbf{P}_{\sigma(\xi+1)}\hat{\mathbf{B}}_{\sigma(\xi)}v(k,l). \tag{11.83}$$

By resorting to the relationships (11.73)–(11.83), it yields from (11.72) that

$$\chi^{+\mathrm{T}}(k,l)\mathbf{P}_{\sigma(\xi+1)}\chi^+(k,l)$$
$$\leq\chi^{\mathrm{T}}(k,l)\Big[4\mathbf{A}_{\sigma(\xi)}^{\mathrm{T}}(k,l)\mathbf{P}_{\sigma(\xi+1)}\mathbf{A}_{\sigma(\xi)}(k,l)$$
$$+ 10\varepsilon^2\pi_{\sigma(\xi),\sigma(\xi+1)}\check{\mathbf{I}} + 10\bar{\varepsilon}^2\psi_{\sigma(\xi),\sigma(\xi+1)}\check{\mathbf{I}}\Big]\chi(k,l)$$
$$+ 4v^{\mathrm{T}}(k,l)\hat{\mathbf{B}}_{\sigma(\xi)}^{\mathrm{T}}\mathbf{P}_{\sigma(\xi+1)}\hat{\mathbf{B}}_{\sigma(\xi)}v(k,l)$$

Main Results

$$+ 5\mathrm{sgn}^\mathrm{T}(s(k,l))\mathbf{F}_{\sigma(\xi)}^\mathrm{T}(k,l)\mathbf{P}_{\sigma(\xi+1)}\mathbf{F}_{\sigma(\xi)}(k,l)\mathrm{sgn}(s(k,l))$$
$$+ 2\chi^\mathrm{T}(k,l)\mathbf{A}_{\sigma(\xi)}^\mathrm{T}(k,l)\mathbf{P}_{\sigma(\xi+1)}\hat{\mathbf{B}}_{\sigma(\xi)}v(k,l)$$
$$+ 10\left(\pi_{\sigma(\xi),\sigma(\xi+1)} + \psi_{\sigma(\xi),\sigma(\xi+1)}\|B^\mathrm{T}XB\|^2\right)d^2. \tag{11.84}$$

Meanwhile, by means of the actual control input (11.62), it gets

$$u^\mathrm{T}(k,l)Q_{\sigma(\xi)}u(k,l)$$
$$=x^\mathrm{T}(k,l)K_{\sigma(\xi)}^\mathrm{T}(B^\mathrm{T}XB)^{-1}\Phi_{\sigma(\xi)}Q_{\sigma(\xi)}\Phi_{\sigma(\xi)}(B^\mathrm{T}XB)^{-1}K_{\sigma(\xi)}x(k,l)$$
$$+ \hbar^\mathrm{T}(k,l)\left(B^\mathrm{T}XB\right)^{-1}\Phi_{\sigma(\xi)}Q_{\sigma(\xi)}\Phi_{\sigma(\xi)}\left(B^\mathrm{T}XB\right)^{-1}\hbar(k,l)$$
$$+ v^\mathrm{T}(k,l)\left(I - \Phi_{\sigma(\xi)}\right)Q_{\sigma(\xi)}\left(I - \Phi_{\sigma(\xi)}\right)v(k,l)$$
$$- 2x^\mathrm{T}(k,l)K_{\sigma(\xi)}^\mathrm{T}(B^\mathrm{T}XB)^{-1}\Phi_{\sigma(\xi)}Q_{\sigma(\xi)}\Phi_{\sigma(\xi)}\left(B^\mathrm{T}XB\right)^{-1}\hbar(k,l)$$
$$- 2x^\mathrm{T}(k,l)K_{\sigma(\xi)}^\mathrm{T}(B^\mathrm{T}XB)^{-1}\Phi_{\sigma(\xi)}Q_{\sigma(\xi)}\left(I - \Phi_{\sigma(\xi)}\right)v(k,l)$$
$$+ 2\hbar^\mathrm{T}(k,l)\left(B^\mathrm{T}XB\right)^{-1}\Phi_{\sigma(\xi)}Q_{\sigma(\xi)}\left(I - \Phi_{\sigma(\xi)}\right)v(k,l). \tag{11.85}$$

It can be obtained that

$$\hbar^\mathrm{T}(k,l)\left(B^\mathrm{T}XB\right)^{-1}\Phi_{\sigma(\xi)}Q_{\sigma(\xi)}\Phi_{\sigma(\xi)}\left(B^\mathrm{T}XB\right)^{-1}\hbar(k,l)$$
$$\leq 2m\gamma_{1\sigma(\xi)}^2\|\left(B^\mathrm{T}XB\right)^{-1}\Phi_{\sigma(\xi)}Q_{\sigma(\xi)}\Phi_{\sigma(\xi)}\left(B^\mathrm{T}XB\right)^{-1}\|\cdot\|s(k,l)\|$$
$$+ 2\lambda^\mathrm{T}(k,l)\left(B^\mathrm{T}XB\right)^{-1}\Phi_{\sigma(\xi)}Q_{\sigma(\xi)}\Phi_{\sigma(\xi)}\left(B^\mathrm{T}XB\right)^{-1}\lambda(k,l), \tag{11.86}$$
$$- 2x^\mathrm{T}(k,l)K_{\sigma(\xi)}^\mathrm{T}(B^\mathrm{T}XB)^{-1}\Phi_{\sigma(\xi)}Q_{\sigma(\xi)}\Phi_{\sigma(\xi)}\left(B^\mathrm{T}XB\right)^{-1}\hbar(k,l)$$
$$\leq x^\mathrm{T}(k,l)K_{\sigma(\xi)}^\mathrm{T}(B^\mathrm{T}XB)^{-1}\Phi_{\sigma(\xi)}Q_{\sigma(\xi)}\Phi_{\sigma(\xi)}(B^\mathrm{T}XB)^{-1}K_{\sigma(\xi)}x(k,l)$$
$$+ 2m\gamma_{1\sigma(\xi)}^2\|\left(B^\mathrm{T}XB\right)^{-1}\Phi_{\sigma(\xi)}Q_{\sigma(\xi)}\Phi_{\sigma(\xi)}\left(B^\mathrm{T}XB\right)^{-1}\|\cdot\|s(k,l)\|$$
$$+ 2\lambda^\mathrm{T}(k,l)\left(B^\mathrm{T}XB\right)^{-1}\Phi_{\sigma(\xi)}Q_{\sigma(\xi)}\Phi_{\sigma(\xi)}\left(B^\mathrm{T}XB\right)^{-1}\lambda(k,l), \tag{11.87}$$
$$2\hbar^\mathrm{T}(k,l)\left(B^\mathrm{T}XB\right)^{-1}\Phi_{\sigma(\xi)}Q_{\sigma(\xi)}\left(I - \Phi_{\sigma(\xi)}\right)v(k,l)$$
$$\leq v^\mathrm{T}(k,l)\left(I - \Phi_{\sigma(\xi)}\right)Q_{\sigma(\xi)}\left(I - \Phi_{\sigma(\xi)}\right)v(k,l)$$
$$+ 2m\gamma_{1\sigma(\xi)}^2\|\left(B^\mathrm{T}XB\right)^{-1}\Phi_{\sigma(\xi)}Q_{\sigma(\xi)}\Phi_{\sigma(\xi)}\left(B^\mathrm{T}XB\right)^{-1}\|\cdot\|s(k,l)\|$$
$$+ 2\lambda^\mathrm{T}(k,l)\left(B^\mathrm{T}XB\right)^{-1}\Phi_{\sigma(\xi)}Q_{\sigma(\xi)}\Phi_{\sigma(\xi)}\left(B^\mathrm{T}XB\right)^{-1}\lambda(k,l). \tag{11.88}$$

Applying the relationships (11.86)–(11.88) into (11.85) gives

$$u^\mathrm{T}(k,l)Q_{\sigma(\xi)}u(k,l)$$
$$\leq 2x^\mathrm{T}(k,l)K_{\sigma(\xi)}^\mathrm{T}(B^\mathrm{T}XB)^{-1}\Phi_{\sigma(\xi)}Q_{\sigma(\xi)}\Phi_{\sigma(\xi)}(B^\mathrm{T}XB)^{-1}K_{\sigma(\xi)}x(k,l)$$
$$+ 2v^\mathrm{T}(k,l)\left(I - \Phi_{\sigma(\xi)}\right)Q_{\sigma(\xi)}\left(I - \Phi_{\sigma(\xi)}\right)v(k,l)$$
$$- 2x^\mathrm{T}(k,l)K_{\sigma(\xi)}^\mathrm{T}(B^\mathrm{T}XB)^{-1}\Phi_{\sigma(\xi)}Q_{\sigma(\xi)}\left(I - \Phi_{\sigma(\xi)}\right)v(k,l)$$
$$+ 6m\gamma_{1\sigma(\xi)}^2\|\left(B^\mathrm{T}XB\right)^{-1}\Phi_{\sigma(\xi)}Q_{\sigma(\xi)}$$

$$\times \Phi_{\sigma(\xi)} \left(B^{\mathrm{T}}XB\right)^{-1} \| \cdot \|s(k,l)\|$$
$$+ 6\lambda^{\mathrm{T}}(k,l) \left(B^{\mathrm{T}}XB\right)^{-1} \Phi_{\sigma(\xi)} Q_{\sigma(\xi)} \Phi_{\sigma(\xi)} \left(B^{\mathrm{T}}XB\right)^{-1} \lambda(k,l). \qquad (11.89)$$

Substituting (11.84) and (11.89) into (11.71) and using (11.66), we get

$$\Delta \hat{V}(k,l,\sigma(\xi))$$
$$\leq \chi^{\mathrm{T}}(k,l)\mathbf{X}_{\sigma(\xi),\sigma(\xi+1)}(k,l)\chi(k,l) - \kappa\|\chi(k,l)\|^2$$
$$+ v^{\mathrm{T}}(k,l)\mathbf{Y}_{\sigma(\xi),\sigma(\xi+1)}v(k,l)$$
$$+ 2\chi^{\mathrm{T}}(k,l)\mathbf{A}_{\sigma(\xi)}^{\mathrm{T}}(k,l)\mathbf{P}_{\sigma(\xi+1)}\hat{\mathbf{B}}_{\sigma(\xi)}v(k,l)$$
$$- 2x^{\mathrm{T}}(k,l)K_{\sigma(\xi)}^{\mathrm{T}}(B^{\mathrm{T}}XB)^{-1}\Phi_{\sigma(\xi)}Q_{\sigma(\xi)}\left(I - \Phi_{\sigma(\xi)}\right)v(k,l)$$
$$- u^{\mathrm{T}}(k-1,l)Q_{\sigma(\xi)}^{h}u(k-1,l) - u^{\mathrm{T}}(k,l-1)Q_{\sigma(\xi)}^{v}u(k,l-1)$$
$$+ 5m\|\mathbf{P}_{\sigma(\xi+1)}\|\left(\tilde{\gamma}_{\sigma(\xi)}\|s(k,l)\| + \mathbf{T}_{\sigma(\xi)}^2\right)$$
$$+ 6m\gamma_{1\sigma(\xi)}^2\| \left(B^{\mathrm{T}}XB\right)^{-1}\Phi_{\sigma(\xi)}Q_{\sigma(\xi)}$$
$$\times \Phi_{\sigma(\xi)}\left(B^{\mathrm{T}}XB\right)^{-1}\| \cdot \|s(k,l)\|$$
$$+ 10\left(\pi_{\sigma(\xi),\sigma(\xi+1)} + \psi_{\sigma(\xi),\sigma(\xi+1)}\|B^{\mathrm{T}}XB\|^2\right)d^2$$
$$\leq \tilde{\chi}^{\mathrm{T}}(k,l)\tilde{\mathbf{X}}_{\sigma(\xi),\sigma(\xi+1)}(k,l)\tilde{\chi}(k,l)$$
$$- \kappa\|\chi(k,l)\|^2 + \vec{\gamma}_{\sigma(\xi),\sigma(\xi+1)}\|\chi(k,l)\| + \hat{\mathbf{T}}_{\sigma(\xi)}, \qquad (11.90)$$

where $\tilde{\chi}(k,l) \triangleq \left[\begin{array}{ccc} \chi^{\mathrm{T}}(k,l) & u^{\mathrm{T}}(k-1,l) & u^{\mathrm{T}}(k,l-1) \end{array} \right]^{\mathrm{T}}$.
From (11.90), the condition (11.69) ensures

$$\Delta \hat{V}(k,l,\sigma(\xi)) < - \kappa\|\chi(k,l)\|^2 + \vec{\gamma}_{\sigma(\xi),\sigma(\xi+1)}\|\chi(k,l)\| + \hat{\mathbf{T}}_{\sigma(\xi)}$$
$$= - \kappa\left(\|\chi(k,l)\| - \vec{\kappa}_{\sigma(\xi),\sigma(\xi+1)}\right)\left(\|\chi(k,l)\| + \hat{\kappa}_{\sigma(\xi),\sigma(\xi+1)}\right), \qquad (11.91)$$

where $\vec{\kappa}_{\sigma(\xi),\sigma(\xi+1)}$ *is defined in (11.70), and* $\hat{\kappa}_{\sigma(\xi),\sigma(\xi+1)}$ *is defined as*
$\hat{\kappa}_{\sigma(\xi),\sigma(\xi+1)} = \frac{-\vec{\gamma}_{\sigma(\xi),\sigma(\xi+1)} + \sqrt{\vec{\gamma}_{\sigma(\xi),\sigma(\xi+1)}^2 + 4\kappa\hat{\mathbf{T}}_{\sigma(\xi)}}}{2\kappa}$.
It is clearly seen from (11.91) that when the state trajectories escape the domain $\mathbf{O}$ *defined in (11.70), that is,* $\|\chi(k,l)\| > \bar{\kappa} \geq \vec{\kappa}_{\sigma(\xi),\sigma(\xi+1)}$ *for any* $\sigma(\xi), \sigma(\xi+1) \in \mathcal{M}$, *it yields* $\Delta\hat{V}(k,l,\sigma(\xi)) < 0$. *This just means that the extended state trajectories* $\chi(k,l)$ *outside the domain* $\mathbf{O}$ *is strictly decreasing. Thus, both of the system state* $x(k,l)$ *and the common sliding variable* $s(k,l)$ *are attracted to the domain* $\mathbf{O}$ *by the proposed second-order SMC law (11.56).*

Notice that it is difficult to solve the *nonconvex* condition (11.69) directly. This fact pushes us to propose a solvable approach in the following theorem for the conditions in Theorem 11.3.

Theorem 11.4 *Consider the 2-D system (11.1) with the Round-Robin protocol (11.6)–(11.7) and the token-dependent 2-D second-order SMC law (11.56)*

Main Results 235

with (11.57)–(11.60). For preassigned parameters $\alpha_1 \geq 0$, $\alpha_2 \geq 0$, *a matrix* $X > 0$ *and any* $j \in \mathcal{M}$, *assume that there exist matrices* $\vec{K}_j \in \mathbb{R}^{m \times (n_h + n_v)}$, $\vec{P}_j^h > 0$, $\vec{P}_j^v > 0$, $\vec{Q}_j^h > 0$, $\vec{Q}_j^v > 0$, $\vec{W}_j > 0$, $\vec{Z}_j > 0$, *and scalars* $\vec{\kappa} > 0$, $\nu_j > 0$, $\mu_j > 0$, $\vec{\pi}_{j,j+1} > 0$, $\vec{\psi}_{j,j+1} > 0$ *such that the following coupled LMIs hold:*

$$\begin{bmatrix} -\vec{\pi}_{j,j+1}I & (I - \Phi_j)B^{\mathrm{T}}\vec{\pi}_{j,j+1} \\ \star & -\vec{P}_{j+1} \end{bmatrix} \leq 0, \tag{11.92}$$

$$\begin{bmatrix} -\vec{\psi}_{j,j+1}I & \hat{\mathbf{B}}_j^{\mathrm{T}}\vec{\psi}_{j,j+1} \\ \star & -\vec{P}_{j+1} \end{bmatrix} \leq 0, \tag{11.93}$$

$$\vec{\mathbf{X}}_{j,j+1} \triangleq \begin{bmatrix} \vec{\mathbf{X}}_{j,j+1}^{(11)} & \vec{\mathbf{X}}_{j,j+1}^{(12)} \\ \star & \vec{\mathbf{X}}_{j,j+1}^{(22)} \end{bmatrix} < 0, \tag{11.94}$$

where

$$\vec{\mathbf{X}}_{j,j+1}^{(11)} \triangleq \begin{bmatrix} -\mathrm{diag}\{\vec{\mathbf{P}}_j, \vec{Q}_j^h, \vec{Q}_j^v\} & \mathbb{K}_j \\ \star & \vec{\Sigma}_{j,j+1} \end{bmatrix},$$

$$\mathbb{K}_j \triangleq \begin{bmatrix} \vec{\mathbf{A}}_j^{\mathrm{T}} & \sqrt{3}\vec{\mathbf{A}}_j^{\mathrm{T}} & 0 & \vec{\mathbf{K}}_j^{\mathrm{T}} \\ \alpha_1 \vec{Q}_j^h \hat{\mathbf{B}}_j^{\mathrm{T}} & 0 & \sqrt{3}\alpha_1 \vec{Q}_j^h \hat{\mathbf{B}}_j^{\mathrm{T}} & -\alpha_1 \vec{Q}_j^h(I-\Phi_j) \\ \alpha_2 \vec{Q}_j^v \hat{\mathbf{B}}_j^{\mathrm{T}} & 0 & \sqrt{3}\alpha_2 \vec{Q}_j^v \hat{\mathbf{B}}_j^{\mathrm{T}} & -\alpha_2 \vec{Q}_j^v(I-\Phi_j) \end{bmatrix}$$

$$\begin{matrix} \vec{\mathbf{K}}_j^{\mathrm{T}} & 0 & \sqrt{6}\vec{\mathbf{Z}}_j^{\mathrm{T}} & \sqrt{10}\varepsilon\vec{\mathbf{P}}_j\vec{\mathbf{I}}^{\mathrm{T}} & \sqrt{10}\varepsilon\vec{\mathbf{P}}_j\vec{\mathbf{I}}^{\mathrm{T}} & \vec{\mathbf{P}}_j \\ 0 & \alpha_1 \vec{Q}_j^h(I-\Phi_j) & 0 & 0 & 0 & 0 \\ 0 & \alpha_2 \vec{Q}_j^v(I-\Phi_j) & 0 & 0 & 0 & 0 \end{matrix} \Bigg],$$

$$\vec{\mathbf{A}}_j \triangleq \begin{bmatrix} A\vec{P}_j - B\Phi_j(B^{\mathrm{T}}XB)^{-1}\vec{K}_j & 0 & B\Phi_j(B^{\mathrm{T}}XB)^{-1}\vec{Z}_j \\ -\Phi_j\vec{K}_j & 0 & \Phi_j\vec{Z}_j \\ 0 & 0 & \vec{Z}_j \end{bmatrix},$$

$$\vec{\mathbf{K}}_j \triangleq \begin{bmatrix} \Phi_j(B^{\mathrm{T}}XB)^{-1}\vec{K}_j & 0 & 0 \end{bmatrix}, \vec{\mathbf{Z}}_j \triangleq \begin{bmatrix} 0 & 0 & \Phi_j(B^{\mathrm{T}}XB)^{-1}\vec{Z}_j \end{bmatrix},$$

$$\vec{\Sigma}_{j,j+1} \triangleq -\mathrm{diag}\{\vec{\mathbf{P}}_{j+1}, \vec{\mathbf{P}}_{j+1}, \vec{\mathbf{P}}_{j+1}, 2\mu_{j+1}I, 2\mu_{j+1}I,$$

$$2\mu_{j+1}I, 2\mu_{j+1}I, \vec{\pi}_{j,j+1}I, \vec{\psi}_{j,j+1}I, \vec{\kappa}I\},$$

$$\vec{\mathbf{X}}_{j,j+1}^{(12)} \triangleq \begin{bmatrix} \vec{\mathbb{N}}_j & 0 \\ 0 & \vec{\mathbb{U}}_{j+1} \\ 0 & 0 \end{bmatrix}, \ \vec{\mathbb{N}}_j \triangleq \begin{bmatrix} \vec{\mathbf{N}}_j & 0 \\ 0 & 0 \\ 0 & 0 \\ 0 & \nu_j\vec{\mathbf{M}}_j \\ 0 & \sqrt{3}\nu_j\vec{\mathbf{M}}_j \\ 0 & 0 \end{bmatrix},$$

$$\vec{\mathbf{N}}_j \triangleq \begin{bmatrix} \vec{P}_j N^{\mathrm{T}} \\ 0 \\ 0 \end{bmatrix}, \ \vec{\mathbf{M}}_j \triangleq \begin{bmatrix} \tilde{X}_j M \\ 0 \\ 0 \end{bmatrix},$$

$$\vec{\mathbb{U}}_{j+1} \triangleq \begin{bmatrix} \mu_{j+1}I & \mu_{j+1}I & 0 & 0 & 0 & 0 & 0 & 0 \\ 0 & 0 & \mu_{j+1}I & \mu_{j+1}I & 0 & 0 & 0 & 0 \\ 0 & 0 & 0 & 0 & \mu_{j+1}I & \mu_{j+1}I & 0 & 0 \\ 0 & 0 & 0 & 0 & 0 & 0 & \mu_{j+1}I & \mu_{j+1}I \end{bmatrix},$$

$$\vec{\mathbf{X}}_{j,j+1}^{(22)} \triangleq -\text{diag}\{\nu_j I, \nu_j I, \vec{Q}_{j+1}^h, \vec{Q}_{j+1}^v, \vec{Q}_{j+1}^h, \vec{Q}_{j+1}^v, \vec{Q}_{j+1}^h, \vec{Q}_{j+1}^v\},$$

with $\vec{P}_{m+1}^h = \vec{P}_1^h$, $\vec{P}_{m+1}^v = \vec{P}_1^v$, $\vec{Q}_{m+1}^h = \vec{Q}_1^h$, $\vec{Q}_{m+1}^v = \vec{Q}_1^v$, $\vec{W}_{m+1} = \vec{W}_1$, $\vec{Z}_{m+1} = \vec{Z}_1$, $\mu_{m+1} = \mu_1$, $\vec{\pi}_{m,m+1} = \vec{\pi}_{m,1}$, and $\vec{\psi}_{m,m+1} = \vec{\psi}_{m,1}$. Then, both the state trajectories of the closed-loop system (11.63) and the sliding function of (11.64) are driven into the domain $\mathbf{O}$ around the origin by the 2-D second-order SMC law $\bar{u}(k,l)$ in (11.56) with the gain matrices $K_j = \vec{K}_j \cdot \text{diag}\left\{\vec{P}_j^h, \vec{P}_j^v\right\}^{-1}$, $j \in \mathcal{M}$.

Proof *Along with similar lines as in Theorem 11.2, the proof of this theorem can be attained readily by using Schur complement and the relationship (11.55).*

Remark 11.5 *Under the Round-Robin protocol (11.6)–(11.7), Theorems 11.2 and 11.4 provide two sufficient yet solvable conditions for the 2-D SMC design problems of the uncertain Roesser model (11.1) with first- and second-order sliding mode, respectively. The desired 2-D SMC laws in form of (11.15) and (11.56) can be obtained via solving coupled LMIs (11.49)–(11.53) and (11.92)–(11.94), respectively. Our proposed 2-D first- and second-order SMC design schemes exhibit the following distinct features: 1) a new 2-D common sliding surface is constructed properly to relax the assumptions on the input matrix as in [2, 171, 192]; 2) to cope with the impacts from the introduced Round-Robin scheduling mechanism and the ZOHs, the designed 2-D first- and second-order SMC laws depend on the periodic scheduler $\sigma(\xi)$; and 3) the ultimate boundedness of the state trajectories of the resulted closed-loop system and the controlled common sliding variable are ensured simultaneously by the proposed 2-D first- and second-order SMC design schemes via some token-dependent Lyapunov-like functions.*

11.3   Solving Algorithms and Examples

In this section, two optimized solving algorithms are first designed for Theorems 11.2 and 11.4. Furthermore, two examples will be provided to illustrate the advantages and the effectiveness of the proposed robust 2-D first- and second-order SMC approaches under Round-Robin protocol.

Solving Algorithms and Examples 237

11.3.1 Solving algorithms

One can observe from Theorems 11.2 and 11.4 that the computation of the obtained sufficient conditions relies on the choices of the preassigned parameters $\alpha_1 \geq 0$, $\alpha_2 \geq 0$ with $\alpha_1 + \alpha_2 = 1$ in (11.8) and the matrix $X > 0$ in (11.13). Besides, it is found that the resultant 2-D SMC performance reflects on the ultimate bounds Ω in (11.26) and $\mathbf{O}$ in (11.70) for first- and second-order sliding mode, respectively. To enhance the practical applications, the following two minimization problems are considered for the proposed 2-D SMC design schemes:

$$\min_{\alpha_1, \alpha_2, X} \hat{\rho}$$

$$\text{subject to: } \alpha_1 \geq 0, \ \alpha_2 \geq 0, \ \alpha_1 + \alpha_2 = 1, \ X > 0,$$

$$\text{and coupled LMIs (11.49)–(11.53)}, \tag{11.95}$$

and

$$\min_{\alpha_1, \alpha_2, X} \bar{\kappa}$$

$$\text{subject to: } \alpha_1 \geq 0, \ \alpha_2 \geq 0, \ \alpha_1 + \alpha_2 = 1, \ X > 0,$$

$$\text{and coupled LMIs (11.92)–(11.94)}. \tag{11.96}$$

Actually, it is difficult to solve the above two optimization problems readily due to matrix $X > 0$ involving $n_h + n_v$ positive elements in diagonal and other $\frac{(n_h + n_v)(n_h + n_v - 1)}{2}$ real elements. In what follows, we provide two solving algorithms to overcome the above searching obstacle:

- Let $X = \varsigma I > 0$ and $\varsigma \triangleq \frac{\varrho}{1 - \varrho}$. Clearly, it has $\varsigma \in (0, +\infty)$ if and only if $\varrho \in (0, 1)$. Now, the optimized design parameters can be found by a simple two-loop linear searching for α_1 and ϱ with two chosen small steps.

- Although the above approach is easy to implement, some conservatism may be leaded in solving. To this end, the binary-based genetic algorithm (GA) as shown in [134, 143] can be formulated to solve the optimization problems (11.95) and (11.96).

11.3.2 Example 1: 2-D SMC without Round-Robin protocol

This example is conducted to give a fair comparison between the proposed 2-D common sliding function approach in this chapter and the existing results in [2,171,192]. To this end, we set the scheduling signal $\sigma(\xi) = 1$ for all $\xi \in \mathbb{N}^+$ and the updating matrix be $\Phi_1 = I$ in the scheduling mechanism (11.11). That is to say, the actual control input $u(k, l)$ reduces to be $u(k, l) = \bar{u}(k, l)$, i.e. the case without Round-Robin protocol.

Now, the following stationary random field model in image processing [3] is concerned:

$$\eta(k + 1, l + 1) = a_1 \eta(k, l + 1) + a_2 \eta(k + 1, l) - a_1 a_2 \eta(k, l), \tag{11.97}$$

where $k = 0, 1, \ldots$ and $j = 0, 1, \ldots$ are the vertical and horizontal position variable, $\eta(k, l)$ is the state of the random field at special coordinates (k, l), and $a_1^2 < 1$ and $a_2^2 < 1$ with a_1 and a_2 representing the vertical and horizontal correlations of the random field, respectively. Let $x^h(k, l) = \eta(k, l + 1) - a_2\eta(k, l)$ and $x^v(k, l) = \eta(k, l)$. Then, equation (11.97) can be represented by the Roesser model (11.1) with $A = \begin{bmatrix} a_1 & 0 \\ 1 & a_2 \end{bmatrix}$ and $a_1 = 0.9$, $a_2 = 0.9$. Besides, other parameters in (11.1) are supposed to be

$$B_1 = 0, B_2 = 1, M = \begin{bmatrix} 0.3 \\ -0.4 \end{bmatrix}, N = \begin{bmatrix} 0.4 & 0.2 \end{bmatrix},$$

$$Z(k, l) = \sin(0.2(k + l + 1)),$$

$$f(x(k, l)) = 0.2\sqrt{|x^h(k, l)x^v(k, l)|} + 0.01.$$

Thus, the scalars in (11.2) can be chosen as $\varepsilon = 0.2$ and $d = 0.01$. Under the initial and boundary conditions (11.3) with $\phi^h(0, l) = 0.6$, $\phi^v(k, 0) = -0.7$, $z_1 = 43$, and $z_2 = 32$, it is easily found that the horizontal and vertical state trajectories in the open-loop case are unstable.

In the case without Round-Robin protocol, our purpose is to design a stabilizing 2-D SMC law for the uncertain stationary random field model. Notice that the existing methods in [2,171,192] cannot be employed to design a stabilizing 2-D SMC law for this unstable system since the assumptions therein on the input matrix B are not satisfied. Now, by employing the proposed 2-D common sliding function, the following 2-D SMC schemes are designed:

•**First-Order SMC Case.** A linear searching algorithm for matrix $X = \frac{nt}{1-nt}I$ ($n \in \mathbb{N}^+$) with a small step $t = 0.001$ is applied to solve Theorem 11.2. Fig. 11.2 depicts the evolution of the upper bound $\hat{\rho}$ for the sliding domain Ω versus the scalar n, in which the following optimized gain matrix is found at $n = 17$:

$$X = 0.0173I, \ K_1 = \begin{bmatrix} 0.0175 & 0.0162 \end{bmatrix}.$$

Then, the 2-D first-order SMC law $\bar{u}(k, l)$ in form of (11.15) is designed as $\bar{u}(k, l) = -1.0116x^h(k, l) - 0.9364x^v(k, l) - (0.3789\|x(k, l)\| + 0.01) \cdot \mathrm{sgn}(s(k, l))$. Under the obtained 2-D first-order SMC law, the horizontal and vertical state trajectories $x^h(k, l)$ and $x^v(k, l)$ are plotted in Fig. 11.3. The evolutions of the 2-D common sliding variable $s(k, l)$ and the 2-D first-order SMC input $\bar{u}(k, l)$ are shown in Fig. 11.4.

•**Second-Order SMC Case.** Theorem 11.4 is solved via a linear searching algorithm for matrix $X = \frac{nt}{1-nt}I$ ($n \in \mathbb{N}^+$) with a small step $t = 0.001$, in which the feasible domain is found at $n \in [332, 435]$. For the given parameters in (11.58)–(11.60) are $\gamma_{11} = \gamma_{21}^h = \gamma_{21}^v = 0.02$, $\beta_1 = \beta_2 = 0.5$ and the sampling periods $T^h = T^v = 10^{-4}$, the evolution of the upper bound $\bar{\kappa}$ for the sliding domain $\mathbf{O}$ versus the scalar n is shown in Fig. 11.5. It is clear that the following optimized gain matrix is attained at $n = 435$:

$$X = 0.7699I, \ K_1 = \begin{bmatrix} 0.8166 & 0.7356 \end{bmatrix}.$$

Solving Algorithms and Examples

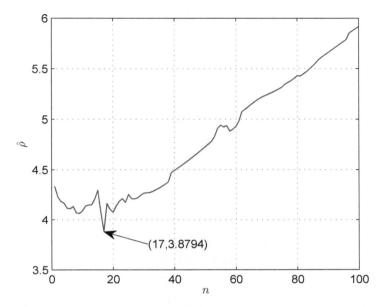

FIGURE 11.2: $\hat{\rho}$ vs. n (First-Order SMC Case in Example 1).

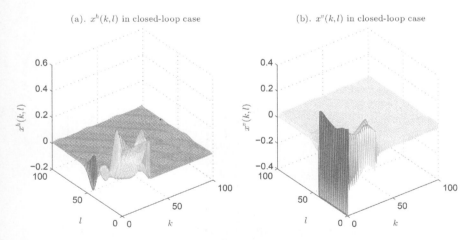

FIGURE 11.3: Horizontal and vertical state trajectories in closed-loop case of Example 1 (First-Order SMC Case).

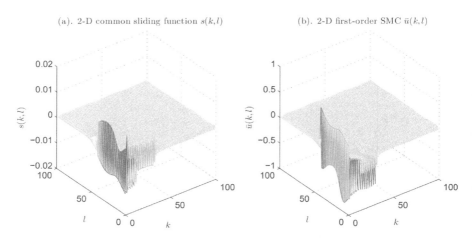

FIGURE 11.4: Evolutions of 2-D common sliding variable and 2-D first-order SMC input in Example 1 (First-Order SMC Case).

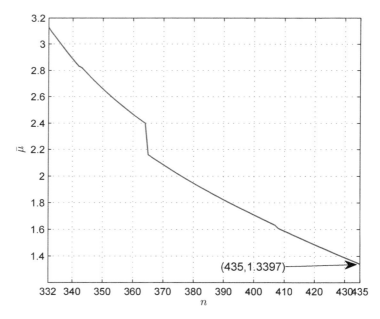

FIGURE 11.5: $\hat{\rho}$ vs. n (Second-Order SMC Case in Example 1).

Solving Algorithms and Examples

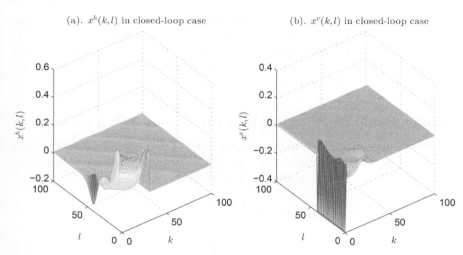

FIGURE 11.6: Horizontal and vertical state trajectories in closed-loop case of Example 1 (Second-Order SMC Case).

Now, the 2-D second-order SMC law $\bar{u}(k,l)$ in form of (11.56) is computed as $\bar{u}(k,l) = -1.0607x^h(k,l) - 0.9554x^v(k,l) + 1.2989\hbar(k,l)$, where $\hbar(k,l)$ is defined in (11.57). Fig. 11.6 plots the horizontal and vertical state trajectories $x^h(k,l)$ and $x^v(k,l)$ under the obtained 2-D second-order SMC law. Meanwhile, Fig. 11.7 shows the evolutions of the 2-D common sliding variable $s(k,l)$ and the 2-D second-order SMC input $\bar{u}(k,l)$.

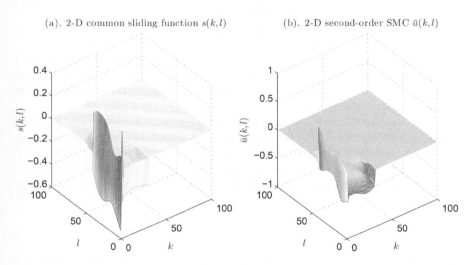

FIGURE 11.7: Evolutions of 2-D common sliding variable and 2-D second-order SMC input in Example 1 (Second-Order SMC Case).

242   *2-D SMC Under RRP*

One can conclude from this example that the proposed 2-D common sliding function approach in this chapter shows a certain advantage over the 2-D directional sliding function approach as in [2, 171, 192]. Besides, it is observed from Figs. 11.2 and 11.5 that the 2-D second-order SMC design may render a less quasi-sliding domain than the 2-D first-order SMC case.

11.3.3   Example 2: 2-D SMC with Round-Robin protocol

As one of famous partial differential equations (PDEs), the following Darboux equation can find real applications in many dynamical processes, e.g., gas absorption, water stream heating and air drying:

$$\frac{\partial^2 s(x,t)}{\partial x \partial t} = a_0 s(x,t) + a_1 \frac{\partial s(x,t)}{\partial t} + a_2 \frac{\partial s(x,t)}{\partial t} + bf(x,t),$$

where $s(x,t)$ is an unknown space function at $x \in [0, x_f]$ and time $t \in [0, \infty)$; a_0, a_1, a_2 and b are real coefficients, $f(x,t)$ is the input function. By adopting a similar transformation technique as in [38] to the PDE, a modified uncertain Roesser model in form of (11.1) can be obtained with considering the possible uncertainties and external disturbance:

$$A = \begin{bmatrix} 1 & 0 \\ 0 & 1 \end{bmatrix} + \begin{bmatrix} \Delta x & 0 \\ 0 & \Delta t \end{bmatrix} \begin{bmatrix} a_1 & a_1 a_2 + a_0 \\ 1 & a_2 \end{bmatrix},$$

$$B_1 = \begin{bmatrix} 0.1 & 0 \end{bmatrix}, B_2 = \begin{bmatrix} 0 & 0.1 \end{bmatrix}, M = \begin{bmatrix} -0.2 & 0.01 \end{bmatrix}^{\mathrm{T}},$$

$$N = \begin{bmatrix} 0.1 & 0.3 \end{bmatrix}, Y(k,l) = \cos(0.4(k + l + 1)),$$

$$f(x(k,l)) = \begin{bmatrix} 0.1 \sin(x^h(k,l)) \\ 0.1 \sin(|x^v(k,l)|) \end{bmatrix} + \begin{bmatrix} \frac{d}{\sqrt{2}} \\ \frac{d}{\sqrt{2}} \end{bmatrix},$$

with $\Delta x = 0.1$, $\Delta t = 0.1$, $a_0 = 0.2$, $a_1 = -0.3$, $a_2 = -0.1$, and $d = 0.01$. Under the initial and boundary conditions (11.3) with $\phi^h(0,l) = 0.3$, $\phi^v(k,0) = -0.4$, $z_1 = 32$ and $z_2 = 43$, it is easily to verify that the uncertain 2-D system without control is unstable.

Now, we consider to design a remote 2-D SMC scheme to stabilize the unstable 2-D Roesser model over a limited communication network, in which the 2-D Round-Robin protocol (11.6)–(11.8) is employed to schedule the communication resource. The following optimized first- and second-order SMC strategies are obtained via solving minimization problems (11.95)–(11.96) by GA[1]:

---

[1]GA settings in Example 2 are: population size $N_c = 80$; maximum of generations $T_{\max} = 100$; single-point crossover probability: $p_c = 0.8$; single-bit mutation probability $p_m = 0.05$; bounds of the diagonal elements in X and the scalars α_1 are $\bar{t} = 1$ and $\underline{t} = 0.0001$ with lengths of binary strings 10; bounds of the non-diagonal elements in X are $\bar{t} = 2$ and $\underline{t} = -2$ with lengths of binary strings 11.

Solving Algorithms and Examples

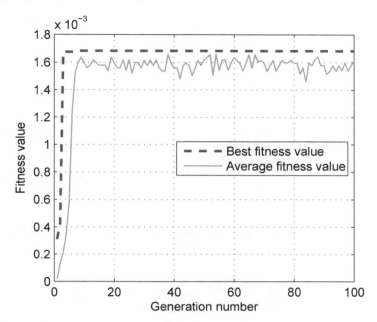

FIGURE 11.8: Evolution of fitness value in each generation (First-Order SMC Case).

•**First-Order Sliding Mode Case.** For the optimization problem (11.95), GA finds the following optimized solutions:

$$\alpha_1 = 0.5602, \alpha_2 = 0.4398, X = \begin{bmatrix} 0.5396 & -0.0576 \\ -0.0576 & 0.5191 \end{bmatrix},$$

$$K_1 = \begin{bmatrix} -1.4584 & 19.9803 \\ 13.1842 & -179.9200 \end{bmatrix}, K_2 = \begin{bmatrix} -582.4253 & 32.9851 \\ 62.2174 & -3.5183 \end{bmatrix}.$$

Fig. 11.8 shows the best and average fitness values of each generation in running of GA, in which fitness function is defined as **Fitness** $\triangleq \frac{1}{\rho}$.

By using the above solutions, the 2-D first-order SMC law (11.15) renders the horizontal and vertical state trajectories $x^h(k,l)$ and $x^v(k,l)$ in the closed-loop case to be stable under the 2-D Round-Robin protocol as depicted in Fig. 11.9. Denote $s(k,l) \triangleq \begin{bmatrix} s_1(k,l) & s_2(k,l) \end{bmatrix}^T$ and $u(k,l) \triangleq \begin{bmatrix} u_1(k,l) & u_2(k,l) \end{bmatrix}^T$. The evolutions of 2-D sliding variables $s_1(k,l)$ and $s_2(k,l)$ are shown in Fig. 11.10. Fig. 11.11 plots the evolutions of the actuators $u_1(k,l)$ and $u_2(k,l)$, respectively.

•**Second-Order Sliding Mode Case.** For the optimization problem (11.96), GA gives the following optimized solutions:

$$\alpha_1 = 0.7967, \alpha_2 = 0.2033, X = \begin{bmatrix} 0.9462 & -0.0870 \\ -0.0870 & 0.7302 \end{bmatrix},$$

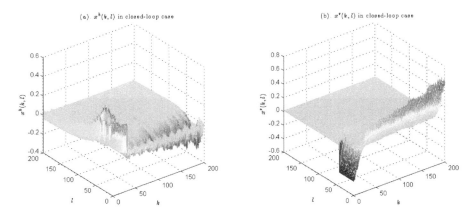

FIGURE 11.9: The state trajectories of the closed-loop system in Example 2 (First-Order SMC Case).

$$K_1 = \begin{bmatrix} 0.0091 & -0.0169 \\ -0.0118 & 0.1436 \end{bmatrix}, K_2 = \begin{bmatrix} 0.0680 & -0.0044 \\ -0.0057 & 0.0060 \end{bmatrix}.$$

By choosing the fitness function as **Fitness** $\triangleq \frac{1}{\bar{\kappa}}$, the best and average fitness values of each generation in running of GA is presented in Fig. 11.12. Under the 2-D Round-Robin protocol (11.6)–(11.8) and the 2-D second-order SMC law (11.56) with the above optimized solutions and $\gamma_{11} = \gamma_{21}^h = \gamma_{21}^v = 0.02$, $\beta_1 = \beta_2 = 0.5$ as well as the sampling periods $T^h = T^v = 10^{-4}$, the horizontal and vertical state trajectories $x^h(k,l)$ and $x^v(k,l)$ in the closed-loop case are given in Fig. 11.13. Fig. 11.14 depicts the evolutions of the

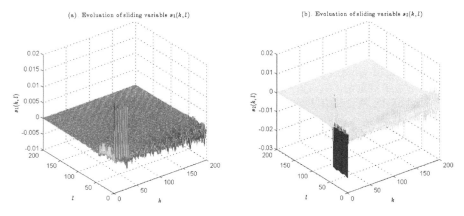

FIGURE 11.10: Evolutions of sliding variables in Example 2 (First-Order SMC Case).

Solving Algorithms and Examples 245

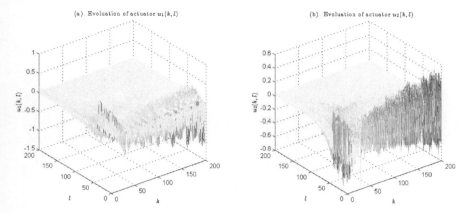

FIGURE 11.11: Evolutions of actuators in Example 2 (First-Order SMC Case).

sliding variables $s_1(k,l)$ and $s_2(k,l)$. The responses of the actuators $u_1(k,l)$ and $u_2(k,l)$ are shown in Fig. 11.15.

The simulation results shown in Figs. 11.8–11.15 have verified the effectiveness of the proposed 2-D first- and second-order SMC design approaches under

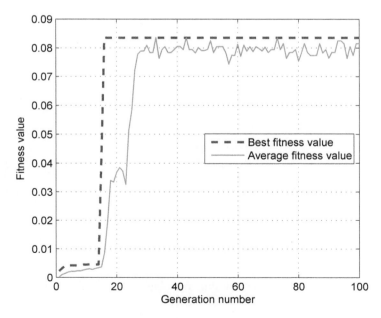

FIGURE 11.12: Evolution of fitness value in each generation (Second-Order SMC Case).

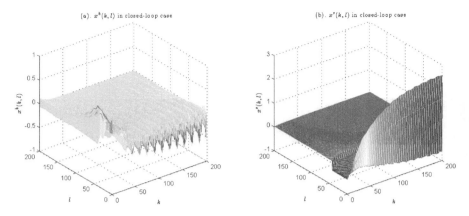

FIGURE 11.13: The state trajectories of the closed-loop system in Example 2 (Second-Order SMC Case).

Round-Robin protocol scheduling. Besides, one can also find from Figs. 11.8 and 11.12 that the 2-D second-order SMC approach would provide more robustness than the 2-D first-order SMC case in eliminating the matched external disturbance $f(x(k,l))$.

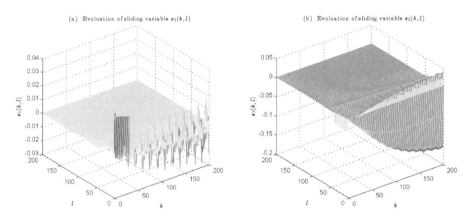

FIGURE 11.14: Evolutions of sliding variables in Example 2 (Second-Order SMC Case).

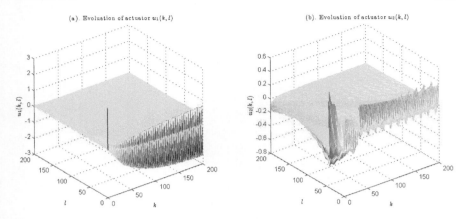

FIGURE 11.15: Evolutions of actuators in Example 2 (Second-Order SMC Case).

11.4 Conclusion

In this chapter, we have studied the 2-D SMC problem for an uncertain Roesser model under the Round-Robin protocol scheduling. By introducing the notion of "time instant" for 2-D systems, the Round-Robin protocol has been applied to determine which actuator obtains access to the C/A communication network and a set of ZOHs has been employed to keep the values in other actuators unchanged until the new value received. To cope with the impacts from the periodic scheduling signal and the ZOHs, token-dependent 2-D first- and second-order SMC laws have been designed, respectively, based on a new 2-D common sliding function. By resorting to the token-dependent Lyapunov-like functions, sufficient conditions have been established to guarantee the ultimate boundedness of both the controlled state trajectories and sliding variable. The advantage and effectiveness of the obtained results by applying 2-D common sliding function have been demonstrated by two simulation examples.

12

Conclusions and Future Topics

12.1 Conclusions

In this book, the recent advances on networked sliding mode control (SMC) problems have been investigated for one-dimensional (1-D) and two-dimensional (2-D) networked control systems (NCSs) under various communication protocols, which include static/dynamic/periodic event-triggered mechanism, stochastic communication protocol, Round-Robin protocol, weighted try-once-discard protocol, multiple-packet transmission protocol, and redundant channel transmission protocol. The super-twisting algorithm and the extended-state-observer-based SMC scheme are involved in this book for suppressing chattering.

The SMC designs for 1-D and 2-D NCSs are studied, respectively, in Part I (Chapters 2–9) and Part II (Chapters 10–11). In each chapter of Parts I and II, the problem formulations are first considered with introducing the key idea of the used communication protocols. Then, some novel sliding functions and SMC schemes are properly designed by considering the scheduling effects from the communication protocols. Furthermore, by drawing on a variety of theories and methodologies such as Lyapunov function, finite-time stability, and linear matrix inequalities, the sufficient conditions for guaranteeing the reachability to the ideal sliding surface and the desired stability of the sliding mode dynamics are derived simultaneously. It is shown that the communication protocols improve the usage of communication network at the cost of weakening some disturbance rejection performances of the SMC schemes. The satisfied SMC schemes are solved via some convex optimization algorithms as well as genetic algorithm. Finally, the proposed results are illustrated in some numerical examples and a practical experiment of permanent magnet synchronous motor speed regulation system. The recent progress on SMC problems for 1-D NCSs has been surveyed in Chapter 1.

This book has established a unified theoretical framework for analysis and synthesis of networked SMC problems in the context of several different sorts of communication protocols. However, the obtained results are still quite limited. Some of the possible future reseal topics are listed in the sequel.

DOI: 10.1201/9781003309499-12

250   *Conclusions and Future Topics*

12.2   Future Topics

- This book just addressed the networked SMC under stochastic communication protocol, Round-Robin protocol, weighted try-once-discard protocol, multiple-packet transmission protocol, and the redundant channel transmission protocol in the framework of discrete-time sliding mode. To date, the problems of the continuous-time sliding mode designs under the above mentioned communication protocols still remain open.

- Actually, some other important communication protocols are not concerned in this book, such as, coding-decoding communication protocols [79,170] and encryption-decryption protocols [50, 222], which serves our future research topics.

- Up to now, the corresponding study of the networked higher-order SMC has been relatively scattered despite certain limited results such as [124, 145], and there still exist a huge gap towards the investigation on the networked higher-order SMC under various communication protocols, let alone the case where 2-D NCSs [182] are considered.

- It is well recognized that how to reduce the conservatism is a key challenge in designing networked SMC for multi-input-multi-output NCSs, in which some nonconvex and nonlinear conditions are inevitably existed. In this book, we make the first attempt to handle the problem by using genetic algorithm so as to render a "smarter" networked SMC [196, 197]. Nonetheless, this issue still need to be explored in the future.

Bibliography

[1] E. H. Abed. Singularly perturbed Hopf Bifurcation, *IEEE Transactions on Circuits and Systems*, vol. CAS-32, no. 12, pp. 1270–1280, 1985.

[2] H. Adloo, P. Karimaghaee and A. S. Sarvestani. An extension of sliding mode control design for the 2-D systems in Roesser model. In *Joint 48th IEEE Conference on Decision and Control and 28th Chinese Control Conference*, Shanghai, China, 2009, pp. 7753–7758.

[3] C. K. Ahn, P. Shi and M. V. Basin. Two-dimensional dissipative control and filtering for Roesser model. *IEEE Transactions on Automatic Control*, vol. 60, no. 7, pp. 1745–1759, 2015.

[4] A. Argha, L. Li and S. W. Su. A new approach to applying discrete sliding mode control to 2D systems. In *Proceedings of 52nd IEEE Annual Conference on Decision and Control*, Florence, Italy, 2013, pp. 3584–3589.

[5] A. Argha, L. Li and S. W. Su. Sliding mode stabilisation of networked systems with consecutive data packet dropouts using only accessible information, *International Journal of Systems Science*, vol. 48, no. 6, pp. 1291–1300, 2016.

[6] A. Bartoszewicz and P. Latosiński. Discrete time sliding mode control with reduced switching-a new reaching law approach, *International Journal of Robust and Nonlinear Control*, vol. 26, no. 1, pp. 47–68, 2016.

[7] M. V. Basin and P. C. Rodríguez-Ramírez. Sliding mode controller design for stochastic polynomial systems with unmeasured states, *IEEE Transactions on Industrial Electronics*, vol. 61, no. 1, pp. 387–396, 2014.

[8] M. V. Basin and P. C. Rodríguez-Ramírez. Sliding mode controller design for linear stochastic systems with unknown parameters, *Journal of the Franklin Institute*, vol. 351, no. 4, pp. 2243–2260, 2014.

[9] A. K. Behera, K. Abhisek, and B. Bandyopadhyay. Event-triggered sliding mode control for a class of nonlinear systems. *International Journal of Control*, vol. 89, no. 9, pp. 1916–1931, 2016.

[10] A. Benzaouia, A. Hmamed, F. Tadeo and A. E. L. Hajjaji. Stabilisation of discrete 2D time switching systems by state feedback control,

International Journal of Systems Science, vol. 42, no. 3, pp. 479–487, 2011.

[11] S. P. Bhat and D. S. Bernstein. Finite-time stability of continuous autonomous systems. *SIAM Journal on Control and Optimization*, vol. 38, no. 3, pp. 751–766, 2006.

[12] P. Bolzern, P. Colaneri and G. De Nicolao. Design of stabilizing strategies for discrete-time dual switching linear systems, *Automatica*, vol. 69, pp. 93–100, 2016.

[13] D. P. Borgers, R. Postoyan, A. Anta, P. Tabuada, D. Nešić and W. P. M. H. Heemels. Periodic event-triggered control of nonlinear systems using overapproximation techniques, *Automatica*, vol. 94, pp. 81–87, 2018.

[14] M. S. Boudellioua, K. Gałkowski and E. Rogers. Equivalence of wave linear repetitive processes and the singular 2-D Roesser state-space model, *Multidimensional Systems and Signal Processing*, 2019, DOI: 10.1007/s11045-019-00654-7.

[15] E. K. Boukas and Z. K. Liu. Robust H_∞ control of discrete-time Markovian jump linear systems with mode-dependent time-delays, *IEEE Transactions on Automatic Control*, vol. 46, no. 12, pp. 1918–1924, 2001.

[16] R. Caballero-Águila, A. Hermoso-Carazo and J. Linares-Pérez. Optimal state estimation for networked systems with random parameter matrices, correlated noises and delayed measurements, *International Journal of General Systems*, vol. 44, no. 2, pp. 142–154, 2015.

[17] R. Canahuire and A. L. Serpa. Reduced order H_∞ controller design for vibration control using genetic algorithms. *Journal of Vibration and Control*, vol. 23, no. 10, pp. 1693–1707, 2017.

[18] S. Cao, J. Liu and Y. Yi. Non-singular terminal sliding mode adaptive control of permanent magnet synchronous motor based on a disturbance observer, *The Journal of Engineering*, vol. 2019, no. 15, pp. 629–634, 2019.

[19] Z. Cao, Y. Niu, H.-K. Lam and J. Zhao. Sliding mode control of Markovian jump fuzzy systems: A dynamic event-triggered method. *IEEE Transactions on Fuzzy Systems*, DOI: 10.1109/TFUZZ.2020.3009729, 2020 (in press).

[20] Z. Cao, Y. Niu and J. Song. Finite-time sliding mode control of Markovian jump cyber-physical systems against randomly occurring injection attacks, *IEEE Transactions on Automatic Control*, vol. 65, no. 3, pp. 1264–1271, 2020.

Bibliography 253

[21] Z. Cao, Y. Niu and Y. Zou. Adaptive neural sliding mode control for singular semi-Markovian jump systems against actuator attacks, *IEEE Transactions on Systems, Man, and Cybernetics: Systems*, 2019, DOI: 10.1109/TSMC.2019.2898428.

[22] G. Cena, S. Scanzio and A. Valenzano. Experimental characterization of redundant channels in industrial Wi-Fi networks, in *IEEE World Conference on Factory Communication Systems (WFCS)*, 2016.

[23] Y. Chen, Z. Wang, Y. Yuan and P. Date. Distributed H_∞ filtering for switched stochastic delayed systems over sensor networks with fading measurements, *IEEE Transactions on Cybernetics*, vol. 50, no. 1, pp. 2–14, Jan. 2020.

[24] D. Christmann, R. Gotzhein, S. Siegmuund and F. Wirth. Realization of Try-Once-Discard in wireless multi-hop networks, *IEEE Transactions on Industrial Informatics*, vol. 10, no. 1, pp. 17–26, 2014.

[25] A. E. F. Clementi, A. Monti and R. Silvestri. Round Robin is optimal for fault-tolerant broadcasting on wireless networks, *Journal of Parallel and Distributed Computing*, vol. 64, no. 1, pp. 89–96, 2004.

[26] E. F. Colet, and L. M. Fridman. *Advances in Variable Structure and Sliding Mode Control*, Springer, 2006.

[27] O. L. V. Costa, M. D. Fragoso and M. G. Todorov. A detector-based approach for the H_2 control of Markov jump linear systems with partial information, *IEEE Transactions on Automatic Control*, vol. 60, no. 5, pp. 1219–1234, 2015.

[28] M. Cucuzzella and A. Ferrara. Practical second order sliding modes in single-loop networked control of nonlinear systems. *Automatica*, vol. 89, pp. 235–240, 2018.

[29] M. Dawa, G. Kaddoum and Z. Sattar. A generalized lower bound on the bit error rate of DCSK systems over multi-path Rayleigh fading channels, *IEEE Transactions on Circuits and Systems II: Express Briefs*, vol. 65, no. 3, pp. 321–325, 2018.

[30] A. Dhawan and H. Kar. An LMI approach to robust optimal guaranteed cost control of 2-D discrete systems described by the Roesser model, *Signal Processing*, vol. 90, no. 9, pp. 2648–2654, 2010.

[31] D. Ding, Z. Wang, B. Shen and H. Shu. State-saturated H_∞ filtering with randomly occurring nonlinearities and packet dropouts: the finite-horizon case, *International Journal of Robust and Nonlinear Control*, vol. 23, no. 16, pp. 1803–1821, 2013.

[32] D. Ding, Z. Wang, J. Lam and B. Shen. Finite-horizon H_∞ control for discrete time-varying systems with randomly occurring nonlinearities and fading measurements, *IEEE Transactions on Automatic Control*, vol. 60, no. 9, pp. 2488–2493, 2015.

[33] J. Dong and G. H. Yang. H_∞ control for fast sampling discrete-time singularly perturbed systems, *Automatica*, vol. 44, no. 5, pp. 1385–1393, 2008.

[34] M. C. F. Donkers and W. P. M. H. Heemels. Output-based event-triggered control with guaranteed L_∞-gain and improved and decentralized event-triggering, *IEEE Transactions on Automatic Control*, vol. 57, no. 6, pp. 1362–1376, 2012.

[35] M. C. F. Donkers, W. P. M. H. Heemels, D. Bernardini, A. Bemporad and V. Shneer. Stability analysis of stochastic networked control systems, *Automatica*, vol. 48, no. 4, pp. 917–925, 2012.

[36] M. C. F. Donkers, W. P. M. H. Heemels, N. van de Wouw and L. Hetel. Stability analysis of networked control systems using a switched linear systems approach, *IEEE Transactions on Automatic Control*, vol. 56, no. 9, pp. 2101–2115, 2011.

[37] H. Du, J. Lam and K. Y. Sze. Non-fragile output feedback H_∞ vehicle suspension control using genetic algorithm, *Engineering Applications of Artificial Intelligence*, vol. 16, no. 7–8, pp. 667–680, 2003.

[38] C. Du, L. Xie and C. Zhang. H_∞ control and robust stabilization of two-dimensional systems in Roesser models. *Automatica*, vol. 37, no. 2, pp. 205–211, 2001.

[39] M. Dymkov, K. Galkowski, E. Rogers, V. Dymkou and S. Dymkou. Modeling and control of a sorption process using 2D systems theory. *The 2011 International Workshop on Multidimensional (nD) Systems*, pp. 1–6, 2011.

[40] C. Edwards, A. Akoachere and S. K. Spurgeon. Sliding-mode output feedback controller design using linear matrix inequalities, *IEEE Transactions on Automatic Control*, vol. 46, no. 1, pp. 115–119, 2001.

[41] S. V. Emel'yanov. Method of designing complex control algorithm using an error and its first time derivative only (in Russian). *Automation and Remote Control*, vol. 18, no. 10, 1957.

[42] S. V. Emel'yanov, I. A. Burovoi and et. al. Mathematical models of process in technology and development of variable structure control system (in Russian). *Metallurgy (Moscow)*, vol. 18, no. 7, 1964.

Bibliography 255

[43] S. V. Emel'yanov and V. A. Taran. On a class of variable structure control systems (in Russian). *Proceedings of USSR Academy of Sciences, Energy and Automation*, no. 3, 1962.

[44] N. Elia. Remote stabilization over fading channels, *Systems & Control Letter*, vol. 54, no. 3, pp. 237–249, 2005.

[45] X. Fan and Z. Wang. Event-triggered integral sliding mode control for linear systems with disturbance, *Systems & Control Letters*, vol. 138, 104669, 2020.

[46] Z. Fei, S. Shi, C. Zhao and L. Wu. Asynchronous control for 2-D switched systems with mode-dependent average dwell time, *Automatica*, vol. 79, pp. 198–206, 2017.

[47] J. Feng and F. Hao. Event-triggered nonsingular fast terminal sliding mode control for nonlinear dynamical systems. In *Chinese Intelligent Systems Conference*, pp. 150–158, 2019.

[48] Y. Feng, X. Yu, and Z. Man. Non-singular adaptive terminal sliding mode control of rigid manipulators. *Automatica*, vol. 38, no. 12, pp. 2159–2167, 2002.

[49] Y. Gao, B. Sun and G. Lu. Passivity-based integral sliding-mode control of uncertain singularly perturbed systems, *IEEE Transactions on Circuits and Systems–II: Express Briefs*, vol. 58, no. 6, pp. 386–390, 2011.

[50] C. Gao, Z. Wang, X. He and H. Dong. Fault-tolerant consensus control for multi-agent systems: An encryption-decryption scheme, *IEEE Transactions on Automatic Control*, DOI: 10.1109/TAC.2021.3079407, 2021.

[51] W. Gao, Y. Wang and A. Homaifa. Discrete-time variable structure control systems, *IEEE Transactions on Industrial Electronics*, vol. 42, no. 2, pp. 117–122, 1995.

[52] K. Galkowski and E. Rogers. Control systems analysis for the Fornasini-Marchesini 2D systems model-progress after four decades. *International Journal of Control*, vol. 91, no. 12, pp. 2801–2822, 2018.

[53] H. Geng, Y. Liang and X. Zhang. The linear minimum mean square error observer for multi-rate sensor fusion with missing measurements, *IET Control Theory & Applications*, vol. 8, no. 14, pp. 175–183, 2014.

[54] A. Girard. Dynamic triggering mechanisms for event-triggered control, *IEEE Transactions on Automatic Control*, vol. 60, no. 7, pp. 1992–1997, 2015.

[55] C. A. C. Gonzaga and O. L. V. Costa. Stochastic stabilization and induced ℓ_2-gain for discrete-time Markov jump Lur'e systems with control saturation, *Automatica*, vol. 50, no. 9, pp. 2397–2404, 2014.

[56] W. P. M. H. Heemels, M. C. F. Donkers and A. R. Teel. Periodic event-triggered control for linear systems, *IEEE Transactions on Automatic Control*, vol. 58, no. 4, pp. 847–861, 2013.

[57] W. P. M. H. Heemels, K. H. Johansson and P. Tabuada. An introduction to event-triggered and self-triggered control, In *2012 IEEE 51st IEEE Conference on Decision and Control (CDC)*, pp. 3270–3285, 2012.

[58] W. P. M. H. Heemels, A. R. Teel, N. Van de Wouw and D. Nesic. Networked control systems with communication constraints: Tradeoffs between transmission intervals, delays and performance, *IEEE Transactions on Automatic control*, vol. 55, no. 8, pp. 1781–1796, 2010.

[59] J. P. Hespana, P. Naghshtabrizi and Y. G. Xu. A survey of recent results in networked control systems, *Proceedings of the IEEE*, vol. 95, no. 1, pp. 138–162, 2007.

[60] R. A. Horn and C. R. Johnson. *Matrix Analysis (Second Edition)*, Cambridge University Press, Cambridge, 2013.

[61] J. Hu, Z. Wang, Y. Niu and H. Gao. Sliding mode control for uncertain discrete-time systems with Markovian jumping parameters and mixed delays, *Journal of the Franklin Institute*, vol. 351, no. 4, pp. 2185–2202, 2014.

[62] J. Hu, Z. Wang, H. Gao and L. K. Stergioulas. Robust sliding mode control for discrete stochastic systems with mixed time delays, randomly occurring uncertainties, and randomly occurring nonlinearities, *IEEE Transactions on Industrial Electronics*, vol. 59, no. 7, pp. 3008–3015, 2012.

[63] S. Hu, and W.-Y. Yan. Stability of networked control systems under a multiple-packet transmission policy. *IEEE Transactions on Automatic Control*, vol. 53, no. 7, pp. 1706–1711, 2018.

[64] S. Hu, D. Yue, X. Yin, X. Xie and Y. Ma. Adaptive event-triggered control for nonlinear discrete-time systems, *International Journal of Robust and Nonlinear Control*, vol. 26, no. 18, pp. 4104–4125, 2016.

[65] P. Ignaciuk and A. Bartoszewicz. Discrete-time sliding-mode congestion control in multisource communication networks with time-varying delay, *IEEE Transactions on Control Systems Technology*, vol. 19, no. 4, pp. 852–867, 2011.

Bibliography

[66] G. P. Incremona and A. Ferrara. Adaptive model-based event-triggered sliding mode control. *International Journal of Adaptive Control and Signal Processing*, vol. 30, no. 8–10, pp. 1298–1316, 2016.

[67] U. Itkis. *Control Systems of Variable Structure*. Wiley, New York, 1976.

[68] X. Ji, T. Liu, Y. Sun and H. Su. Stability analysis and controller synthesis for discrete linear time-delay systems with state saturation nonlinearities, *International Journal of Systems Science*, vol. 42, no. 3, pp. 397–406, 2011.

[69] C. Y. Jung, H. Y. Hwang, D. K. Sung and G. U. Hwang. Enhanced Markov chain model and throughput analysis of the slotted CSMA/CA for IEEE 802.15. 4 under unsaturated traffic conditions, *IEEE Transactions on Vehicular Technology*, vol. 58, no. 1, pp. 473–478, 2009.

[70] H. Kando and T. Iwazumi. Multirate digital control design of an optimal regulator via singular perturbation theory, *International Journal of Control*, vol. 44, no. 6, pp. 1555-1578, 1986.

[71] J. Lam and H. K. Tam. Robust output feedback pole assignment using genetic algorithm. *Proceedings of the Institution of Mechanical Engineers, Part I: Journal of Systems and Control Engineering*, vol. 214, no. 5, pp. 327–334, 2000.

[72] L. B. Le, E. Hossain and A. S. Alfa. Service differentiation in multirate wireless networks with weighted round-robin scheduling and ARQ-based error control, *IEEE Transactions on Communications*, vol. 54, no. 2, pp. 208–215, 2006.

[73] A. Levant. Sliding order and sliding accuracy in sliding mode control. *International Journal of Control*, vol. 58, no. 6, pp. 1247–1263, 1993.

[74] A. Levant, and L. Fridman. Higher order sliding modes. *International Journal of Robust and Nonlinear Control*, vol. 18, pp. 381–384, 2008.

[75] X. Li, H. Gao and C. Wang. Generalized Kalman-Yakubovich-Popov lemma for 2-D FM LSS model, *IEEE Transactions on Automatic Control*, vol. 57, no. 12, pp. 3090–3103, 2012.

[76] J. Li, Y. Niu, and J. Song. Finite-time boundedness of sliding mode control under periodic event-triggered strategy. *International Journal of Robust and Nonlinear Control*, vol. 31, no. 2, pp. 623–639, 2021.

[77] Q. Li, B. Shen, Y. Liu and F. E. Alsaadi. Event-triggered H_∞ state estimation for discrete-time stochastic genetic regulatory networks with Markovian jumping parameters and time-varying delays, *Neurocomputing*, vol. 174, pp. 912–920, 2016.

[78] Q. Li, B. Shen, Z. Wang, T. Huang and J. Luo. Synchronization control for a class of discrete time-delay complex dynamical networks: A dynamic event-triggered approach, *IEEE Transactions on Cybernetics*, DOI: 10.1109/TCYB.2018.2818941, 2018.

[79] J. Li, Z. Wang, H. Dong and X. Yi. Outlier-resistant observer-based control for a class of networked systems under encodingCdecoding mechanism, *IEEE Systems Journal*, DOI: 10.1109/JSYST.2020.3044238, 2020.

[80] H. Li, J. Wang and P. Shi. Output-feedback based sliding mode control for fuzzy systems with actuator saturation, *IEEE Transactions on Fuzzy Systems*, vol. 24, no. 6, pp. 1282–1293, 2016.

[81] B. Li, Z. Wang, Q. L. Han. Input-to-state stabilization of delayed differential systems with exogenous disturbances: The event-triggered case, *IEEE Transactions on Systems, Man, and Cybernetics: Systems*, DOI: 10.1109/TSMC.2017.2719960, 2017.

[82] H. Li, H. Yang, F. Sun, and Y. Xia. Sliding-mode predictive control of networked control systems under a multiple-packet transmission policy. *IEEE Transactions on Industrial Electronics*, vol. 61, no. 11, pp. 6234–6243, 2014.

[83] J. Li, Q. Zhang, X. G. Yan and S. Spurgeon. Observer-based fuzzy integral sliding mode control for nonlinear descriptor systems, *IEEE Transactions on Fuzzy Systems*, vol. 26, no. 5, pp. 2818–2832, 2018.

[84] S. Li, M. Zhou and X. Yu. Design and implementation of terminal sliding mode control method for PMSM speed regulation system. *IEEE Transactions on Industrial Informatics*, vol. 9, no. 4, pp. 1879–1891, 2012.

[85] K. Liu, E. Fridman and L. Hetel. Stability and L_2-gain analysis of networked control systems under Round-Robin scheduling: A time-delay approach, *Systems & Control Letters*, vol. 61, no. 5, pp. 666–675, 2012.

[86] H. Liu, Z. Wang, B. Shen and H. Dong. Event-triggered H_∞ state estimation for delayed stochastic memristive neural networks with missing measurements: The discrete time case, *IEEE Transactions on Neural Networks and Learning Systems*, vol. 29, no. 8, pp. 3726–3737, 2018.

[87] W. Liu, Z. Wang, Y. Yuan, N. Zeng, K. Hone and X. Liu. A novel sigmoid-function-based adaptive weighted particle swarm optimizer, *IEEE Transactions on Cybernetics*, 2019, DOI: 10.1109/TCYB.2019.2925015.

[88] J. Liu, L. Wu, C. Wu, W. Luo and L. G. Franquelo. Event-triggering dissipative control of switched stochastic systems via sliding mode. *Automatica*, vol. 103, pp. 261-273, 2019.

Bibliography

[89] B. Lian, Q. Zhang and J. Li. Integrated sliding mode control and neural networks based packet disordering prediction for nonlinear networked control systems. *IEEE Transactions on Neural Networks and Learning Systems*, vol. 30, no. 8, pp. 2324–2335, 2019.

[90] B. Lian, Q. Zhang and J. Li. Sliding mode control and sampling rate strategy for networked control systems with packet disordering via Markov chain prediction. *ISA Transactions*, vol. 83, pp. 1–12, 2018.

[91] K. Liu and E. Fridman. Discrete-time network-based control under scheduling and actuator constraints, *International Journal of Robust and Nonlinear Control*, vol. 25, no. 12, pp. 1816–1830, 2015.

[92] K. Liu, E. Fridman and K. H. Johansson. Networked control with stochastic scheduling, *IEEE Transactions on Automatic Control*, vol. 60, no. 11, pp. 3071–3076, 2015.

[93] K. Liu, E. Fridman, K. H. Johansson and Y. Xia. Quantized control under Round-Robin communication protocol, *IEEE Transactions on Industrial Electronics*, vol. 63, no. 7, pp. 4461–4471, 2016.

[94] J. Liu, Y. Gao, Y. Yin, J. Wang, W. Luo and G. Sun. *Sliding Mode Control Methodology in the Applications of Industrial Power Systems*, Springer, 2020.

[95] Y. Liu, Y. Niu, J. Lam and B. Zhang. Sliding mode control for uncertain switched systems with partial actuator faults, *Asian Journal of Control*, vol. 16, no. 5, pp. 1–10, 2014.

[96] W. Liu, Z. Wang, H. Dai and M. Naz. Dynamic output feedback control for fast sampling discrete-time singularly perturbed systems, *IET Control Theory & Applications*, vol. 10, no. 15, pp. 1782–1788, 2016.

[97] M. Liu, L. Zhang, P. Shi and Y. Zhao. Sliding mode control of continuous-time Markovian jump systems with digital data transmission, *Automatica*, vol. 80, pp. 200–209, 2017.

[98] J. Ludwiger, M. Steinberger and M. Horn. Spatially distributed networked sliding mode control, *IEEE Control Systems Letters*, vol. 3, no. 4, pp. 972–977, 2019.

[99] J. Ludwiger, M. Steinberger, M. Horn, G. Kubin and A. Ferrara. Discrete time sliding mode control strategies for buffered networked systems, *in 2018 IEEE Conference on Decision and Control (CDC)*, pp. 6735–6740, 2018.

[100] W. Marszalek and Z. Trzaska. Mixed-mode oscillations in a modified Chua's circuit, *Circuits, Systems and Signal Processing*, vol. 29, no. 6, pp. 1075–1087, 2010.

Bibliography

[101] L. Ma, Z. Wang, Y. Bo and Z. Guo. Robust H_∞ sliding mode control for nonlinear stochastic systems with multiple data packet losses, *International Journal of Robust and Nonlinear Control*, vol. 22, no. 5, pp. 473–491, 2012.

[102] L. Ma, Z. Wang and H. K. Lam. Event-triggered mean-square consensus control for time-varying stochastic multi-agent system with sensor saturations, *IEEE Transactions on Automatic Control*, vol. 62, no. 7, pp. 3524–3531, 2017.

[103] Z. Man, A. Paplinski and H. Wu. A robust MIMO terminal sliding mode control scheme for rigid robotic manipulators. *IEEE Transactions on Automatic Control*, vol. 39, no. 12, pp. 2464–2469, 1994.

[104] Z. Man and X. Yu. Terminal sliding mode control of MIMO linear systems. *IEEE Transactions on Circuits and Systems I: Fundamental Theory and Applications*, vol. 44, no. 11, pp. 1065–1070, 1997.

[105] D. Mehdi, E. K. Boukas and O. Bachelier. Static output feedback design for uncertain linear discrete-time system, *IMA Journal of Mathematical Control and Information*, vol. 21, no. 1, pp. 1–13, 2004.

[106] X. Meng and T. Chen. Event detection and control co-design of sampled-data systems. *International Journal of Control*, vol. 87, no. 4, pp. 777–786, 2014.

[107] A. R. Mesquita, J. P. Hespanha and G. N. Nair. Redundant data transmission in control/estimation over lossy networks, *Automatica*, vol. 48, no. 8, pp. 1612–1620, 2012.

[108] Z. Ning, L. Zhang, J. de. J. Rubio and X. Yin. Asynchronous filtering for discrete-time fuzzy affine systems with variable quantization density, *IEEE Transactions on Cybernetics*, vol. 47, no. 1, pp. 153–164, 2017.

[109] Y. Niu and D. W. C. Ho. Design of sliding mode control subject to packet losses, *IEEE Transactions on Automatic Control*, vol. 55, no. 11, pp. 2623–2628, 2010.

[110] Y. Niu, D. W. C. Ho and Z. Wang. Improved sliding mode control for discrete-time systems via reaching law, *IET Control Theory & Applications*, vol. 4, no. 11, pp. 2245–2251, 2010.

[111] Y. Niu, Z. Wang and X. Wang. Robust sliding mode design for uncertain stochastic systems based on H_∞ control method, *Optimal Control Applications and Methods*, vol. 31, no. 2, pp. 93–104, 2010.

[112] M. Ogura, A. Cetinkaya, T. Hayakawa and V. M. Preciado. State-feedback control of Markov jump linear systems with hidden Markov mode observation, *Automatica*, vol. 89, pp. 65–72, 2018.

Bibliography

[113] R. C. L. F. Oliveira, A. N. Vargas, J. B. R. do Val and P. L. D. Peres. Mode-independent H_2-control of a DC motor modelled as a Markov jump linear system, *IEEE Transactions on Control Systems Technology*, vol. 22, no. 5, pp. 1915–1919, 2014.

[114] P. Pakshin, J. Emelianova, K. Gałkowski and E. Rogers. Stabilization of two-dimensional nonlinear systems described by Fornasini–Marchesini and Roesser Models. *SIAM Journal on Control and Optimization*, vol. 56, no. 5, pp. 3848–3866, 2018.

[115] J. Qiu, Y. Wei, H. R. Karimi and H. Gao. Reliable control of discrete-time piecewise-affine time-delay systems via output feedback, *IEEE Transactions on Reliability*, vol. 67, no. 1, pp. 79–91, 2018.

[116] L. Qu, W. Qiao and L. Qu. An enhanced linear active disturbance rejection rotor position sensorless control for permanent magnet synchronous motors, *IEEE Transactions on Power Electronics*, vol. 35, no. 6, pp. 6175–6184, 2020.

[117] L. Qu, W. Qiao and L. Qu. Active-disturbance-rejection-based sliding-mode current control for permanent-magnet synchronous motors, *IEEE Transactions on Power Electronics*, DOI: 10.1109/TPEL.2020.3003666, 2020 (in press).

[118] D. E. Quevedo, A. Ahlén and K. H. Johansson. State estimation over sensor networks with correlated wireless fading channels, *IEEE Transactions on Automatic Control*, vol. 58, no. 3, pp. 581–593, 2013.

[119] W. Perruquetti and J. P. Barbot. *Sliding Mode Control in Engineering*, CRC Press, 2002.

[120] C. Ren, X. Li, X. Yang and S. Ma. Extended state observer-based sliding mode control of an omnidirectional mobile robot with friction compensation, *IEEE Transactions on Industrial Electronics*, vol. 66, no. 12, pp. 9480–9489, 2019.

[121] E. Rogers, K. Galkowski and D. H. Owens. *Control systems theory and applications for linear repetitive processes*. Berlin: Springer-Verlag, 2007.

[122] P. Sadeghi, R. A. Kennedy, P. B. Rapajic and R. Shams. Finite-state Markov modeling of fading channels: A survey of principles and applications, *IEEE Signal Processing Magazine*, vol. 25, no. 5, pp. 57–80, 2008.

[123] I. Salgado, S. Kamal, B. Bandyopadhyay, I. Chairez and L. Fridman. Control of discrete time systems based on recurrent Super-Twisting-like algorithm. *ISA Transactions*, vol. 64, pp. 47–55, 2016.

[124] D. Shah, A. Mehta, K. Patel and A. Bartoszewicz. Event-triggered discrete higher-order SMC for networked control system having network irregularities. *IEEE Transactions on Industrial Informatics*, vol. 16, no. 11, pp. 6837–6847, 2020.

[125] N. K. Sharma, and J. Sivaramakrishnan. *Discrete-Time Higher Order Sliding Mode: The Concept and The Control*, Springer, 2019.

[126] M. Shen, J. H. Park and D. Ye. A separated approach to control of Markov jump nonlinear systems with general transition probabilities, *IEEE Transactions on Cybernetics*, vol. 46, no. 9, pp. 2010–2018, 2016.

[127] B. Shen, Z. Wang, D. Wang and Q. Li. State-saturated recursive filter design for stochastic time-varying nonlinear complex networks under deception attacks. *IEEE Transactions on Neural Networks and Learning Systems*, 2019, DOI: 10.1109/TNNLS.2019.2946290.

[128] B. Shen, Z. Wang, D. Wang and H. Liu. Distributed state-saturated recursive filtering over sensor networks under Round-Robin protocol. *IEEE Transactions on Cybernetics*, 2019, DOI: 10.1109/TCYB.2019.2932460.

[129] Y. Shen, Z. Wang, B. Shen, F. E. Alsaadi and F. E. Alsaadi. Fusion estimation for multi-rate linear repetitive processes under weighted Try-Once-Discard protocol, *Information Fusion*, vol. 55, pp. 281–291, Mar. 2020.

[130] Z. Shu, J. Lam and J. Xiong. Static output-feedback stabilization of discrete-time Markovian jump linear systems: A system augmentation approach, *Automatica*, vol. 46, no. 4, pp. 687–694, 2010.

[131] H. Song, S. C. Chen and Y. Yam. Sliding mode control for discrete-time systems with Markovian packet dropouts, *IEEE Transactions on Cybernetics*, vol. 47, no. 11, pp. 3669–3679, 2017.

[132] J. Song, D. W. C. Ho, and Y. Niu. Model-based event-triggered sliding-mode control for multi-input systems: Performance analysis and optimization. *IEEE Transactions on Cybernetics*, 2020, doi: 10.1109/TCYB.2020.3020253.

[133] J. Song and Y. Niu. Dynamic event-triggered sliding mode control: Dealing with slow sampling singularly perturbed systems. *IEEE Transactions on Circuits and Systems II: Express Briefs*, 2019, DOI: 10.1109/TCSII.2019.2926879.

[134] J. Song and Y. Niu. Co-design of 2D event generator and sliding mode controller for 2D Roesser model via genetic algorithm. *IEEE Transactions on Cybernetics*, 2020, DOI: 10.1109/TCYB.2019.2959139.

Bibliography

[135] J. Song, Y. Niu and H.-K. Lam. Reliable sliding mode control of fast sampling singularly perturbed systems: A redundant channel transmission protocol approach. *IEEE Transactions on Circuits and Systems I: Regular Papers*, vol. 66, no. 11, pp. 4490–4501, 2019.

[136] J. Song, Y. Niu, J. Lam and H.-K. Lam. Fuzzy remote tracking control for randomly varying local nonlinear models under fading and missing measurements, *IEEE Transactions on Fuzzy Systems*, vol. 26, no. 3, pp. 1125–1137, 2018.

[137] J. Song, Y. Niu and J. Xu. An event-triggered approach to sliding mode control of Markovian jump Lur'e systems under hidden mode detections, *IEEE Transactions on Systems, Man, and Cybernetics: Systems*, DOI: 10.1109/TSMC.2018.2847315, 2018.

[138] J. Song, Y. Niu and Y. Zou. Finite-time stabilization via sliding mode control, *IEEE Transactions on Automatic Control*, vol. 62, no. 3, pp. 1478–1483, 2017.

[139] J. Song, Y. Niu and Y. Zou. Asynchronous sliding mode control of Markovian jump systems with time-varying delays and partly accessible mode detection probabilities, *Automatica*, vol. 93, pp. 33–41, 2018.

[140] J. Song, Y. Niu and Y. Zou. A parameter-dependent sliding mode approach for finite-time bounded control of uncertain stochastic systems with randomly varying actuator faults and its application to a parallel active suspension system, *IEEE Transactions on Industrial Electronics*, vol. 65, no. 10, pp. 8124–8132, 2018.

[141] J. Song, Z. Wang and Y. Niu. Static output-feedback sliding mode control under Round-Robin protocol. *International Journal of Robust and Nonlinear Control*, vol. 28, no. 18, pp. 5841-5857, 2018.

[142] J. Song, Z. Wang and Y. Niu. On H_∞ sliding mode control under stochastic communication protocol, *IEEE Transactions on Automatic Control*, vol. 64, no. 5, pp. 2174–2181, 2019.

[143] J. Song, Z. Wang, Y. Niu and H. Dong. Genetic-algorithm-assisted sliding mode control for networked state-saturated systems over hidden Markov fading channels. *IEEE Transactions on Cybernetics*, DOI: 10.1109/TCYB.2020.2980109, 2020.

[144] J. Song, Z. Wang, Y. Niu and J. Hu. Observer-based sliding mode control of state-saturated systems under weighted try-once-discard protocol. *International Journal of Robust and Nonlinear Control*, vol. 30, no. 18, pp. 7991–8006, 2020.

[145] J. Song, W. X. Zheng, and Y. Niu. Self-triggered sliding mode control for networked PMSM speed regulation system: A PSO-optimized super-twisting algorithm. *IEEE Transactions on Industrial Electronics*, 2021, doi: 10.1109/TIE.2021.3050348.

[146] Y. Song, Z. Wang, D. Ding and G. Wei. Robust model predictive control under redundant channel transmission with applications in networked DC motor systems, *International Journal of Robust and Nonlinear Control*, vol. 26, no. 18, pp. 3937–3957, 2016.

[147] F. Stadtmann and O. L. V. Costa. H_2-control of continuous-time hidden Markov jump linear systems, *IEEE Transactions on Automatic Control*, vol. 62. no. 8, pp. 4031–4037, 2017.

[148] M. Steinberger, M. Horn, and L. Fridman. *Variable-Structure Systems and Sliding-Mode Control: From Theory to Practice*, Springer, 2020.

[149] X. Su, X. Liu, P. Shi and R. Yang. Sliding mode control of discrete-time switched systems with repeated scalar nonlinearities, *IEEE Transactions on Automatic Control*, vol. 62, no. 9, pp. 4604–4610, 2017.

[150] Y.-C. Sun and G.-H. Yang. Periodic event-triggered resilient control for cyber-physical systems under denial-of-service attacks, *Journal of the Franklin Institute*, vol. 355, no. 13, pp. 5613–5631, 2018.

[151] M. Tabbara and D. Nešić. Input-output stability of networked control systems with stochastic protocols and channels, *IEEE Transactions on Automatic Control*, vol. 53, no. 5, pp. 1160–1175, 2008.

[152] P. Tabuada. Event-triggered real-time scheduling of stabilizing control tasks, *IEEE Transactions on Automatic Control*, vol. 52, no. 9, pp. 1680-1685, 2007.

[153] V. Ugrinovskii and E. Fridman. A Round-Robin type protocol for distributed estimation with H_∞ consensus, *Systems & Control Letters*, vol. 69, pp. 103–110, 2014.

[154] V. I. Utkin. Variable structure system with sliding mode. *IEEE Transactions on Automatic Control*, vol. 22, no. 2, pp. 212–221, 1977.

[155] V. I. Utkin. Sliding mode control design principles and applications to electric drives. *IEEE Transactions on Industrial Electronics*, vol. 40, no. 1, pp. 23–36, 1993.

[156] V. I. Utkin. Sliding mode control of DC/DC converters, *Journal of the Franklin Institute*, vol. 350, no. 8, pp. 2146–2165, 2013.

[157] V. Utkin, J. Guldner and J. Shi. *Sliding Mode Control in Electro-Mechanical Systems*, CRC press, 2009.

Bibliography 265

[158] V. Utkin, A. Poznyak, Y. V. Orlov and A. Polyakov. *Road Map for Sliding Mode Control Design*, Springer, 2020.

[159] A. N. Vargas, E. F. Costa and J. B. R. do Val. On the control of Markov jump linear systems with no mode observation: Application to a DC motor device, *International Journal of Robust and Nonlinear Control*, vol. 23, no. 10, pp. 1136–1156, 2013.

[160] S. T. Venkataraman and S. Gulati. Control of nonlinear systems using terminal sliding modes. *Transactions of the ASME*, vol. 115, 554–560, 1993.

[161] G. Walsh, H. Ye and L. Bushnell. Stability analysis of networked control systems, *IEEE Transactions on Control Systems Technology*, vol. 10, no. 3, pp. 438–446, 2002.

[162] C. Wen, Z. Wang, Q. Liu and F. E. Alsaadi. Recursive distributed filtering for a class of state-saturated systems with fading measurements and quantization effects, *IEEE Transactions on Systems, Man, and Cybernetics: Systems*, vol. 48, no. 6, pp. 930–941, 2016.

[163] X. Wan, Z. Wang, Q. L. Han and M. Wu. Finite-time H_∞ state estimation for discrete time-delayed genetic regulatory networks under stochastic communication protocols, *IEEE Transactions on Circuits and Systems I-Regular Papers*, DOI: 10.1109/TCSI.2018.2815269, 2018.

[164] X. Wan, Z. Wang, M. Wu and X. Liu. H_∞ state estimation for discrete-time nonlinear singularly perturbed complex networks under the Round-Robin protocol, *IEEE Transactions on Neural Networks and Learning Systems*, vol. 30, no. 2, pp. 415–426, 2019.

[165] J. Wang, X. Du and D. Ding. Event-triggered control of two-dimensional discrete-time systems in Roesser model. In *Proceedings of the 33rd Chinese Control Conference*, Nanjing, China, 2014, pp. 5851–5856.

[166] Y. Wang, Y. Feng, X. Zhang and J. Liang. A new reaching law for antidisturbance sliding-mode control of PMSM speed regulation system, *IEEE Transactions on Power Electronics*, vol. 35, no. 4, pp. 4117–4126, 2020.

[167] F. Wang, Z. Wang, J. Liang and X. Liu. Resilient filtering for linear time-varying repetitive processes under uniform quantizations and Round-Robin protocols. *IEEE Transactions on Circuits and Systems I: Regular Papers*, vol. 65, no. 9, pp. 2992–3004, 2018.

[168] F. Wang, Z. Wang, J. Liang and X. Liu. Resilient state estimation for 2-D time-varying systems with redundant channels: A variance-constrained approach. *IEEE Transactions on Cybernetics*, 2018, DOI: 10.1109/TCYB.2018.2821188.

[169] D. Wang, Z. Wang, B. Shen and Q. Li. H_∞ finite-horizon filtering for complex networks with state saturations: The weighted try-once-discard protocol, *International Journal of Robust and Nonlinear Control*, vol. 29, no. 7, pp. 2096–2111, 2019.

[170] L. Wang, Z. Wang, B. Shen and G. Wei. Recursive filtering with measurement fading: A multiple description coding scheme, *IEEE Transactions on Automatic Control*, DOI: 10.1109/TAC.2020.3034196, 2020.

[171] L. Wu and H. Gao. Sliding mode control of two-dimensional systems in Roesser model. *IET Control Theory & Applications*, vol. 2, no. 4, pp. 352–364, 2008.

[172] L. Wu, Y. Gao, J. Liu and H. Li. Event-triggered sliding mode control of stochastic systems via output feedback, *Automatica*, vol. 82, pp. 79–92, 2017.

[173] L. Wu, R. Yang, P. Shi and X. Su. Stability analysis and stabilization of 2-D switched systems under arbitrary and restricted switchings, *Automatica*, vol. 59, pp. 206–215, 2015.

[174] D. Wu, J. Wu, and S. Chen. Robust H_∞ control for networked control systems with uncertainties and multiple-packet transmission. *IET Control Theory and Applications*, vol. 4, no. 5, pp. 701–709, 2009.

[175] P. Wu, W. Zhang and S. Han. Self-triggered nonsingular terminal sliding mode control. In *2018 Annual American Control Conference (ACC)*, pp. 6513–6520, 2018.

[176] N. Xiao, Y. Niu and L. Xie. State feedback stabilization over finite-state fading channels, *Asian Journal of Control*, vol. 18, no. 3, pp. 1052–1061, 2016.

[177] S. Xu and G. Feng. New results on H_∞ control of discrete singularly perturbed systems, *Automatica*, vol. 45, no. 10, pp. 2339–2343, 2009.

[178] W. Xu, A. K. Junejo, Y. Liu and M. R. Islam. Improved continuous fast terminal sliding mode control with extended state observer for speed regulation of PMSM drive system, *IEEE Transactions on Vehicular Technology*, vol. 68, no. 11, pp. 10465–10476, 2019.

[179] Y. Xu, R. Lu, P. Shi, H. Li and S. Xie. Finite-time distributed state estimation over sensor networks with Robin-Robin protocol and fading channels, *IEEE Transactions on Cybernetics*, vol. 48, no. 1, pp. 336–345, 2018.

[180] X.-G. Yan, S. K. Spurgeon and C. Edwards. Static output feedback sliding mode control for time-varying delay systems with time-delayed nonlinear disturbances, *International Journal of Robust and Nonlinear Control*, vol. 20, no. 7, pp. 777–788, 2010.

Bibliography 267

[181] X.-G. Yan, S. K. Spurgeon and C. Edwards. Memoryless static output feedback sliding mode control for nonlinear systems with delayed disturbances, *IEEE Transactions on Automatic Control*, vol. 59, no. 7, pp. 1906–1912, 2014.

[182] Y. Yan, L. Su, V. Gupta and P. Antsaklis. Analysis of two-dimensional feedback systems over networks using dissipativity, *IEEE Transactions on Automatic Control*, vol. 65, no. 8, pp. 3241–3255, 2020.

[183] L. Xu, S. Yan, Z. Lin and S. Matsushita. A new elementary operation approach to multidimensional realization and LFR uncertainty modeling: The MIMO case, *IEEE Transactions on Circuits and Systems I: Regular Paper*, vol. 59, no. 3, pp. 638–651, 2012.

[184] Y. Yan, S. Yu and C. Sun. Quantization-based event-triggered sliding mode tracking control of mechanical systems, *Information Sciences*, vol. 523, pp. 296–306, 2020.

[185] C. Yang, Z. Che, J. Fu and L. Zhou. Passivity-based integral sliding mode control and ε-bound estimation for uncertain singularly perturbed systems with disturbances, *IEEE Transactions on Circuits and Systems–II: Express Briefs*, DOI: 10.1109/TCSII.2018.2849744, 2018.

[186] C. Yang and Q. Zhang. Multiobjective control for T-S fuzzy singularly perturbed systems, *IEEE Transactions on Fuzzy Systems*, vol. 17, no. 1, 2009.

[187] M. Yang, X. Lang, J. Long and D. Xu. A Flux immunity robust predictive current control with incremental model and extended state observer for PMSM drive. *IEEE Transactions on Power Electronics*, vol. 32, no. 12, pp. 9267–9279, 2017.

[188] J. Yang, J. Sun, W. X. Zheng and S. Li. Periodic event-triggered robust output feedback control for nonlinear uncertain systems with time-varying disturbance, *Automatica*, vol. 94, pp. 324–333, 2018.

[189] L. Yang, and J. Yang. Nonsingular fast terminal sliding-mode control for nonlinear dynamical systems. *International Journal of Robust and Nonlinear Control*, vol. 21, no. 16, pp. 1865–1879, 2011.

[190] Y. Yang, D. Yue and C. Xu. Dynamic event-triggered leader-following consensus control of a class of linear multi-agent systems, *Journal of the Franklin Institute*, DOI: 10.1016/j.jfranklin.2018.08.007, 2018.

[191] R. Yang and W. X. Zheng. Two-dimensional sliding mode control of discrete-time Fornasini-Marchesini systems, *IEEE Transactions on Automatic Control*, vol. 64, no. 9, pp. 3943–3948, 2019.

268   *Bibliography*

[192] R. Yang, W. X. Zheng and Y. Yu. Event-triggered sliding mode control of discrete-time two-dimensional systems in Roesser model. *Automatica*, vol. 114, 108813, 2020.

[193] N. Yeganefar, N. Yeganefar, M. Ghamgui and E. Moulay. Lyapunov theory for 2-D nonlinear Roesser models: Application to asymptotic and exponential stability. *IEEE Transactions on Automatic Control*, vol. 58, no. 5, pp. 1299–1304, 2013.

[194] K.-Y. You and L.-H. Xie. Survey of recent progress of networked control systems, *Acta Automatica Sinica*, vol. 39, no. 2, pp. 101–117, 2013.

[195] K. D. Young, V. I. Utkin and U. Ozguner. A control engineer's guide to sliding mode control. *IEEE Transactions on Control Systems Technology*, vol. 7, no. 3, pp. 328–342, 1999.

[196] X. Yu and O. Kaynak. Sliding-mode control with soft computing: A survey, *IEEE Transactions on Industrial Electronics*, vol. 56, no. 9, pp. 3275–3285, 2009.

[197] X. Yu and O. Kaynak. Sliding mode control made smarter: A computational intelligence perspective. *IEEE Systems, Man, and Cybernetics Magazine*, vol. 3, no. 2, pp. 31–34, 2017.

[198] X. Yu and Z. Man. Model reference adaptive control systems with terminal sliding modes. *International Journal of Control*, vol. 64, no. 6, pp. 1165–1176, 1996.

[199] X. Yu, and Z. Man. Fast terminal sliding-mode control design for nonlinear dynamical systems. *IEEE Transactions on Circuits and Systems I: Fundamental Theory and Applications*, vol. 49, no. 2, pp. 261–264, 2002.

[200] Y. Yuan, Z. Wang, Y. Yu, L. Guo and H. Yang. Active disturbance rejection control for a pneumatic motion platform subject to actuator saturation: An extended state observer approach. *Automatica*, vol. 107, pp. 353–361, 2019.

[201] M. Zak. Terminal attractors for addressable memory in neural network. *Physics Letters A*, vol. 133, no. 1-2, pp. 18–22, 1988.

[202] L. Zhang, H. Gao and O. Kaynak. Network-induced constraints in networked control systems-A survey, *IEEE Transactions on Industrial Informatics*, vol. 9, no. 1, pp. 403–416, 2013.

[203] X.-M. Zhang, Q.-L. Han and X. Yu. Survey on recent advances in networked control systems, *IEEE Transactions on Industrial Informatics*, vol. 12, no. 5, pp. 1740–1752, 2016.

Bibliography

[204] X.-M. Zhang, Q.-L. Han and B.-L. Zhang. An overview and deep investigation on sampled-data-based event-triggered control and filtering for networked systems, *IEEE Transactions on Industrial Informatics*, vol. 13, no. 1, pp. 4–16, 2017.

[205] J. Zhang, X. Liu, Y. Xia, Z. Zuo and Y. Wang. Disturbance observer-based integral sliding-mode control for systems with mismatched disturbances. *IEEE Transactions on Industrial Electronics*, vol. 63, no. 11, pp. 7040–7048, 2016.

[206] H. Zhang, Q. Liu, J. Zhang, S. Chen and C. Zhang. Speed regulation of permanent magnet synchronous motor using event triggered sliding mode control. *Mathematical Problems in Engineering*, vol. 2018, Article ID: 2394535, 2018.

[207] Z. Zhang and Y. Niu. Adaptive sliding mode control for interval type-2 stochastic fuzzy systems subject to actuator failures. *International Journal of Systems Science*, vol. 49, no. 15, pp. 3169–3181, 2018.

[208] Z. Zhang, Y. Niu and J. Song. Input-to-state stabilization of interval type-2 fuzzy systems subject to cyber attacks: An observer-based adaptive sliding mode approach. *IEEE Transactions on Fuzzy Systems*, vol. 28, no. 1, pp. 190–203, 2020.

[209] L. Zhang, Z. Ning and Z. Wang. Distributed filtering for fuzzy time-delay systems with packet dropouts and redundant channels, *IEEE Transactions on Systems, Man, and Cybernetics: Systems*, vol. 46, no. 4, pp. 559–572, 2016.

[210] H. Zhang, J. Wang and Y. Shi. Robust H_∞ sliding-mode control for Markovian jump systems subject to intermittent observations and partially known transition probabilities, *Systems & Control Letters*, vol. 62, no. 12, pp. 1114–1124, 2013.

[211] J. Zhang and Y. Xia. Design of static output feedback sliding mode control for uncertain linear systems, *IEEE Transactions on Industrial Electronics*, vol. 57, no. 6, pp. 2161–2170, 2010.

[212] S. Zhang and V. Vittal. Wide-area control resiliency using redundant communication paths, *IEEE Transactions on Power Systems*, vol. 29, no. 5, pp. 2189–2199, 2014.

[213] B.-C. Zheng and G. Yang. Quantized output feedback stabilization of uncertain systems with input nonlinearities via sliding mode control, *International Journal of Robust and Nonlinear Control*, vol. 24, no. 2, pp. 228–246, 2014.

[214] B.-C. Zheng, X. Yu and Y. Xue. Quantized feedback sliding-mode control: An event-triggered approach, *Automatica*, vol. 91, pp. 126–135, 2018.

[215] X.-L. Zhu, and G.-H. Yang. State feedback controller design of networked control systems with multiple-packet transmission. *International Journal of Control*, vol. 82, no. 1, pp. 86–94, 2009.

[216] Y. Zhu, L. Zhang and W. X. Zheng. Distributed H_∞ filtering for a class of discrete-time Markov jump Lur'e systems with redundant channels, *IEEE Transactions on Industrial Electronics*, vol. 63, no. 3, pp. 1876–1885, 2016.

[217] L. Zou, Z. Wang and H. Gao. Observer-based H_∞ control of networked systems with stochastic communication protocol: The finite-horizon case, *Automatica*, vol. 63, pp. 366–373, 2016.

[218] L. Zou, Z. Wang, H. Gao and X. Liu. State estimation for discrete-time dynamical networks with time-varying delays and stochastic disturbances under the Round-Robin protocol, *IEEE Transactions on Neural Networks and Learning Systems*, vol. 28, no. 5, pp. 1139–1151, 2017.

[219] L. Zou, Z. Wang and H. Gao. Set-membership filtering for time-varying systems with mixed time-delays under Round-Robin and Weighted Try-Once-Discard protocols, *Automatica*, vol. 74, pp. 341–348, 2016.

[220] L. Zou, Z. Wang, Q.-L. Han and D. Zhou. Moving horizon estimation for networked time-delay systems under Round-Robin protocol, *IEEE Transactions on Automatic Control*, vol. 64, no. 12, pp. 5191–5198, 2019.

[221] L. Zou, Z. Wang, Q. L. Han and D. Zhou. Ultimate boundedness control for networked systems with try-once-discard protocol and uniform quantization effects, *IEEE Transactions on Automatic Control*, vol. 62, no. 12, pp. 6582–6588, 2017.

[222] L. Zou, Z. Wang, J. Hu, Y. Liu and X. Liu. Communication-protocol-based analysis and synthesis of networked systems: Progress, prospects and challenges, *International Journal of Systems Science*, DOI: 10.1080/00207721.2021.1917721, 2021.

Index

H_∞, 26, 82
2-D Lyapunov function, 199
2-D event generators, 209
2-D sliding function, 195, 217
2-D super-twisting-like algorithm, 227

asynchronous sliding mode control, 79

chattering, 7, 144, 173, 228
communication protocol, 11, 250
coupled LMIs, 91, 153, 226, 235

dynamic event-triggered, 11, 103

equivalent control law, 3, 106, 147, 165, 166, 195, 198, 218
event-triggered, 11, 81, 193
event-triggered asynchronous SMC, 83
extended-state-observer, 8, 161

fast-sampling SPS, 117

genetic algorithm, 153, 161, 193, 237

hidden Markov fading channels., 143
hidden Markov model, 79

LMIs, 33, 35, 53, 93, 105, 107, 110, 123, 127
Lur'e-type Lyapunov functional, 84

Markov chain, 24, 80
Markov fading channels, 139
Markovian jump systems, 79
mean-square exponentially ultimately bounded, 121

multiple-packet transmission protocol, 13, 140

networked control systems, 10, 79
nonsingular terminal sliding function, 7, 165

observer-based SMC, 65
operational amplifier circuit, 131
optimization problem, 35, 44, 52, 94, 121, 153, 201, 237
output-feedback sliding function, 120

periodic event-triggered, 11, 167
permanent magnet synchronous motor, 1, 161

quasi-sliding mode, 9, 87, 106, 148

reachability, 4, 27, 49, 69, 86, 105, 126, 147, 170, 193, 213
reaching law, 196
redundant channel transmission protocol, 13, 117
Roesser model, 9, 194, 214
Round-Robin protocol, 12, 41, 213

separation strategy, 52
singularly perturbed system, 101, 117
sliding function, 25, 44, 65, 82, 103, 143, 165
sliding mode control, 1, 23, 41, 61, 101, 117, 139, 193, 213
sliding variable, 3, 58, 59, 70, 71, 77, 96, 98, 104, 105, 110, 115, 128, 136, 171, 196, 208, 228, 238
slow-sampling SPS, 102

state-saturated, 62, 139, 140
state-saturated observer, 65
static output-feedback, 41
stochastic communication protocol, 12, 23
stochastic Lyapunov function, 27, 154
super-twisting algorithm, 227

terminal sliding mode control, 5, 161

token-dependent Lyapunov function, 28, 31, 46
two-dimensional, 9, 193, 213

weighted Try-Once-Discard protocol, 13, 61

Zeno phenomenon, 12, 167